T0179808

Point Processes and Jump Diffusions

The theory of marked point processes on the real line is of great and increasing importance in areas such as insurance mathematics, queuing theory and financial economics. However, the theory is often viewed as technically and conceptually difficult and has proved to be a block for PhD students looking to enter the area.

This book gives an intuitive picture of the central concepts as well as the deeper results, while presenting the mathematical theory in a rigorous fashion and discussing applications in filtering theory and financial economics. Consequently, readers will get a deep understanding of the theory and how to use it. A number of exercises of differing levels of difficulty are included, providing opportunities to put new ideas into practice. Graduate students in mathematics, finance and economics will gain a good working knowledge of point-process theory, allowing them to progress to independent research.

Tomas Björk was Professor Emeritus of Mathematical Finance at the Stockholm School of Economics and previously worked at the Mathematics Department of the Royal Institute of Technology, Stockholm. Björk served as co-editor of *Mathematical Finance*, on the editorial board for *Finance and Stochastics* and several other journals, and was President of the Bachelier Finance Society. He was particularly known for his research on point-process-driven forward-rate models, finite-dimensional realizations of infinite-dimensional SDEs, and time-inconsistent control theory. He was the author of the well-known textbook *Arbitrage Theory in Continuous Time* (1998), now in its fourth edition.

Tomas Björk 1947–2021

Point Processes and Jump Diffusions

An Introduction with Finance Applications

TOMAS BJÖRK

Stockholm School of Economics

CAMBRIDGE
UNIVERSITY PRESS

CAMBRIDGE
UNIVERSITY PRESS

University Printing House, Cambridge CB2 8BS, United Kingdom

One Liberty Plaza, 20th Floor, New York, NY 10006, USA

477 Williamstown Road, Port Melbourne, VIC 3207, Australia

314–321, 3rd Floor, Plot 3, Splendor Forum, Jasola District Centre, New Delhi – 110025, India

103 Penang Road, #05-06/07, Visioncrest Commercial, Singapore 238467

Cambridge University Press is part of the University of Cambridge.

It furthers the University's mission by disseminating knowledge in the pursuit of
education, learning, and research at the highest international levels of excellence.

www.cambridge.org
Information on this title: www.cambridge.org/9781316518670
DOI: 10.1017/9781009002127

First published 2021

Printed in the United Kingdom by TJ Books Limited, Padstow Cornwall

A catalogue record for this publication is available from the British Library.

ISBN 978-1-316-51867-0 Hardback

Contents

Preface

This text is intended as an introductory overview of stochastic calculus for marked point processes and jump diffusions with applications to filtering, stochastic control and finance.

The pedagogical approach is that I have tried to provide as much intuition as possible while being reasonably precise. In more concrete terms this means that most major results are preceded by a heuristic argument, often including "infinitesimal" arguments, which leads up to the formulation of a conjecture. As an example, when discussing measure transformations, some rather simple arguments will lead us to conjecture the formulation of the Girsanov theorem. This conjecture is then given a formal proof. Most results are given full formal proofs, apart from the fact that I go lightly on technical details like integrability and regularity assumptions. This is done in order to increase readability and to highlight the main ideas in the proofs. For more technical points the reader is referred to the specialist literature. A few results, where the proofs are really hard, are presented without proofs, but often with a heuristic argument to make them believable. The reader is again referred to the literature.

The dependence structure of the text is that Parts II–V on control, filtering and finance build on Part I. Parts II–V can, however, be read independently of each other.

The reader is assumed to be familiar with basic measure theory and stochastic calculus for Wiener processes. No previous knowledge of finance is assumed.

Acknowledgements
I am very grateful for very helpful comments from Alexander Aurell, Rasmus Damsbo, Matt P. Dziubinski, Nicky van Foreest, Mia Hinnerich, Johan Kjaer, Timo Koski, Mariana Khapko and Agatha Murgoci.

Warm thanks go to Thaleia Zariphopoulou, for valuable help with the book, and for care and friendship during a very difficult time for me. It meant a lot.

I am also very grateful to the staff at Cambridge University Press for the help with producing the book. Special thanks are due to David Tranah who, many years ago, encouraged me to write this book. He has shown a large amount of patience with me during the years, and his help in finishing the book has been invaluable.

Part I

Point Processes

1   Counting Processes

In this chapter we introduce the concept of a counting process. For references to the literature on counting processes and more general point processes, see the Notes at the end of the chapter.

1.1   Generalities and the Poisson Process

Our main objective is to study point processes on the real positive line, and the simplest type of a point process is a *counting process*. The formal definition is as follows.

Definition 1.1   A random process $\{N_t; \ t \in R_+\}$ is a **counting process** if it satisfies the following conditions.

1. The trajectories of N are, with probability one, right continuous and piecewise constant.
2. The process starts at zero, so

$$N_0 = 0.$$

3. For each t

$$\Delta N_t = 0, \quad \text{or} \quad \Delta N_t = 1.$$

   with probability one. Here ΔN_t denotes the jump size of N at time t, or more formally

$$\Delta N_t = N_t - N_{t-}.$$

In more pedestrian terms, the process N starts at $N_0 = 0$ and stays at the level 0 until some random time T_1 when it jumps to $N_{T_1} = 1$. It then stays at level 1 until the another random time T_2 when it jumps to the value $N_{T_2} = 2$ etc. We will refer to the random times $\{T_n; \ n = 1, 2, \ldots\}$ as the **jump times** of N. Counting processes are often used to model situations where some sort of well-specified **events** are occurring randomly in time. A typical example of an event could be the arrival of a new customer at a queue, an earthquake in a well-specified geographical area, or a company going bankrupt. The interpretation is then that N_t denotes the number of events that have occurred in the time interval $[0, t]$. Thus N_t could be the number of customers who have arrived at a certain queue during the interval $[0, t]$ etc. With this interpretation, the jump times $\{T_n; \ n = 1, 2, \ldots\}$ are often also referred to as the **event times** of the process N.

Before we go on to the general theory of counting processes, we will study the **Poisson**

process in some detail. The Poisson process is the single most important of all counting processes, and among counting processes it occupies very much the same place that the Wiener process does among diffusion processes. We start with some elementary facts concerning the Poisson distribution.

Definition 1.2 A random variable X is said to have a **Poisson distribution** with parameter α if it takes values among the natural numbers, and the probability distribution has the form

$$P(X = n) = e^{-\alpha} \frac{\alpha^n}{n!}, \quad n = 0, 1, 2, \ldots$$

We will often write this as $X \sim \mathrm{Po}(\alpha)$.

We recall that, for any random variable X, its **characteristic function** φ_X is defined by

$$\varphi_X(u) = E\left[e^{iuX}\right], \quad u \in R,$$

where i is the imaginary unit. We also recall that the distribution of X is completely determined by φ_X. We will need the following well-known result concerning the Poisson distribution.

Proposition 1.3 *Let X be $\mathrm{Po}(\alpha)$. Then the characteristic function is given by*

$$\varphi_X(u) = e^{\alpha\left(e^{iu} - 1\right)}.$$

The mean and variance are given by

$$E[X] = \alpha, \quad \mathrm{Var}(X) = \alpha.$$

Proof This is left as an exercise.   □

We now leave the Poisson distribution and go on to the Poisson process.

Definition 1.4 Let $(\Omega, \mathcal{F}, P)$ be a probability space with a given filtration $\mathbf{F} = \{\mathcal{F}_t\}_{t \geq 0}$, and let λ be a real number with $\lambda > 0$. A counting process N is a **Poisson process with intensity** λ **with respect to the filtration F** if it satisfies the following conditions.

1. N is adapted to **F**.
2. For all $s \leq t$ the random variable $N_t - N_s$ is independent of $\mathcal{F}_s$.
3. For all $s \leq t$, the conditional distribution of the increment $N_t - N_s$ is given by

$$P(N_t - N_s = n \,|\, \mathcal{F}_s) = e^{-\lambda(t-s)} \frac{\lambda^n(t-s)^n}{n!}, \quad n = 0, 1, 2, \ldots \tag{1.1}$$

In concrete terms this says that the increment $N_t - N_s$ is Poisson with parameter $\lambda(T-s)$ and independent of $\mathcal{F}_s$. In the definition above we encounter the somewhat forbidding looking formula (1.1). As it turns out, there is another way of characterizing the Poisson process, which is much easier to handle than distributional specification above. This alternative characterization is done in terms of the "infinitesimal characteristics" of the process, and we now go on to discuss this.

1.2   Infinitesimal Characteristics

One of the main ideas in modern process theory is that the "true nature" of a process is revealed by its "infinitesimal characteristics". For a diffusion process the infinitesimal characteristics are the drift and the diffusion terms. For a counting process, the natural infinitesimal object is the "predictable conditional jump probability per unit time", and informally we define this as

$$\frac{P\left(dN_t = 1 \,|\, \mathcal{F}_{t-}\right)}{dt}.$$

The increment process dN is informally interpreted as

$$dN_t = N_t - N_{t-dt},$$

and the sigma algebra $\mathcal{F}_{t-}$ is defined by

$$\mathcal{F}_{t-} = \bigvee_{0 \le s < t} \mathcal{F}_s \tag{1.2}$$

The reason why we define dN_t as $N_t - N_{t-dt}$ instead of $N_{t+dt} - N_t$ is that we want the increment process dN to be adapted. The term "predictable" will be very important later on, and will be given a precise mathematical definition. We also note that the increment dN_t only takes two possible values, namely $dN_t = 0$ or $dN_t = 1$ depending on whether or not an event has occurred at time t. We can thus write the conditional jump probability as an expected value, namely as

$$P\left(dN_t = 1 \,|\, \mathcal{F}_{t-}\right) = E^P\left[dN_t |\, \mathcal{F}_{t-}\right].$$

Suppose now that N is a Poisson process with intensity λ, and that h is a small real number. According to the definition we then have

$$P\left(N_t - N_{t-h} = 1 \,|\, \mathcal{F}_{t-h}\right) = e^{-\lambda h} \lambda h.$$

Expanding the exponential we thus have

$$P\left(N_t - N_{t-h} = 1 \,|\, \mathcal{F}_{t-h}\right) = \lambda h \sum_{n=0}^{\infty} \frac{(-\lambda h)^n}{n!}.$$

As h becomes "infinitesimally small" the higher-order terms can be neglected and as a formal limit when $h \to dt$ we obtain

$$P\left(dN_t = 1 \,|\, \mathcal{F}_{t-}\right) = \lambda dt, \tag{1.3}$$

or equivalently

$$E^P\left[dN_t |\, \mathcal{F}_{t-}\right] = \lambda dt. \tag{1.4}$$

This entire discussion has obviously been very informal, but nevertheless the formula (1.4) has a great intuitive value. It says that we can interpret the parameter λ as the **conditional jump intensity**. In other words, λ is the (conditional) expected number of jumps per unit of time. The point of this is twofold.

- The concept of a conditional jump intensity is easy to interpret intuitively, and it can also easily be generalized to a large class of counting processes.
- The distribution of a counting process is completely determined by its conditional jump intensity, and equation (1.4) is *much* simpler than that equation (1.1).

The main project of this text is to develop a mathematically rigorous theory of counting processes, building on the intuitively appealing concept of a conditional jump intensity. As the archetypical example we will of course use the Poisson process, and to start with we need to reformulate the nice but very informal relation (1.4) to something more mathematically precise. To do this we start by noting (again informally) that if we subtract the conditional expected number of jumps λdt from the actual number of jumps dN_t then the difference

$$dN_t - \lambda dt,$$

should have zero conditional mean. The implication of this is that we are led to conjecture that if we define the process M by

$$\begin{cases} dM_t &= dN_t - \lambda dt, \\ M_0 &= 0, \end{cases}$$

or, equivalently, on integrated form as

$$M_t = N_t - \lambda t,$$

then M should be a martingale. This conjecture is in fact true.

Proposition 1.5 *Assume that N is an **F**- Poisson process with intensity λ. Then the process M, defined by*

$$M_t = N_t - \lambda t, \tag{1.5}$$

*is an **F**-martingale.*

Proof We have to show that $E\left[N_t - N_s \mid \mathcal{F}_s\right] = \lambda(t - s)$. This however follows directly from the fact that the conditional distribution of $N_t - N_s$, given $\mathcal{F}_s$, is Poisson with parameter $\lambda(t - s)$.   □

This somewhat trivial result is much more important than it looks like at first sight. It is in fact the natural starting point of the "martingale approach" to counting processes. Indeed, as we will see below, the martingale property of M above, is not only a *consequence* of the fact that N is a Poisson process but, in fact, the martingale property *characterizes* the Poisson process within the class of counting processes. More precisely, we will show below that if N is an arbitrary counting process and if the process M, defined above is a martingale, then this *implies* that N must be Poisson with intensity λ. This is a huge technical step forward in the theory of counting processes, the reason being that it is often relatively easy to check the martingale property of M, whereas it is typically a very hard task to check that the conditional distribution of the increments of N is given by (1.1).

Furthermore, it turns out that a very big class of counting processes can be characterized by a corresponding martingale property and this fact, coupled with a (very simple form of) stochastic differential calculus for counting processes, will provide us with a very powerful toolbox for a fairly advanced study of counting processes on filtered probability spaces.

To develop this theory we need to carry out the following program.

1. Assuming that a process A is of bounded variation, we need to develop a theory of stochastic integrals of the form

$$\int_0^t h_s \, dA_s,$$

   where the integrand h should be required to have some nice measurability property.
2. In particular, if M is a martingale of bounded variation, we would like to know under what conditions a process X of the form

$$X_t = \int_0^t h_s \, dM_s,$$

   is a martingale. Is it for example enough that h is adapted? (Compare with the Wiener case).
3. Develop a differential calculus for stochastic integrals of the type above. In particular we would like to derive an extension of the Itô formula to the counting process case.
4. Use the theory to study general counting processes in terms of their martingale properties.
5. Given a Wiener process W, we recall that there exists a powerful martingale representation theorem which says that (for the internal filtration) *every* martingale X can be written as $X_t = X_0 + \int_0^t h_s \, dW_s$. Does there exist a corresponding theory for counting processes?
6. Study how the conditional jump intensity will change under an absolutely continuous change of measure. Does there exist a Girsanov theory for counting processes?
7. Finally we want to apply the theory above in order to study more concrete problems, like optimal control, and arbitrage theory for economies where asset prices are driven by jump diffusions.

1.3   Notes

The textbook Brémaud (1981) is a classic in the field. The monograph Last & Brandt (1995) contains an almost encyclopedic study of (marked) point processes. In Cont & Tankov (2003) the reader will find an in depth study of Lévy processes and their applications to finance. For general semimartingale theory see Cohen & Elliott (2015), Jacod & Shiryaev (1987), or Protter (2004).

2 Stochastic Integrals and Differentials

2.1 Integrators of Bounded Variation

In this section, the main object is to develop a stochastic integration theory for integrals of the form

$$\int_0^t h_s \, dA_s,$$

where A is a process of bounded variation. In a typical application, the integrator A could for example be given by

$$A_t = N_t - \lambda t,$$

where N is a Poisson process with intensity λ, and in particular we will investigate under what conditions the process X defined by

$$X_t = \int_0^t h_s \, [dN_s - \lambda ds],$$

is a martingale. Apart from this, we also need to develop a stochastic differential calculus for processes of this kind, derive the relevant Itô formula, and to study stochastic differential equations, driven by counting processes.

Before we embark on this program, the following two points are worth mentioning.

- Compared to the definition of the usual Itô integral for Wiener processes, the integration theory for point processes is quite simple. Since all integrators will be of bounded variation, the integrals can be defined pathwise, as opposed to the Itô integral which has to be defined as an L^2 limit.

- On the other hand, compared to the Itô integral, where the natural requirement is that the integrands are adapted, the point-process integration theory requires much more delicate measurability properties of the integrands. In particular we need to understand the fundamental concept of a **predictable process**.

In order to get a feeling for the predictability concept, and its relation to martingale theory, we will start by giving a brief recapitulation of discrete-time stochastic integration theory.

2.2   Discrete-Time Stochastic Integrals

In this section we briefly discuss the simplest type of stochastic integration, namely integration of discrete-time processes. This will serve as an introduction to the more complicated continuous-time theory later on, and it is also important in its own right. We start by defining the discrete stochastic integral.

Definition 2.1   Consider a probability space $(\Omega, \mathcal{F}, P)$, equipped with a discrete-time filtration $\mathbf{F} = \{\mathcal{F}_n\}_{n=0}^{\infty}$.

- For any random process Y, the **increment process** ΔY is defined by

$$(\Delta Y)_n = Y_n - Y_{n-1}, \tag{2.1}$$

  with the convention $Y_{-1} = 0$. For simplicity of notation we will sometimes denote $(\Delta Y)_n$ by ΔY_n.
- For any two processes X and Y, the **discrete stochastic integral** process $X \star Y$ is defined by

$$(X \star Y)_n = \sum_{k=0}^{n} X_k (\Delta Y)_k. \tag{2.2}$$

Instead of $(X \star Y)_n$ we will sometimes write $\int_0^n X_s \, dY_s$.

The reason why we define ΔY by "backward increments" as $(\Delta Y)_n = Y_n - Y_{n-1}$, instead of "forward increments" $(\Delta Y)_n = Y_{n+1} - Y_n$, is that by using backwards increments the process ΔY is adapted whenever Y is adapted.

From standard Itô integration theory we recall that if W is a Wiener process and if h is a square-integrable adapted process, then the integral process Z, given by

$$Z_t = \int_0^t h_s \, dW_s,$$

is a martingale. It is therefore natural to expect that a similar result would hold for the discrete-time integral, but this is not the case. Indeed, as we will see below, the correct measurability concept is that of a *predictable process* rather than that of an adapted process.

Definition 2.2

- A random process X is **F-adapted** if, for each n, X_n is $\mathcal{F}_n$-measurable.
- A random process X is **F-predictable** if, for each n, X_n is $\mathcal{F}_{n-1}$-measurable. Here we use the convention $\mathcal{F}_{-1} = \mathcal{F}_0$.

We note that a predictable process is "known one step ahead in time".

The main result for stochastic integrals is that when you integrate a *predictable process* X with respect to a martingale M, then the result is a new martingale.

Proposition 2.3   *Assume that the filtered probability space $(\Omega, \mathcal{F}, P, \mathbf{F})$ carries the processes X and M, where X is predictable, M is a martingale and $X_n(\Delta M)_n \in L^1$ for each n. Then the stochastic integral $X \star M$ is a martingale.*

Proof   We recall that in discrete time, a process Z is a martingale if and only if

$$E[\Delta Z_n | \mathcal{F}_{n-1}] = 0, \quad n = 0, 1, \ldots$$

Thus, defining Z as

$$Z_n = \sum_{k=0}^{n} X_k \Delta M_k,$$

it follows that

$$\Delta Z_n = X_n \Delta M_n$$

and we obtain

$$E[\Delta Z_n | \mathcal{F}_{n-1}] = E[X_n \Delta M_n | \mathcal{F}_{n-1}] = X_n E[\Delta M_n | \mathcal{F}_{n-1}] = 0.$$

In the second equality we used the fact that X is predictable, and in the third equality we used the martingale property of M.   □

2.3   Stochastic Integrals in Continuous Time

We now go back to continuous time and assume that we are given a filtered probability space $(\Omega, \mathcal{F}, P, \mathbf{F})$. Before going on to define the new stochastic integral we need to define a number of measurability properties for random processes, and in particular we need to define the discrete-time version of the predictability concept.

Definition 2.4

- A random process X is said to be **cadlag** (continu à droite, limites à gauche) if the trajectories are right continuous with left hand limits, with probability one.
- The class of adapted cadlag processes A with $A_0 = 0$, such that the trajectories of A are of finite variation on the interval $[0, T]$ is denoted by $\mathcal{V}_T$. Such a process is said to be of **finite variation on** $[0, T]$, and will thus satisfy the condition

$$\int_0^T |dA_t| < \infty \quad P\text{-a.s.}$$

- We denote by $\mathcal{A}_T$ the class of processes in $\mathcal{V}_T$ such that

$$E\left[\int_0^T |dA_t|\right] < \infty.$$

Such a process is said to be of **integrable variation on** $[0, T]$.
- The class of processes belonging to $\mathcal{V}_T$ for all $T < \infty$ is denoted by $\mathcal{V}$. Such a process is said to be of **finite variation**.
- The class of processes belonging to $\mathcal{A}_T$ for all $T < \infty$ is denoted by $\mathcal{A}$. Such a process is said to be of **integrable variation**.

Remark   Note that both the cadlag property, and the property of being adapted, are parts of the definition of $\mathcal{V}_T$ and $\mathcal{A}_T$.

We now come to the two main measurability properties of random processes. Before we go on to the definitions, we recall that a random process X on the time interval R_+ is a mapping

$$X : \Omega \times R_+ \rightarrow R,$$

where the value of X at time t, for the elementary outcome $\omega \in \Omega$ is denoted either by $X(t, \omega)$ or by $X_t(\omega)$.

Definition 2.5   The **optional** σ-algebra on $R_+ \times \Omega$ is generated by all processes Y of the form

$$Y_t(\omega) = Z(\omega)I\{r \leq t < s\},\qquad(2.3)$$

where I is the indicator function, r and s are fixed real numbers, and Z is an $\mathcal{F}_r$ measurable random variable. A process X which, viewed as a mapping $X : \Omega \times R_+ \rightarrow R$, is measurable with respect to the optional σ-algebra is said to be an **optional process**.

The definition above is perhaps somewhat forbidding when you meet it the first time. Note however, that every generator process Y above is adapted and cadlag, and we have in fact the following result, the proof of which is nontrivial and omitted.

Proposition 2.6   *The optional σ-algebra is generated by the class of adapted cadlag processes.*

In particular it is clear that every process of finite variation, and every adapted process with continuous trajectories is optional. The optional measurability concept is in fact "the correct one" instead of the usual concept of a process being adapted. The difference between an adapted process and an optional one is that optionality for a process X implies a joint measurability property in (t, ω), whereas X being adapted only implies that the mapping $X_t : \Omega \rightarrow R$ is $\mathcal{F}_t$ measurable in ω for each fixed t. For "practical" purposes, the difference between an adapted process and an optional process is very small and the reader may, without great risk, interpret the term "optional" as "adapted". The main point of the optionality property is the following result, which shows that optionality is preserved under stochastic integration.

Proposition 2.7   *Assume that A is of finite variation and that h is an optional process satisfying the condition*

$$\int_0^t |h_s||dA_s| < \infty, \quad \text{for all } t.$$

Then the following assertions hold.

- *The process $X = h \star A$ defined, for each ω, by*

$$X_t(\omega) = \int_0^t h_s(\omega)dA_s(\omega),$$

   is well defined, for almost each ω, as a Lebesgue–Stieltjes integral.
- *The process X is cadlag and optional, so in particular it is adapted.*

- *If h also satisfies the condition*

$$E\left[\int_0^t |h_s||dA_s|\right] < \infty, \quad \textit{for all } t,$$

  then X is of integrable variation.

Proof The proposition is easy to prove if h is generator process of the form (2.3). The general case can then be proved by approximating h by a linear combination of generator processes, or by using a monotone class argument.   □

Remark   Note again that since A is of finite variation it is, by definition, optional. If we only require that h is adapted and A of finite variation (and thus adapted), then this would *not* guarantee that X is adapted.

2.4   Stochastic Integrals and Martingales

Suppose that M is a martingale of integrable variation. We now turn to the question under which conditions on the integrand h, a stochastic process of the form

$$X_t = \int_0^t h_s dM_s,$$

is itself a martingale. With the Wiener theory fresh in the memory, one is perhaps led to conjecture that it is enough to require that h (apart from obvious integrability properties) is adapted, or perhaps optional. This conjecture is, however, *not* correct and it is easy to construct a counterexample.

Example   Let Z be a non-trivial random variable with

$$E[Z] = 0, \quad E[Z^2] < \infty,$$

and define the process M by

$$M_t = \begin{cases} 0, & 0 \le t < 1, \\ Z, & t \ge 1. \end{cases}$$

If we define the filtration **F** by $\mathcal{F}_t = \sigma\{M_s; s \le t\}$, then it is easy to see that M is a martingale of integrable variation. In particular, M is optional, so let us define the integrand h as $h = M$. If we now define the process X by

$$X_t = \int_0^t M_s dM_s,$$

then it is clear that the integrator M has a point mass of size Z at $t = 1$. In particular we have $X_1 = M_1 \Delta M_1 = Z^2$, and we immediately obtain

$$X_t = \begin{cases} 0, & 0 \le t < 1, \\ Z^2, & t \ge 1. \end{cases}$$

From this it is clear that X is a non-decreasing process, so in particular it is *not* a martingale.

Note, however, that if we define h as $h_t = M_{t-}$ then X will be a martingale (why?). As we will see, it is not a coincidence that this choice of h is *left*-continuous.

It is clear from this example that we must demand more than mere optionality from the integrand h in order to ensure that that the stochastic integral $h \star M$ is a martingale. From the discrete-time theory we recall that if M is a martingale and if h is *predictable*, then $h \star M$ is a martingale. We also recall that predictability of h in discrete time means that $h_n \in \mathcal{F}_{n-1}$ and now the question is how to generalize this concept to continuous time.

The obvious idea is of course to say that a continuous-time process h is predictable if $h_t \in \mathcal{F}_{t-}$ for all $t \in R_+$, and in order to see if this is a good idea we now give some informal and heuristic arguments. Let us thus assume that M is a martingale of bounded variation, that $h_t \in \mathcal{F}_{t-}$ for all $t \in R_+$, and that all necessary integrability conditions are satisfied. We then define the process X by

$$X_t = \int_0^t h_s dM_s,$$

and we now want to check if X is a martingale. Loosely speaking, and comparing with discrete-time theory, we expect the process X to be a martingale if and only if

$$E\left[dX_t | \mathcal{F}_{t-}\right] = 0,$$

for all t. By definition we have

$$dX_t = h_t dM_t,$$

so we obtain

$$E\left[dX_t | \mathcal{F}_{t-}\right] = E\left[h_t dM_t | \mathcal{F}_{t-}\right].$$

Since $h_t \in \mathcal{F}_{t-}$, we can pull this term outside the expectation, and since M is a martingale we have $E\left[dM_t | \mathcal{F}_{t-}\right] = 0$, so we obtain

$$E\left[dX_t | \mathcal{F}_{t-}\right] = h_t E\left[dM_t | \mathcal{F}_{t-}\right] = 0,$$

thus "proving" that X is a martingale.

This very informal argument is very encouraging, but it turns out that the requirement $h_t \in \mathcal{F}_{t-}$ is not quite good enough for our purposes, the main reason being that, for each fixed t, it is a measurability argument in the ω variable only. In particular the requirement $h_t \in \mathcal{F}_{t-}$ has the weakness that it does not guarantee that the process X is adapted. We thus need to refine the simple idea above, and it turns out that the following definition is exactly what we need.

Definition 2.8 Given a filtered probability space $(\Omega, \mathcal{F}, P, \mathbf{F})$, we define the **F-predictable** σ-algebra Σ_P on $R_+ \times \Omega$ as the σ-algebra generated by all processes Y of the form

$$Y_t(\omega) = Z(\omega)I\{r < t \le s\}, \tag{2.4}$$

where r and s are real numbers and the random variable Z is $\mathcal{F}_r$- measurable. A process

X which is measurable with respect to the predictable σ-algebra is said to be an **F-predictable process**.

This definition is the natural generalization of the predictability concept from discrete-time theory, and it is extremely important to notice that all the generator processes Y above are *left*-continuous and adapted. It is also possible to show the following result, the proof of which is omitted.

Proposition 2.9 *The predictable σ-algebra is also generated by the class of left-continuous adapted processes.*

In particular, this result implies that every adapted left-continuous process is predictable, and a very important special case of a predictable process is obtained if we start with an adapted cadlag process X and then define a new process Y by

$$Y_t = X_{t-}.$$

Since Y is left-continuous and adapted it will certainly be predictable, and most of the predictable processes that we will meet in "practice" are in fact of this form.

Remark   The working mathematician can, without great risk, interpret the term "predictable" as either "adapted and left-continuous" or as "$\mathcal{F}_{t-}$-adapted".

We can now state the main results of this section.

Proposition 2.10 *Assume that M is a martingale of bounded variation and that h is a predictable process satisfying the condition*

$$E\left[\int_0^t |h_s||dM_s|\right] < \infty, \tag{2.5}$$

for all $t \geq 0$. Then the process X defined by

$$X_t = \int_0^t h_s \, dM_s,$$

is a martingale.

Proof   It is very easy to show that if h is a generator process of the form (2.4) then X is a martingale. The general result then follows by a (non-trivial) approximation argument in the form of a monotone class argument.   □

We will also need the following result, which shows how the predictability property is inherited by stochastic integration.

Proposition 2.11 *Let A be a predictable process of bounded variation (so in particular A is cadlag) and let h be a predictable process satisfying*

$$E\left[\int_0^t |h_s||dA_s|\right] < \infty, \tag{2.6}$$

for all $t \geq 0$. Then the integral process

$$X_t = \int_0^t h_s \, dA_s$$

is predictable.

Proof The result is obvious when h is a generator process of the form (2.4). The general result follows by an approximation argument.   □

We finish this section with a useful lemma.

Lemma 2.12 *Assume that X is optional and cadlag, and define the process Y by $Y_t = X_{t-}$. Then Y is predictable and for any optional process h we have*

$$\int_0^t h_s X_s \, ds = \int_0^t h_s Y_s \, ds,$$

for all $t \geq 0$.

Proof The predictability follows from the fact that Y is left-continuous and adapted. Since X is cadlag, X and Y will (for a fixed trajectory) only differ on a finite number of points, and since we are integrating with respect to Lebesgue measure the integrals will coincide.   □

2.5   The Itô Formula

Given the standard setting of a filtered probability space, let us consider an optional cadlag process X. If X can be represented on the form

$$X_t = X_0 + A_t + \int_0^t \sigma_s \, dW_s, \quad t \in R_+, \tag{2.7}$$

where the process, A, is of bounded variation, W is a Wiener process and σ is an optional process, then we say that X has a **stochastic differential** and we write

$$dX_t = dA_t + \sigma_t \, dW_t. \tag{2.8}$$

In our applications, the process A will always be of the form

$$dA_t = \mu_t \, dt + \beta_t \, dN_t, \tag{2.9}$$

where μ and β are predictable and N is a counting process, but in principle we allow A to be an arbitrary process of bounded variation (and thus cadlag and adapted). As in the standard Wiener case, it is important to note that the differential expression (2.8) is, by definition, nothing other than a shorthand notation for the integral expression (2.7).

The first question to ask is whether there exists an Itô formula for processes of this kind. In other words, let X have a stochastic differential of the form (2.8), let $F(t, x)$ be a given a smooth function, and define the process Z by

$$Z_t = F(t, X_t). \tag{2.10}$$

The question is now whether Z has a stochastic differential and, if so, what it looks like. This question is answered within general semimartingale theory, but since that theory is outside the scope of the present text we will, for the moment, only discuss the simpler case when X has the differential

$$dX_t = \mu_t dt + \sigma_t dW_t + \beta_t dN_t. \tag{2.11}$$

Now, *between the jumps* of N the process X will have the dynamics

$$dX_t = \mu_t dt + \sigma_t dW_t,$$

and this is of course handled by the standard Itô formula.

$$dZ_t = \left\{ \frac{\partial F}{\partial t}(t, X_t) + \mu_t \frac{\partial F}{\partial x}(t, X_t) + \frac{1}{2}\sigma_t^2 \frac{\partial^2 F}{\partial x^2}(t, X_t) \right\} dt + \sigma_t \frac{\partial F}{\partial x}(t, X_t) dW_t.$$

On the other hand, *at a jump time* t, the process N has a jump size of $\Delta N_t = N_t - N_{t-} = 1$ which implies that the process X will have a jump of size

$$\Delta X_t = \beta_t \Delta N_t = \beta_t.$$

Since $Z_t = F(t, X_t)$, the induced jump of Z is given by

$$\Delta Z_t = F(t, X_t) - F(t-, X_{t-}).$$

Furthermore, since $X_t = X_{t-} + \Delta X_t = X_{t-} + \beta_t$, we obtain

$$\Delta Z_t = F(t-, X_{t-} + \beta_t) - F(t-, X_{t-}).$$

Because F is assumed to be smooth we can also write this as

$$\Delta Z_t = F(t, X_{t-} + \beta_t) - F(t, X_{t-}).$$

If we note that $dN_t = 1$ at a jump time and that $dN_t = 0$ at times of no jumps, we can summarize our findings as follows.

Proposition 2.13 (The Itô formula)   *Assume that X has dynamics of the form*

$$dX_t = \mu_t dt + \sigma_t dW_t + \beta_t dN_t, \tag{2.12}$$

where μ and σ are optional, β is predictable and W is a Wiener process. Let F be a $C^{1,2}$ function. Then the following Itô formula holds.

$$dF(t, X_t) = \left\{ \frac{\partial F}{\partial t}(t, X_t) + \mu_t \frac{\partial F}{\partial x}(t, X_t) + \frac{1}{2}\sigma_t^2 \frac{\partial^2 F}{\partial x^2}(t, X_t) \right\} dt$$
$$+ \sigma_t \frac{\partial F}{\partial x}(t, X_t) dW_t$$
$$+ \{ F(t, X_{t-} + \beta_t) - F(t, X_{t-}) \} dN_t. \tag{2.13}$$

We see that this is just the standard Itô formula, with the added term

$$\{ F(t, X_{t-} + \beta_t) - F(t, X_{t-}) \} dN_t.$$

If t is not a jump time of N, then $dN_t = 0$ so the jump term disappears. If, on the other hand, N has a jump at time t, then $dN_t = 1$ and the jump term $F(t, X_{t-} + \beta_t) - F(t, X_{t-})$ is added.

We can now compare this version of the Itô formula to what we get by doing a naive and straightforward Taylor expansion at $t-$. The first-order terms are

$$\frac{\partial F}{\partial t}(t-, X_{t-})dt + \frac{\partial F}{\partial x}(t-, X_{t-})dX_t,$$

which, by smoothness of F and Lemma 2.12, can be written as

$$\frac{\partial F}{\partial t}(t, X_t)dt + \frac{\partial F}{\partial x}(t, X_{t-})dX_t.$$

By substituting (2.12) we obtain

$$\frac{\partial F}{\partial t}(t, X_t)dt + \frac{\partial F}{\partial x}(t, X_{t-})dX_t$$
$$= \left\{\frac{\partial F}{\partial t}(t, X_t) + \mu_t\frac{\partial F}{\partial x}(t, X_t)\right\}dt + \sigma_t\frac{\partial F}{\partial x}(t, X_t)dW_t + \frac{\partial F}{\partial x}(t, X_{t-})\beta_t dN_t$$
$$= \left\{\frac{\partial F}{\partial t}(t, X_t) + \mu_t\frac{\partial F}{\partial x}(t, X_t)\right\}dt + \sigma_t\frac{\partial F}{\partial x}(t, X_t)dW_t + \frac{\partial F}{\partial x}(t, X_{t-})\Delta X_t,$$

where we have used the fact that $\beta_t dN_t = \Delta X_t$. Comparing this expression to the Itô formula above, and writing $\{F(t, X_{t-} + \beta_t) - F(t, X_{t-})\}dN_t = \Delta F(t, X_t)$, we can rewrite the Itô formula as

$$dF(t, X_t) = \frac{\partial F}{\partial t}(t, X_t)dt + \frac{\partial F}{\partial x}(t, X_{t-})dX_t + \frac{1}{2}\sigma_t^2\frac{\partial^2 F}{\partial x^2}(t, X_t)dt$$
$$+ \left\{\Delta F(t, X_t) - \frac{\partial F}{\partial x}(t, X_{t-})\Delta X_t\right\}.$$

We formulate this as a proposition.

Theorem 2.14 (The Itô formula) *Assume that X has the dynamics*

$$dX_t = \mu_t dt + \sigma_t dW_t + \beta_t dN_t. \tag{2.14}$$

Assume furthermore that F is a $C^{1,2}$ function. Then the following holds:

$$dF(t, X_t) = \frac{\partial F}{\partial t}(t, X_t)dt + \frac{\partial F}{\partial x}(t, X_{t-})dX_t + \frac{1}{2}\sigma_t^2\frac{\partial^2 F}{\partial x^2}(t, X_t)dt$$
$$+ \left\{\Delta F(t, X_t) - \frac{\partial F}{\partial x}(t, X_{t-})\Delta X_t\right\}. \tag{2.15}$$

Remark Note the evaluation of X at X_{t-} in the term $\frac{\partial F}{\partial x}(t, X_{t-})dX_t$. This implies that the process $\frac{\partial F}{\partial t}(t, X_{t-})$ is *predictable*, and thus that any martingale component in X will be integrated to a new martingale.

2.6 Multidimensional Counting Processes

So far we have only considered a single counting process, but now we introduce the multidimensional case.

Definition 2.15   A **k-dimensional counting process** is a vector $(N^1, \ldots, N^k)$ of counting processes.

If $(N^1, \ldots, N^k)$ is a k-dimensional counting process, then it *can* happen that different components have common jumps, so $\Delta N_t^i \cdot \Delta N_t^j$ is not necessarily equal to zero. As we will see in many cases below, the case when a multidimensional counting process is without common jumps is considerably easier to handle than the case with common jumps. We will thus give these processes a name.

Definition 2.16   A **k-variate point process** is a k-dimensional counting process *without common jumps*.

If $(N^1, \ldots, N^k)$ is a multivariate point process (so no common jumps) then we may define a process N by

$$N_t = \sum_{i=1}^{k} N_t^i. \tag{2.16}$$

Since there are no common jumps for $(N^1, \ldots, N^k)$ we see that N is indeed a counting process, so we denote the jump times of N by $T_1, T_2, \ldots$. We can now define a sequence of random variables $\{Z_n\}_n$, where Z_n can take values in $\{1, 2 \ldots, k\}$ as follows

$$Z_n = \sum_{i=1}^{k} i \cdot \Delta N_{T_n}^i. \tag{2.17}$$

In more pedestrian terms we have the following situation.

- Consider T_n, i.e. the jump time number n of the process N.
- Because of the definition of N and the absence of common jumps, exactly one of the processes $N^1, \ldots, N^k$ will have a jump at T_n.
- If the process N^i has a jump at T_n then $Z_n = i$.

We can recover the processes $N^1, \ldots, N^k$ from N and $\{Z_n\}_n$ by the formula

$$N_t^i = \sum_{n=1}^{\infty} I\{T_n \le t\} I\{Z_n = i\}. \tag{2.18}$$

We thus see that a k-variate point process can be described either by the vector process $(N^1, \ldots, N^k)$, or by the double sequence $\{(T_n, Z_n)\}_{n=1}^{\infty}$. We state this as a rather trivial result.

Remark 2.17   Consider a k-variate point process $N^1, \ldots, N^k$. For every $\omega \in \Omega$, the trajectories $t \longmapsto (N_t^1(\omega), \ldots, N_t^k(\omega))$ are uniquely determined by the collection of points

$$\{(T_n(\omega), Z_n(\omega)\}_{n=1}^{\infty},$$

where we note the following facts.

- For every n we have $(T_n(\omega), Z_n(\omega)) \in R_+ \times E$, where $E = \{1, 2, \ldots, k\}$.
- For every t and every ω there is a most one point $(T_n(\omega), Z_n(\omega))$ with $T_n(\omega) = t$. This is because a k-variate point process has no common jumps.

2.7 The Multivariate Itô Formula

We now have the following multivariate extension of the Itô formula. The proof is almost identical to the scalar case.

Proposition 2.18 (The multivariate Itô formula)   *Assume that X has dynamics of the form*

$$dX_t = \mu_t dt + \sigma_t dW_t + \sum_{i=1}^{k} \beta_t^i dN_t^i, \tag{2.19}$$

where $(N^1, \ldots, N^k)$ is a k-dimensional counting process, μ and σ are optinonal, $\beta^1, \ldots, \beta^k$ are predictable and W is a Wiener process. Let F be a $C^{1,2}$ function. Then the following holds.

- *In the general case when there may be common jumps, we have the Itô formula*

$$dF(t, X_t) = \frac{\partial F}{\partial t}(t, X_t)dt + \frac{\partial F}{\partial x}(t, X_{t-})dX_t + \frac{1}{2}\sigma_t^2 \frac{\partial^2 F}{\partial x^2}(t, X_t)dt$$
$$+ \left\{ \Delta F(t, X_t) - \frac{\partial F}{\partial x}(t, X_{t-})\Delta X_t \right\}. \tag{2.20}$$

- *If $(N^1, \ldots, N^k)$ is a multivariate point process (and thus without without common jumps), then the Itô formula can be written as*

$$dF(t, X_t) = \left\{ \frac{\partial F}{\partial t}(t, X_t) + \mu_t \frac{\partial F}{\partial x}(t, X_t) + \frac{1}{2}\sigma_t^2 \frac{\partial^2 F}{\partial x^2}(t, X_t) \right\} dt$$
$$+ \sigma_t \frac{\partial F}{\partial x}(t, X_t)dW_t$$
$$+ \sum_{i=1}^{k} \left\{ F(t, X_{t-} + \beta_t^i) - F(t, X_{t-}) \right\} dN_t^i. \tag{2.21}$$

Note that in the case without common jumps, at most one of the terms dN_t^i will be different from zero.

2.8 The Multidimensional Itô Formula

Let us finally study the multidimensional case, where we consider a vector process $X = (X^1, \ldots, X^n)^\star$, where * denotes transpose. The component X^i has a stochastic differential of the form

$$dX_t^i = \mu_{it} dt + \sum_{j=1}^{d} \sigma_{ijt} dW_t + \sum_{j=1}^{k} \beta_{ijt} dN_t^j,$$

where $W^1, \ldots, W^d$ are d *independent* Wiener processes and $N^1, \ldots, N^k$ are counting processes. We use the notation

$$
\mu = \begin{bmatrix} \mu_1 \\ \vdots \\ \mu_n \end{bmatrix}, \quad W = \begin{bmatrix} W^1 \\ \vdots \\ W^d \end{bmatrix}, \quad N = \begin{bmatrix} N^1 \\ \vdots \\ N^k \end{bmatrix},
$$

$$
\sigma = \begin{bmatrix} \sigma_{11} & \sigma_{12} & \cdots & \sigma_{1d} \\ \sigma_{21} & \sigma_{22} & \cdots & \sigma_{2d} \\ \vdots & \vdots & \ddots & \vdots \\ \sigma_{n1} & \sigma_{n2} & \cdots & \sigma_{nd} \end{bmatrix}, \quad \beta = \begin{bmatrix} \beta_{11} & \beta_{12} & \cdots & \beta_{1k} \\ \beta_{21} & \beta_{22} & \cdots & \beta_{2k} \\ \vdots & \vdots & \ddots & \vdots \\ \beta_{n1} & \beta_{n2} & \cdots & \beta_{nk} \end{bmatrix}.
$$

Then we may write the X-dynamics as

$$
dX_t = \mu_t dt + \sigma_t dW_t + \beta_t dN_t.
$$

We now have the following Itô formula.

Theorem 2.19 (Itô's formula)   *With notation as above, let the n-dimensional process X have dynamics given by*

$$
dX_t = \mu_t dt + \sigma_t dW_t + \beta_t dN_t,
$$

where μ and σ are optional and β is predictable. Let $F : R_+ \times R^n \to R$ be a $C^{1,2}$ mapping. Then the following hold.

- *The process $F(t, X_t)$ has a stochastic differential given by*

$$
dF(t, X_t) = \left\{ \frac{\partial F}{\partial t} + \sum_{i=1}^{n} \mu_i \frac{\partial F}{\partial x_i} + \frac{1}{2} \sum_{i,j=1}^{n} C_{ij} \frac{\partial^2 F}{\partial x_i \partial x_j} \right\} dt + \sum_{i=1}^{n} \frac{\partial F}{\partial x_i} \sigma_i dW_t
$$
$$
+ \{ F(t, X_{t-} + \beta_t \Delta N_t) - F(t-, X_{t-}) \}. \tag{2.22}
$$

  where σ_i is the ith row of σ and the matrix C is defined by

$$
C = \sigma\sigma^\star,
$$

  where $\star$ again denotes transpose.
- *Alternatively, the differential can be written as*

$$
dF(t, X_t) = \frac{\partial F}{\partial t} dt + \sum_{i=1}^{n} \frac{\partial F}{\partial x_i} dX_t^i + \frac{1}{2} \sum_{i,j=1}^{n} C_{ij} \frac{\partial^2 F}{\partial x_i \partial x_j} dt,
$$
$$
+ \left\{ \Delta F(t, X_t) - \sum_{i=1}^{n} \frac{\partial F}{\partial x_i} \Delta X_t^i \right\}. \tag{2.23}
$$

- *If the counting processes have no common jumps, then the Itô formula can be written as*

$$dF(t, X_t) = \left\{ \frac{\partial F}{\partial t} + \sum_{i=1}^{n} \mu_i \frac{\partial F}{\partial x_i} + \frac{1}{2} \sum_{i,j=1}^{n} C_{ij} \frac{\partial^2 F}{\partial x_i \partial x_j} \right\} dt + \sum_{i=1}^{n} \frac{\partial F}{\partial x_i} \sigma_i dW_t$$

$$+ \sum_{j=1}^{k} \left\{ F(t, X_{t-} + \beta_t^j) - F(t-, X_{t-}) \right\} dN_t^j, \qquad (2.24)$$

where β^j is the jth column of the matrix β.

There is one important special case of the Itô formula for processes of bounded variation, namely the product differentiation formula.

Proposition 2.20 *Assume that X and Y are processes of bounded variation (i.e. with no Wiener component). Then the following holds.*

$$d(X_t Y_t) = X_{t-} dY_t + Y_{t-} dX_t + \Delta X_t \Delta Y_t. \qquad (2.25)$$

Proof The proof is left to the reader.   □

2.9   Stochastic Differential Equations

In this section we will apply the Itô formula in order to study stochastic differential equations driven by a counting process. This turns out to be a bit delicate, and there are some serious potential dangers, so we start with a simple example without a driving Wiener process. Let us thus consider a counting process N, a real number x_0 and two real-valued functions $\mu : R \rightarrow R$ and $\beta : R \rightarrow R$.

An obvious task now is to investigate under what conditions on μ and β the SDE

$$\begin{cases} dX_t &= \mu(X_t)dt + \beta(X_t)dN_t, \\ X_0 &= x_0, \end{cases} \qquad (2.26)$$

has an adapted cadlag solution. A very natural, but naive, conjecture is that (2.26) will always possess a solution, as long as μ and β are "nice enough" (such as for example Lipschitz and linear growth). This, however, is *wrong*, and it is very important to understand the following fact.

$$\boxed{\textit{The SDE (2.26) is fundamentally ill-posed}}$$

To understand why this is so, let us consider the dynamics of X at a jump time t of the counting process N. Suppose therefore that N has a jump at time t. Then the X-dynamics becomes

$$\Delta X_t = X_t - X_{t-} = \beta(X_t)dN_t = \beta(X_t), \qquad (2.27)$$

which we can write as

$$X_t = X_{t-} + \beta(X_t). \qquad (2.28)$$

The problem with this formula is that it describes a *non-causal* dynamics. The natural way of modeling the X-dynamics is of course to model it as being generated "causally" by the N-process, in the sense that at a jump time t, the jump size ΔX_t should be uniquely determined by X_{t-} and by dN_t. In (2.27) however, we see that if we are standing at $t-$, the jump size ΔX_t is determined by X_t; i.e., by the value of X *after* the jump. In particular we see that at a jump time t, the value of X_t (given X_{t-}) is being implicitly determined by the non-linear equation (2.28).

By writing down a seemingly innocent expression like (2.26), one may in fact easily end up with completely nonsensical equations for which there is no solution. Indeed, consider for example the simple case when $\alpha \equiv 0$, $\beta(x) = x$ and $x_0 = 1$. We then have the SDE

$$\begin{cases} dX_t &= X_t dN_t, \\ X_0 &= 1. \end{cases}$$

This does not look particularly strange, but at a jump time t, equation (2.28) will now have the form

$$X_t = X_{t-} + X_t,$$

which implies that

$$X_{t-} = 0.$$

This however, is inconsistent with the initial condition $X_0 = 1$ (why?) so the SDE does not have a solution.

From the discussion above it should be clear that the correct way of writing an SDE driven by a counting process is to formulate it as

$$\begin{cases} dX_t &= \mu(X_{t-})dt + \beta(X_{t-})dN_t, \\ X_0 &= x_0, \end{cases}$$

where of course $\mu(X_{t-})$ can be replaced by $\mu(X_t)$ in the dt term. In fact, we have the following result.

Proposition 2.21 *Assume that the ODE*

$$\begin{cases} \dfrac{dX_t}{dt} &= \mu(X_t), \\ X_0 &= x_0, \end{cases}$$

has a unique global solution for every choice of x_0 and let $\beta : R \to R$ be an arbitrarily chosen function. Then the SDE

$$\begin{cases} dX_t &= \mu(X_{t-})dt + \beta(X_{t-})dN_t, \\ X_0 &= x_0, \end{cases}$$

has a unique global solution.

Proof   We have the following concrete algorithm.

1. Denote the jump times of N by $T_1, T_2, \ldots$.

2. For every fixed ω, solve the ODE

$$\begin{cases} \dfrac{dX_t}{dt} &= \mu(X_t), \\ X_0 &= x_0, \end{cases}$$

on the half-open interval $[0, T_1)$. In particular we have now determined the value of X_{T_1-}.

3. Calculate the value of X_{T_1} by the formula

$$X_{T_1} = X_{T_1-} + \beta(X_{T_1-}).$$

4. Given X_{T_1} from the previous step, solve the ODE

$$\frac{dX_t}{dt} = \mu(X_t)$$

on the interval $[T_1, T_2)$. This will give us X_{T_2-}.

5. Compute X_{T_2} by the formula

$$X_{T_2} = X_{T_2-} + \beta(X_{T_2-}).$$

6.

Continue by induction. □

We illustrate this methodology by solving a concrete SDE, namely the counting process analogue to geometrical Brownian motion

$$\begin{cases} dX_t &= \alpha X_{t-} dt + \beta X_{t-} dN_t, \\ X_0 &= x_0, \end{cases} \tag{2.29}$$

where α and β are real numbers.

To solve this SDE we note that up to the first jump time T_1 we have the ODE

$$\begin{cases} \dfrac{dX_t}{dt} &= \alpha X_t, \\ X_0 &= x_0, \end{cases}$$

with the exponential solution

$$X_t = e^{\alpha t} x_0,$$

so in particular we have $X_{T_1-} = e^{\alpha T_1} x_0$. The jump size at T_1 is given by

$$\Delta X_{T_1} = \beta X_{T_1-},$$

so we have

$$X_{T_1} = X_{T_1-} + \beta X_{T_1-} = (1+\beta) X_{T_1-} = (1+\beta) e^{\alpha T_1} x_0.$$

We now solve the ODE

$$\begin{cases} \dfrac{dX_t}{dt} &= \alpha X_t, \\ X_{T_1} &= (1+\beta) e^{\alpha T_1} x_0, \end{cases}$$

for $t \in [T_1, T_2)$ to obtain

$$X_t = e^{\alpha(t-T_1)}(1+\beta)e^{\alpha T_1}x_0 = e^{\alpha t}(1+\beta)x_0$$

and in particular

$$X_{T_2-} = (1+\beta)e^{\alpha T_2}x_0.$$

As before, the the jump condition gives us

$$X_{T_2} = X_{T_2-} + \beta X_{T_2-} = (1+\beta)X_{T_2-} = (1+\beta)^2 e^{\alpha T_2}x_0.$$

Continuing in this way we see that the solution is given by

$$X_t = x_0(1+\beta)^{N_t}e^{\alpha t}.$$

We may in fact generalize this result as follows.

Proposition 2.22 *Assume that X satisfies the SDE*

$$\begin{cases} dX_t &=& \alpha_t X_{t-}dt + \beta_t X_{t-}dN_t, \\ X_0 &=& x_0, \end{cases}$$

where α and β are predictable processes. Then X can be represented as

$$X_t = x_0 e^{\int_0^t \alpha_s ds} \prod_{T_n \leq t} (1+\beta_{T_n}).$$

If $\beta > -1$, this can also be written as

$$X_t = x_0 e^{\int_0^t \alpha_s ds + \int_0^t \ln(1+\beta_s)dN_s}.$$

2.10   The Watanabe Characterization Theorem

The aim in this section is to prove the Watanabe characterization theorem for the Poisson process. Before we do this, we make a slight extension of the definition of a Poisson process.

Definition 2.23 Let $(\Omega, \mathcal{F}, P)$ be a probability space with a given filtration $\mathbf{F} = \{\mathcal{F}_t\}_{t \geq 0}$ and let $t \rightarrow \lambda_t$ be a deterministic function of time. A counting process N is a **Poisson process with intensity function λ with respect to the filtration F** if it satisfies the following conditions.

1. N is adapted to **F**.
2. For all $s \leq t$ the random variable $N_t - N_s$ is independent of $\mathcal{F}_s$.
3. For all $s \leq t$, the conditional distribution of the increment $N_t - N_s$ is given by

$$P(N_t - N_s = n|\mathcal{F}_s) = e^{-\Lambda_{s,t}}\frac{(\Lambda_{s,t})^n}{n!}, \quad n = 0,1,2,\ldots, \quad (2.30)$$

where

$$\Lambda_{s,t} = \int_s^t \lambda_u du. \quad (2.31)$$

We have the following easy result concerning the characteristic function for a Poisson process.

Lemma 2.24   *For a Poisson process as above, the following holds for all $s < t$:*

$$E\left[e^{iu(N_t - N_s)}\middle|\mathcal{F}_s\right] = e^{\Lambda_{s,t}(e^{iu}-1)} \qquad (2.32)$$

With the definition above it is easy to see that the process X defined by

$$N_t - \int_0^T \lambda_s ds,$$

is an **F**-martingale. The Watanabe theorem says that this martingale property of X is not only a consequence of, but in fact characterizes, the Poisson process among the class of counting processes.

Theorem 2.25 (The Watanabe Characterization Theorem)   *Assume that N is a counting process and that $t \to \lambda_t$ is a deterministic function. Assume furthermore that the process M, defined by*

$$M_t = N_t - \int_0^t \lambda_s ds, \qquad (2.33)$$

*is an **F**-martingale. Then N is Poisson with respect to **F** with intensity function λ.*

Proof   Using a slight extension of Proposition 1.3, it is enough to show that

$$E\left[e^{iu(N_t - N_s)}\middle|\mathcal{F}_s\right] = e^{\Lambda_{s,t}(e^{iu}-1)}, \qquad (2.34)$$

with Λ as in (2.31). We start by proving this for the simpler case when $s = 0$. We thus want to show that

$$E\left[e^{iuN_t}\right] = \exp\left\{\Lambda_{0,t}\left(e^{iu} - 1\right)\right\}.$$

It is now natural to define the process Z by

$$Z_t = e^{iuN_t},$$

and an application of the Itô formula of Proposition 2.13 immediately gives us

$$dZ_t = \left\{e^{iu(N_{t-}+1)} - e^{iuN_{t-}}\right\} dN_t = e^{iuN_{t-}}\left\{e^{iu} - 1\right\} dN_t.$$

We now use the relation

$$dN_t = \lambda_t dt + dM_t,$$

where the martingale M is defined by (2.33) to obtain

$$dZ_t = Z_{t-}\left\{e^{iu} - 1\right\}\lambda_t dt + Z_{t-}\left\{e^{iu} - 1\right\} dM_t.$$

Integrating this over $[0,t]$ we obtain (using the fact that $Z_0 = 1$)

$$Z_t = 1 + \left\{e^{iu} - 1\right\}\int_0^t Z_{s-}\lambda_s ds + \left\{e^{iu} - 1\right\}\int_0^t Z_{s-}\lambda_s dM_s.$$

Since M is a martingale and the integrand Z_{s-} is predictable (left-continuity!) the dM integral is also a martingale so, after taking expectations, we obtain

$$E\left[Z_t\right] = 1 + \left\{e^{iu} - 1\right\} \int_0^t E\left[Z_{s-}\right] \lambda_s ds.$$

Let us now, for a fixed u, define the deterministic function y by

$$y_t = E\left[e^{iuN_t}\right] = E\left[Z_t\right].$$

We thus have

$$y_t = 1 + \left\{e^{iu} - 1\right\} \int_0^t y_{s-} \lambda_s ds,$$

and since we are integrating over Lebesgue measure we can (why?) write this as

$$y_t = 1 + \left\{e^{iu} - 1\right\} \int_0^t y_s \lambda_s ds.$$

Taking the derivative with respect to t we obtain the ODE

$$\begin{cases} \dfrac{dy_t}{dt} &= y_t \left(e^{iu} - 1\right) \lambda_t, \\ y_0 &= 1, \end{cases}$$

with the solution

$$y_t = e^{\left(e^{iu} - 1\right) \int_0^t \lambda_s ds}.$$

This proves (2.34) for the special case when $s = 0$. For the general case, it is clearly enough to show (why?) that

$$E\left[I_A e^{iu(N_t - N_s)}\right] = E\left[I_A\right] e^{\Lambda_{s,t}\left(e^{iu} - 1\right)},$$

for every $A \in \mathcal{F}_s$. To do this we now define, for fixed s, u and A, the process Z on the time interval $[s, \infty)$ by

$$Z_t = I_A e^{iu(N_t - N_s)},$$

and basically copy the argument above.   □

We end this section with a an easy but useful result.

Proposition 2.26 *Assume that N is Poisson with intensity λ and that the predictable process g is integrable or non-negative. Then the following holds:*

$$E\left[\int_0^t g_s dN_s\right] = \int_0^t E\left[g_s\right] \lambda_s ds. \tag{2.35}$$

Proof Since $N_t - \int_0^t \lambda_s ds$ is a martingale and g is predictable, we have

$$E\left[\int_0^t g_s \left\{dN_s - \lambda_s ds\right\}\right] = 0.$$

The Fubini theorem does the rest.   □

2.11   The Laplace Functional

We recall that, for a random variable X, the characteristic function

$$\varphi_X(u) = E\left[e^{iuX}\right]$$

completely characterizes the distribution of X. For a counting process, the same role is played by the Laplace functional. With this in mind, let N be a counting process and denote by $\mathcal{R}$ all non-negative Borel functions $f : R \rightarrow R_+$.

Definition 2.27   The **Laplace functional** is a mapping

$$\mathbf{L}^N : \mathcal{R} \rightarrow R$$

defined by

$$\mathbf{L}^N(f) = E\left[e^{-\int_0^\infty f_t \, dN_t}\right], \tag{2.36}$$

so

$$\mathbf{L}^N(f) = E\left[e^{-\sum_{n=1}^\infty f(T_n)}\right],$$

where $\{T_n\}_{n=1}^\infty$ are the jump times of N.

For a multivariate counting process $N = (N^1, \ldots, N^k)$ we define the Laplace functional $\mathbf{L}^N : \mathcal{R}^k \rightarrow R$ by the formula

$$\mathbf{L}^N(f) = E\left[e^{-\sum_{i=1}^k \int_0^\infty f_t^i \, dN_t^i}\right], \tag{2.37}$$

which in vector notation reads as

$$\mathbf{L}^N(f) = E\left[e^{-\int_0^\infty f_t \, dN_t}\right].$$

In what follows we will often also use the notation

$$\mathbf{L}_t^N(f) = E\left[e^{-\int_0^t f_s \, dN_s}\right]. \tag{2.38}$$

We now have some easy results concerning the distribution of a counting process. For any interval I of the type $I = (s, t]$ we use the notation

$$\Delta N_I = N_t - N_s,$$

and for any counting process N we denote the distribution of N by $\mathcal{L}(N)$.

Proposition 2.28   *Consider two counting processes N and M. The following statements are equivalent.*

(a) $\mathcal{L}(N) = \mathcal{L}(M)$.

(b) $\mathcal{L}(N_{t_1}, \ldots, N_{t_n}) = \mathcal{L}(M_{t_1}, \ldots, M_{t_n})$, *for all finite collections of points $t_1 < t_2 \cdots < t_n$.*

(c) $\mathcal{L}(\Delta N_{I_1}, \ldots, \Delta N_{I_n}) = \mathcal{L}(\Delta M_{I_1}, \ldots, \Delta M_{I_n})$, *for all finite sets of non overlapping intervals $I_1, \ldots, I_n$.*

(d) $\mathbf{L}^N(f) = \mathbf{L}^M(f)$, *for all $f \in \mathcal{R}$.*

In particular, the distribution of a counting process N is completely determined by the Laplace functional $\mathbf{L}^N$*, and two counting processes N and M are independent if and only if*

$$\mathbf{L}^{N,M}(f,g) = \mathbf{L}^N(f) \cdot \mathbf{L}^M(g)$$

for all $f, g \in \mathcal{R}$.

Proof The equivalence between (a) and (b) is standard and the equivalence between (b) and (c) is obvious. By letting f be the indicator $f(t) = I_E(t)$ where $E = \bigcup_{i=1}^N I_i$, it is clear that (d) implies (c), and (d) follows from (c) by a monotone class argument. The final statements are obvious. $\square$

As an example we compute the Laplace functional of a Poisson process with intensity function λ. To this end we define, for any $f \in \mathcal{R}$, the process X by

$$X_t = e^{-\int_0^t f_s \, dN_s}$$

From Itô's formula we obtain

$$dX_t = X_{t-}\left(e^{-f_t} - 1\right) dN_t,$$
$$X_0 = 1$$

which we can write as

$$dX_t = X_{t-}\left(e^{-f_t} - 1\right)\{dN_t - \lambda_t dt\} + X_{t-}\left(e^{-f_t} - 1\right)\lambda_t dt.$$

Since f and λ are deterministic, the integrands are predictable, so the first term is a martingale increment. Integrating, taking expectations and using the notation $y_t = E[X_t]$, we obtain

$$y_t = 1 + \int_0^t y_{s-}\left(e^{-f_s} - 1\right)\lambda_s \, ds.$$

Since we are integrating against Lebesgue measure, we can write this as

$$y_t = 1 + \int_0^t y_s \left(e^{-f_s} - 1\right)\lambda_s \, ds$$

which can be written as the ODE

$$\dot{y}_t = y_t \left(e^{-f_t} - 1\right)\lambda_t,$$
$$y_0 = 1$$

with solution

$$y_t = e^{\int_0^t (e^{-f_s} - 1)\lambda_s \, ds}.$$

We have thus proved the following result.

Proposition 2.29 *If N is Poisson with intensity function* λ*, then the Laplace functional is given by*

$$\mathbf{L}^N(f) = e^{-\int_0^\infty (1 - e^{-f(t)})\lambda_t \, dt}. \tag{2.39}$$

2.12 Exercises

Exercise 2.1 Show that the SDE

$$
\begin{cases}
dX_t &= aX_t dt + \beta dN_t, \\
X_0 &= x_0.
\end{cases}
$$

where a, β and x_0 are real numbers, and N is a counting process, has the solution

$$
X_t = e^{at} x_0 + \beta \int_0^t e^{a(t-s)} dN_s.
$$

Exercise 2.2 Consider the SDE of the previous exercise and assume that N is Poisson with constant intensity λ. Compute $E[X_t]$.

Exercise 2.3 Consider the following SDEs, where N^x and N^y are counting processes without common jumps, and where the parameters $\alpha_X, \alpha_Y, \beta_X, \beta_Y$ are known constants.

$$
dX_t = \alpha_X X_t dt + \beta_X X_{t-} dN_t^x,
$$
$$
dY_t = \alpha_Y Y_t dt + \beta_Y Y_{t-} dN_t^y.
$$

Define the process Z by $Z_t = X_t Y_t$. Then Z will satisfy an SDE. Find this SDE and compute $E[Z_t]$ in the case when N^x and N^y are Poisson with intensities λ_x and λ_y.

Exercise 2.4 Consider the SDEs of the previous exercise. Define the process Z by $Z_t = X_t/Y_t$. Then Z will satisfy an SDE. Find this SDE and compute $E[Z_t]$ in the case when N^x and N^y are Poisson with intensities λ_x and λ_y.

Exercise 2.5 Consider two discrete-time processes X and Y. Prove the product formula

$$
\Delta(XY)_n = X_{n-1}\Delta Y_n + Y_{n-1}\Delta X_n + \Delta X_n \Delta Y_n.
$$

Exercise 2.6 Consider two discrete-time processes X and Y which are both of bounded variation (i.e. they have no driving Wiener process). Use the Itô formula to prove the product formula

$$
d(XY)_t = X_{t-} dY_t + Y_{t-} dX_t + \Delta X_t \Delta Y_t.
$$

As usual, $\Delta X_t = X_t - X_{t-}$ etc.

3   More on Poisson Processes

In this chapter we will use stochastic calculus and the Watanabe theorem in order to derive some well-known results concerning the Poisson process.

3.1   Independence and Absence of Common Jumps

In this section we study the relation between independence and absence of common jumps for two Poisson processes.

Proposition 3.1   *Consider a filtered probability space* $(\Omega, \mathcal{F}, P, \mathbf{F})$. *Assume that N and M are* $(P, \mathbf{F})$-*Poisson processes with intensities* λ *and* μ *respectively. Assume furthermore that N and M are independent. Then they have no common jumps, i.e. we have*

$$\Delta N_t \cdot \Delta M_t = 0, \quad P - a.s. \quad \text{for all } t. \tag{3.1}$$

Proof   It is enough to show that, for every T we have

$$E\left[\sum_{t \leq T} \Delta N_t \cdot \Delta M_t\right] = 0, \tag{3.2}$$

where the sum P-a.s has a finite number of terms.

In order to prove this we choose an arbitrary T and consider the process $X_t = N_t \cdot M_t$. We then compute $E[N_T \cdot M_T]$ in two different ways and compare the results.

By the assumed independence we trivially have

$$E[N_T \cdot M_T] = E[N_T] \cdot E[M_T] = \lambda \mu T^2. \tag{3.3}$$

On the other hand, we can apply the Itô formula to obtain

$$dX_t = N_{t-}dM_t + M_{t-}dN_t + \Delta N_t \cdot \Delta M_t,$$

and we can write this as

$$dX_t = N_{t-}[dM_t - \mu dt] + M_{t-}[dN_t - \lambda dt] + N_{t-}\mu dt + M_{t-}\lambda dt + \Delta N_t \cdot \Delta M_t.$$

In the first two terms we have a predictable process multiplying a martingale differential, so both these terms represent martingale differentials. If we integrate the formula over $[0, t]$ and take expectations, the martingale parts will vanish and we obtain

$$E[N_T \cdot M_T] = E\left[\int_0^T [\mu N_{t-} + \lambda M_{t-}] \, dt\right] + E\left[\sum_{t \leq T} \Delta N_t \cdot \Delta M_t\right].$$

Using Fubini we have

$$E\left[\int_0^T [\mu N_{t-} + \lambda M_{t-}] \, dt\right] = \int_0^T \{\mu E \, [N_{t-}] + \lambda E \, [M_{t-}]\} \, dt$$

$$= \int_0^T \{\mu \lambda t + \lambda \mu t\} \, dt = \lambda \mu T^2.$$

We have thus shown that

$$E \, [N_T \cdot M_T] = \lambda \mu T^2 + E\left[\sum_{t \leq T} \Delta N_t \cdot \Delta M_t\right]$$

and, comparing this with (3.3), gives us (3.2).   □

We have thus proved that, for Poisson processes, independence implies absence of common jumps and from an intuitive point of view, this is not at all surprising. What is much more surprising is that there is a converse result.

Proposition 3.2   *Consider a filtered probability space* $(\Omega, \mathcal{F}, P, \mathbf{F})$. *Assume that N and M are* $(P, \mathbf{F})$*-Poisson processes with intensities* λ *and* μ *respectively. Assume furthermore that N and M have no common jumps. Then N and M are independent.*

Proof   For any non-negative functions $f, g : R \to R$ we define the Laplace functionals $\mathbf{L}_t^N(f)$, $\mathbf{L}_t^M(g)$, and $\mathbf{L}_t^{N,M}(f,g)$ by

$$\mathbf{L}_t^N(f) = E\left[e^{-\int_0^t f_s \, dN_s}\right],$$

$$\mathbf{L}_t^M(g) = E\left[e^{-\int_0^t g_s \, dM_s}\right],$$

$$\mathbf{L}_t^{N,M}(f,g) = E\left[e^{-\int_0^t f_s \, dN_s - \int_0^t g_s \, dM_s}\right].$$

In order to prove the proposition it is enough to show that

$$\mathbf{L}_t^{N,M}(f,g) = \mathbf{L}_t^N(f) \cdot \mathbf{L}_t^M(g), \qquad (3.4)$$

for all f, g and t. We will now prove (3.4) by studying the process

$$X_t = e^{-\int_0^t f_s \, dN_s - \int_0^t g_s \, dM_s}$$

for an arbitrary choice of f and g. We will use the Itô formula and in order to do this we define the processes Y and Z by

$$Y_t = \int_0^t f_s \, dN_s, \qquad Z_t = \int_0^t g_s \, dM_s$$

with stochastic differentials given by

$$dY_t = f_t \, dN_t, \qquad dZ_t = g_t \, dM_t.$$

We thus have

$$\mathbf{L}_t^{N,M}(f,g) = E \, [X_t],$$

where

$$X_t = e^{-(Y_t + Z_t)}.$$

Since Y and Z have no common jumps, we may use version (2.24) of the Itô formula to obtain

$$dX_t = X_{t-}\left(e^{-f_t} - 1\right) dN_t + X_{t-}(e^{-g_t} - 1)\, dM_t.$$

Integrating this gives us

$$X_t = 1 + \int_0^t X_{s-}\left(e^{-f_s} - 1\right) dN_s + \int_0^t X_{s-}(e^{-g_s} - 1)\, dM_s.$$

Taking expectations, using Proposition 2.26 and defining the deterministic function h by $h_t = E[X_t]$, yields

$$h_t = 1 + \int_0^t h_{s-}\left(e^{-f_s} - 1\right) \lambda ds + \int_0^t h_{s-}(e^{-g_s} - 1)\, \mu ds.$$

Since we are integrating with respect to Lebesgue measure we can write this as

$$h_t = 1 + \int_0^t h_s\left(e^{-f_s} - 1\right) \lambda ds + \int_0^t h_s(e^{-g_s} - 1)\, \mu ds.$$

Taking the t-derivative gives us the ODE

$$\dot{h}_t = h_t \left[\lambda\left(e^{-f_t} - 1\right) + \mu(e^{-g_t} - 1)\right],$$
$$h_0 = 1$$

with solution

$$h_t = e^{\lambda \int_0^t (e^{-f_s} - 1)ds + \mu \int_0^t (e^{-g_s} - 1)ds}.$$

We have thus shown that

$$L_t^{N,M}(f,g) = e^{\lambda \int_0^t (e^{-f_s} - 1)ds + \mu \int_0^t (e^{-g_s} - 1)ds}$$

and as special cases we have

$$L_t^N(f) = e^{\lambda \int_0^t (e^{-f_s} - 1)ds}, \qquad L_t^M(g) = e^{\mu \int_0^t (e^{-g_s} - 1)ds}.$$

Since we obviously have

$$L_t^{N,M}(f,g) = L_t^N(f) \cdot L_t^M(g),$$

the proposition is proved.   □

Note that the assumptions of Proposition 3.2 require N and M to be Poisson with respect to the common filtration $\mathbf{F}$. We obviously have $\mathcal{F}_t^N \subseteq \mathcal{F}_t$ and $\mathcal{F}_t^M \subseteq \mathcal{F}_t$, so this assumption is stronger than the assuming M and N to be Poisson with respect to the respective internal filtrations. It is therefore natural to study the situation when we assume no common jumps, but weaken the measurability assumptions by only requiring that N is $\mathbf{F}^N$-Poisson and that M is $\mathbf{F}^M$-Poisson. A natural conjecture is perhaps that these weaker assumptions will also imply independence. This conjecture is in fact not true, as the following counterexample shows.

Consider a Poisson process N with unit intensity, and an independent random variable τ with unit intensity exponential distribution. We view τ as a random point in time, and now define the process M as follows:

$$M_t = \begin{cases} 0 & \text{if } 0 \leq t < \tau, \\ 1 + N_{t-\tau} & \text{if } t \geq \tau. \end{cases}$$

In everyday language this means that if we reset the time and space variables to zero at time τ then M will have exactly the same trajectory as N. It is then more or less obvious that M and N will have no common jumps, but it is also obvious that they are not independent.

3.2 Adding Poisson Processes

In this section and the next we will study the effect of adding and subtracting points in a point process. We start with a simple result on addition.

Proposition 3.3 *Assume that X and Y are independent $(P, \mathbf{F})$-Poisson processes with intensities λ and μ respectively. If the process N is defined by*

$$N_t = X_t + Y_t,$$

then N is $(P, \mathbf{F})$-Poisson with intensity $\lambda + \mu$.

Proof Since X and Y are independent, they have no common jumps. N will thus have jumps of unit size, so N is really a counting process. Using Watanabe we only have to show that the process M defined by $M_t = X_t + Y_t - (\lambda + \mu)t$ is a martingale. This is, however, obvious, since M is the sum of the two martingales $X_t - \lambda t$ and $Y_t - \mu t$. $\quad\square$

3.3 Thinning a Poisson Process

Let N be an arbitrary point process with jump times $\{T_n\}_{n=1}^{\infty}$. Then we can write

$$N_t = \sum_{n=1}^{\infty} I\{T_n \leq t\}.$$

where I is the indicator function. Let furthermore $\{Z_n\}_{n=1}^{\infty}$ be a sequence of (not necessarily i.i.d.) random variables taking the values 0 or 1.

Definition 3.4 The process N^Z defined by

$$N_t^Z = \sum_{n=1}^{\infty} Z_n \cdot I\{T_n \leq t\} \tag{3.5}$$

is called a **thinning** of N.

In everyday language this means that if $Z_n = 1$ we keep the jump time T_n, but if $Z_n = 0$ we delete the jump time T_n.

It is clear that N^Z is a counting process, and we have the following result on independent thinning of a Poisson process.

Proposition 3.5   *Let N be a Poisson process with intensity λ and let $\{Z_n\}_{n=1}^{\infty}$ be an i.i.d. sequence of Bernoulli variables with $P(Z_n = 1) = p$. Assume furthermore that N and $\{Z_n\}_{n=1}^{\infty}$ are independent. Then the thinned process N^Z is an $\mathbf{F}^Z$-Poisson process with intensity $\lambda^Z = p \cdot \lambda$, where $\mathbf{F}^Z$ is shorthand notation for the internal filtration $\mathbf{F}^{N^Z}$.*

Proof   We will use the Laplace functional. By definition we have, for any counting process.

$$\mathbf{L}^{N^Z}(f) = E\left[e^{-\sum_{n=1}^{\infty} f(T_n)Z_n}\right],$$

which we may write as

$$\mathbf{L}^{N^Z}(f) = E\left[\prod_n^{\infty} e^{-f(T_n)Z_n}\right].$$

Conditioning on $\mathcal{F}_{\infty}^N$ and using the independence we obtain

$$\mathbf{L}^{N^Z}(f) = E\left[\prod_{n=1}^{\infty}\left\{pe^{-f(T_n)} + 1 - p\right\}\right] = E\left[e^{\sum_{n=1}^{\infty} \ln\{pe^{-f(T_n)}+1-p\}}\right].$$

Comparing this with the formula

$$L^N(f) = E\left[e^{-\sum_{n=1}^{\infty} f(T_n)}\right]$$

we have thus showed that

$$\mathbf{L}^{N^Z}(f) = \mathbf{L}^N\left(-\ln\left\{pe^{-f} + 1 - p\right\}\right). \tag{3.6}$$

We now recall from Proposition 2.29 that if N is Poisson we have

$$L^N(g) = e^{-\int_0^{\infty}\left(1-e^{-g(t)}\right)\lambda_t\, dt}.$$

Setting $g = -\ln\left\{pe^{-f} + 1 - p\right\}$ gives us

$$\mathbf{L}^{N^Z}(f) = e^{-\int_0^{\infty}\left(1-e^{-f(t)}\right)p\lambda_t\, dt},$$

showing that N^Z is Poisson with intensity $p\lambda$. □

We can in fact improve on this result, so assume again that we have a Poisson process with intensity λ and that we thin it with probability p. Denote the thinned process by N^p. We can then define the process N^q (with $q = 1 - p$) by

$$N_t^q = N_t - N_t^p.$$

Concretely this means that at any jump time T_n we allocate the jump time either to N^p with probability p or to N^q with probability q. Then the following holds.

Proposition 3.6   *With assumptions as above, N^p and N^q are independent Poisson processes with intensities λp and λq respectively.*

Proof That N^p and N^q are Poisson with the given intensities follows from our previous result. Independence then follows from the fact that they (by definition) have no common jumps, and from Proposition 3.2. □

3.4 Marking a Counting Process

In this section we extend the thinning procedure of the previous section. Assume therefore that N is a counting process and that $\{Z_n\}$ are random variables taking values in the set $\{1, 2, \ldots, K\}$.

Definition 3.7 For every $i = 1, \ldots, k$ define the counting process N^i by

$$N_t^i = \sum_{n=1}^{\infty} I\{T_n \leq t\} \cdot I\{Z_n = i\}.$$

The multivariate process N^Z defined by $N^Z = (N^1, \ldots, N^K)$ is then said to be obtained from N by Z-marking.

We have the following nice result for independent marking of a Poisson process.

Proposition 3.8 *Assume that $N^Z = (N^1, \ldots, N^K)$ is obtained from a Poisson process N with intensity λ, where $\{Z_n\}$ are i.i.d. and independent of N and where $P(Z_n = i) = p_i$. Then the following hold.*

- *The process N^i is $\mathbf{F}^{N^Z}$-Poisson with intensity $p_i\lambda$, for all $i = 1, \ldots, K$.*
- *The processes $N^1, \ldots, N^K$ are independent.*

Proof The first item follows easily, by using exactly the same arguments as for Proposition 3.5. Thus the processes $N^1, \ldots, N^K$ are $\mathbf{F}^{N^Z}$-Poisson and, by definition, they have no common jumps. It then follows from Proposition 3.2 that $N^1, \ldots, N^K$ are independent. □

4 Counting Processes with Stochastic Intensities

In this section we will generalize the concept of an intensity from the Poisson case to the case of a fairly general counting process. We consider a filtered probability space $(\Omega, \mathcal{F}, P, \mathbf{F})$ carrying an optional counting process N.

4.1 Definition of Stochastic Intensity

Definition 4.1 Suppose we are given an optional counting process N on the filtered space $(\Omega, \mathcal{F}, P, \mathbf{F})$. Let λ be a non-negative optional random process such that

$$\int_0^t \lambda_s \, ds < \infty, \quad \text{for all } t \geq 0. \tag{4.1}$$

If the condition

$$E\left[\int_0^\infty h_t \lambda_t \, dt\right] = E\left[\int_0^\infty h_t \, dN_t\right] \tag{4.2}$$

holds for every non-negative predictable process h, then we say that N has the **F-intensity** λ.

Note the requirement of predictability of h above. At first sight, this definition may look rather strange, but the intuitive interpretation is easy: Modulo integrability, it says that the difference $dN_t - \lambda_t \, dt$ is a martingale increment. This is clear from the following result.

Proposition 4.2 *Assume that N has the **F**-intensity λ and that N is integrable, in the sense that*

$$E[N_t] < \infty, \quad \text{for all } t \geq 0. \tag{4.3}$$

Then the process M defined by

$$M_t = N_t - \int_0^t \lambda_s \, ds$$

*is an **F**-martingale.*

Proof Fix s and t, with $s < t$, and choose an arbitrary event $A \in \mathcal{F}_s$. If we now define the process h by

$$h_u(\omega) = I_A(\omega) I\{s < u \leq t\},$$

then h is non-negative and predictable (why?). With this choice of h, the relation (4.2) becomes

$$E\left[I_A \int_s^t \lambda_u du\right] = E\left[I_A (N_t - N_s)\right].$$

Because of (4.3) we may now subtract the left-hand side from the right-hand side without any risk of expressions of the type $+\infty - \infty$. The result is

$$E\left[I_A (M_t - M_s)\right] = 0,$$

which shows that M is a martingale. □

Remark The reason why we define the intensity concept by the condition (4.2), rather than by the martingale property of M above, is that (4.2) also covers the case when $E[N_t] = \infty$.

We now have a number of obvious questions to answer.

- Does every counting process have an intensity?
- Is the intensity unique?
- How does the intensity depend on the filtration $\mathbf{F}$?
- What is the intuitive interpretation of λ?

4.2 Existence of an Intensity Process

If N is an optional counting process, then N is of course increasing, so N is (in a trivial way) a local submartingale. We may thus refer to the Doob–Meyer decomposition theorem 7.9 to deduce the existence of an increasing predictable process A with $A_0 = 0$, such that the process M, defined by

$$M_t = N_t - A_t,$$

is a local martingale. In this setting, the process A is referred to as the **compensator** of N. In the general case there is no guarantee that A is even continuous, but if we *assume* that the compensator A is absolutely continuous with respect to Lebesgue measure, then we can write A as

$$A_t = \int_0^t \lambda_s ds,$$

for some non-negative optional process λ, and this λ is of course our intensity process. We thus see that only those counting processes for which the compensator is absolutely continuous will possess an intensity. Furthermore one can show that if a counting process has an intensity, then the distribution of every jump time will have a density with respect to Lebesgue measure. This implies that if we restrict ourselves (as we will do for the rest of the text) to counting processes with intensities then we are basically excluding counting processes with jumps at predetermined points in time.

4.3   Uniqueness

From Definition 4.1 it should be clear that we can *not* expect the intensity process λ to be unique. Indeed, suppose for example that λ is an intensity for N and that λ is cadlag. If we now define the process μ by

$$\mu_t = \lambda_{t-},$$

then it is clear that

$$E\left[\int_0^\infty h_t\lambda_t dt\right] = E\left[\int_0^\infty h_t\mu_t dt\right]$$

for all predictable h, so μ is also an intensity. If, however, we require *predictability*, then we have uniqueness.

Proposition 4.3   *Assume that N has an* **F** *intensity λ^*. Then N will also possess an* **F**-*predictable intensity λ. Furthermore, λ is unique in the sense that if μ is another predictable intensity, then we have*

$$\mu_t(\omega) = \lambda_t(\omega), \quad dPdN_t - a.e.$$

Proof   The formal proof is rather technical and left out. The intuitive idea behind the proof is however very easy to understand. We simply define the process λ by

$$\lambda_t = E\left[\lambda_t^*|\mathcal{F}_{t-}\right],$$

and since λ_t is clearly $\mathcal{F}_{t-}$-measurable for each t, we see that λ is predictable, thus proving the existence of a predictable intensity.

This is in fact, where the formal proof gets technical. The object λ defined above is not really defined as a *bona fide* random process. Instead we have defined λ_t as an equivalence class of random variables for each t, and the problem is to show that we can choose one member of each equivalence class and "glue" these together in such a way that we obtain a predictable process. This is a very technical problem and the reader is referred to to Jacod & Shiryaev (1987).

To show uniqueness $dPdN_t$-a.e., it enough to show that for every predictable non-negative process h we have

$$E\left[\int_0^\infty h_t\lambda_t dN_t\right] = E\left[\int_0^\infty h_t\mu_t dN_t\right].$$

If, in the left-hand side, we use the assumption that N has the intensity μ and on the right-hand side use the fact that N has the intensity λ we see that both sides equal

$$E\left[\int_0^\infty h_t\mu_t\lambda_t dt\right]. \qquad \qquad \square$$

4.4   Interpretation

We now go on the intuitive interpretation of the intensity concept. Let us thus assume that N has the predictable intensity process λ. Modulo integrability, this implies that

$$dN_t - \lambda dt$$

is a martingale increment and heuristically we will thus have

$$E\left[dN_t - \lambda dt \mid \mathcal{F}_{t-}\right] = 0.$$

Since λ is predictable we have $\lambda_t \in \mathcal{F}_{t-}$ so we can move $\lambda_t dt$ outside the expectation and obtain

$$E\left[dN_t \mid \mathcal{F}_{t-}\right] = \lambda_t dt.$$

or, equivalently.

$$P\left(dN_t = 1 \mid \mathcal{F}_{t-}\right) = \lambda_t dt.$$

We thus see that the predictable intensity λ has the interpretation that λ_t is, on the infinitesimal time scale, the conditional probability of a jump per unit time, or, equivalently, the conditional expected number of of jumps per unit time. The difference between this situation and the Poisson case is that we now allow the intensity to be a random process.

Since we know that the predictable intensity is unique, we can summarize the moral of this section so far in the following slogan:

The natural description of the dynamics for a counting process N is in terms of its *predictable intensity* λ, with the interpretation

$$E\left[dN_t \mid \mathcal{F}_{t-}\right] = \lambda_t dt. \tag{4.4}$$

4.5   Dependence on the Filtration

It is important to note that the intensity concept is tied to a particular choice of filtration. If we have two different filtrations **F** and **G**, and a counting process which is optional with respect to to both **F** and **G**, then there is no reason to believe that the **F**-intensity λ^F will coincide with the **G**-intensity λ^G. In the general case there are no interesting relations between λ^F and λ^G, but in the special case when **G** is a sub-filtration of **F**, we have a very precise result.

Proposition 4.4   *Assume that N has the predictable $\mathbf{F}$-intensity λ^F, and that we are given a filtration $\mathbf{G}$ such that*

$$\mathcal{G}_t \subseteq \mathcal{F}_t, \quad \text{for all } t \geq 0.$$

Then there exists a predictable $\mathbf{G}$-intensity λ^G with the property that

$$\lambda_t^G = E\left[\lambda_t^F \mid \mathcal{G}_{t-}\right].$$

Proof Using the intuitive interpretation (4.4) the result follows at once from the calculation

$$\lambda_t^G = E\left[dN_t | \mathcal{G}_{t-}\right] = E\left[E\left[dN_t | \mathcal{F}_{t-}\right]| \mathcal{G}_{t-}\right] = E\left[\lambda_t^F | \mathcal{G}_{t-}\right].$$

A more formal proof is as follows. Let h be an arbitrary non-negative **G**-predictable process. Then h will also be **F**-predictable (why?) and we have

$$E\left[\int_0^\infty h_t\, dN_t\right] = E\left[\int_0^\infty h_t \lambda_t^F\, dt\right] = E\left[\int_0^\infty E\left[h_t \lambda_t^F | \mathcal{G}_{t-}\right] dt\right]$$

$$= E\left[\int_0^\infty h_t E\left[\lambda_t^F | \mathcal{G}_{t-}\right] dt\right] = E\left[\int_0^\infty h_t \lambda_t^G\, dt\right],$$

which shows that λ^G is the predictable **G**-intensity of N. $\hfill\square$

4.6   Conditional Jump Probabilities

Let us consider a k-variate point process $N^1, \ldots, N^k$ on a filtered probability space $(\Omega, \mathcal{F}, P, \mathbf{F})$, where $\mathbf{F} = \mathbf{F}^N$, and denote the **F**-predictable intensity processes by $\lambda^1, \ldots, \lambda^k$. We recall the informal interpretation

$$\lambda_t^i dt = E\left[\Delta N_t^i | \mathcal{F}_{t-}^N\right],$$

so λ^i determines the conditional jump probabilities for the process N^i. We also recall, from Section 2.6, that we can write

$$N_t = \sum_{i=1}^k N_t^i, \tag{4.5}$$

$$Z_n = \sum_{i=1}^k i \cdot \Delta N_{T_n}^i, \tag{4.6}$$

$$N_t^i = \sum_{n=1}^\infty I\{T_n \le t\} I\{Z_n = i\} ., \tag{4.7}$$

where $\{T_n\}_n$ are the jump times of the process N. We denote the predictable intensity of N by λ so we have (why?)

$$\lambda_t = \sum_{i=1}^k \lambda_t^i. \tag{4.8}$$

Suppose now that we are standing at time t and that there is a jump of N at time t. This means that exactly one of the processes $N^1, \ldots, N^k$ has a jump at time t, and our problem is to determine which process this is. Slightly more precisely we would like to determine the probability

$$P\left(dN_t^i = 1 | \mathcal{F}_{t-}^N,\ dN_t = 1\right).$$

In other words, suppose we have observed N on $[0, t)$ and been given the information that there is a jump of N at time t. We then want to compute the probability that it is the process N^i that has generated the jump.

Intuitively this is not so hard. With a certain amount of hand waving, we have

$$P\left(dN_t^i = 1 \,\middle|\, \mathcal{F}_{t-}^N, dN_t = 1\right) = \frac{P\left(dN_t^i = 1 \,\&\, dN_t = 1 \,\middle|\, \mathcal{F}_{t-}^N\right)}{P\left(dN_t = 1 \,\middle|\, \mathcal{F}_{t-}^N\right)}$$

$$= \frac{P\left(dN_t^i = 1 \,\middle|\, \mathcal{F}_{t-}^N\right)}{\lambda_t dt} = \frac{\lambda_t^i dt}{\lambda_t dt} = \frac{\lambda_t^i}{\lambda_t}.$$

The precise formulation of this result is as follows. The proof is rather hard so the reader is referred to Last & Brandt (1995) for a full proof.

Proposition 4.5 *With notation as above we have*

$$P\left(Z_n = i \,\middle|\, \mathcal{F}_{T_n-}^N\right) = \frac{\lambda_{T_n}^i}{\lambda_{T_n}}, \quad \text{on } \{T_n < \infty\}. \tag{4.9}$$

4.7   A Random Time Change

In this section we will show that every counting process possessing a positive intensity can be transformed into a Poisson process by using a random clock. This is the Poisson parallel to a result in Wiener theory which says that every integral with respect to a Wiener process can be viewed as a standard Wiener process under a random clock. Let us thus consider a counting process N with the **F**-intensity process λ, and assume that λ is strictly positive. We now define, for each t, the random variable C_t by

$$C_t = \inf\left\{s \geq 0 : \int_0^s \lambda_u du = t\right\}. \tag{4.10}$$

Because of the assumption that $\lambda > 0$ we see that C_t is uniquely determined by the relation

$$\int_0^{C_t} \lambda_u du = t.$$

The family of random variables $\{C_t : t \geq 0\}$ is referred to as a **random clock**, and we now define the process N_t^C by

$$N_t^C = N_{C_t}.$$

It is easy to see (how?) that C_t is an **F**-stopping time for each fixed t, so we can now define a new filtration **G** by $\mathcal{G}_t = \mathcal{F}_{C_t}$. Since N is **F**-optional it follows that N^C is **G**-optional, and we have the following result.

Proposition 4.6 *The process N^C defined above is a **G**-Poisson process with unit intensity.*

Proof   It is obvious that N^C is a jump process. Referring to the Watanabe theorem we thus have to prove that the process M defined by

$$M_t = N_t^C - t$$

is a **G**-martingale. We know, however, that the process

$$N_t - \int_0^t \lambda_s \, ds$$

is an **F**-martingale, so from the optional sampling theorem it follows that the process

$$N_{C_t} - \int_0^{C_t} \lambda_s \, ds$$

is a **G**-martingale, but by definition we have

$$N_{C_t} = N_t^C, \quad \text{and} \quad \int_0^{C_t} \lambda_s \, ds = t$$

which concludes the proof. $\square$

Remark 4.7   The assumption that λ is strictly positive is not necessary, but it simplifies the proof.

4.8   Exercises

Exercise 4.1   Assume that the k-variate point process $(N^1, \ldots, N^k$ on $(\Omega, \mathcal{F}, P, \mathbf{F})$ has **F**-intensities λ^1, λ^k. Prove that the intensity λ of $N = \sum_i N^i$ is given by

$$\lambda_t = \sum_{i=1}^k \lambda_t^i.$$

5   Martingale Representations and Girsanov Transformations

In this chapter we present two main theoretical workhorses for counting process theory: the martingale representation theorem and the Girsanov theorem. These results will be used over and over again in connection with general counting process theory. They are also fundamental for the analysis of arbitrage-free capital markets.

5.1   The Martingale Representation Theorem

Assume that we are given a filtered probability space $(\Omega, \mathcal{F}, P, \mathbf{F})$ carrying an integrable adapted point process N with $\mathbf{F}$-intensity λ. From Propositions 2.10 and 4.2 we know that, for every choice of a predictable (and sufficiently integrable) process h the process X defined by

$$X_t = \int_0^t h_s \left[dN_s - \lambda_s ds \right] \tag{5.1}$$

is an $\mathbf{F}$-martingale. An interesting question is now to ask whether also the converse statement also is true, i.e. to ask if *every* $\mathbf{F}$-martingale X can be represented on the form (5.1). That this cannot possible be the case is clear from the following counterexample.

Assume for simplicity that N is Poisson with constant intensity, and suppose that the space also carries an independent $\mathbf{F}$-Wiener process W. Then, setting $X = W$, it is clear that X is an $\mathbf{F}$-martingale, but it is also clear that X cannot have the representation (5.1). The reason is of course that X has continuous trajectories, whereas a stochastic integral with respect to the compensated N-process, will have trajectories with jumps. The more informal reason is of course that the Wiener process W "has nothing at all to do with the point process N". In order to have any chance of obtaining a positive result we therefore have to guarantee that the space carries "nothing other than the process N itself". The natural condition is given in the following fundamental result, which is the point process analogue of the corresponding martingale representation result for Wiener processes.

Theorem 5.1 *Assume that N is an integrable point process with intensity λ, and that the filtration is the* internal *one generated by N, i.e.*

$$\mathcal{F}_t = \mathcal{F}_t^N. \tag{5.2}$$

*Then, for every **F**-martingale X there will exist a predictable process h such that*

$$X_t = X_0 + \int_0^t h_s \left[dN_s - \lambda_s ds \right].$$ (5.3)

Furthermore, the process h is unique $dP(\omega)dN_t(\omega) - a.e..$

Proof   This is a deep and difficult result involving a detailed study of the structure of the internal filtration. The reader is referred to Brémaud (1981) or Last & Brandt (1995) for a proof. If N is Poisson then you can basically copy the proof for the Wiener case, but in the general case this does not work.   □

Remark   We remark that this is an abstract existence result. There is generally no concrete characterization of the integrand h.

The result above generalizes immediately to the multidimensional setting, and we can also include a finite number of driving Wiener processes.

Theorem 5.2   *Let $(\Omega, \mathcal{F}, P, \mathbf{F})$ be a filtered probability space carrying k counting processes $N^1, \ldots, N^k$, as well as a standard d-dimensional Wiener process $W^1, \ldots, W^d$. Assume that the filtration **F** is the internal one, i.e.*

$$\mathcal{F}_t = \sigma \left\{ N_s^i, W_s^j; \ i = 1, \ldots, k, \ j = 1, \ldots, \ldots, d; \ s \le t \right\}.$$ (5.4)

Assume furthermore that N^i has the predictable intensity λ^i for $i = 1, \ldots, k$. Then, for every F-martingale X, there will exist predictable processes $h^1, \ldots, h^k$ and $g^1, \ldots, g^d$ such that

$$X_t = X_0 + \sum_{i=1}^{k} h_s^i \left[dN_s^i - \lambda_s^i ds \right] + \sum_{j=1}^{d} \int_0^t g_s^j dW_s^j.$$ (5.5)

5.2   The Girsanov Theorem

We will now study how the intensity λ of a counting process N changes when we transform the original measure P to a new measure $Q \ll P$. As a result we will obtain a counting process version of the standard Girsanov theorem for Wiener processes. At the end of the chapter we apply the theory to maximum likelihood estimation and we also define and prove the existence of Cox processes. To set the scene we consider a given filtered space $(\Omega, \mathcal{F}, P, \mathbf{F})$ carrying an adapted counting process N with **F**-predictable P-intensity λ. We study the process N on a fixed time interval $[0, T]$.

Let us now assume that we change measure from P to Q, where $Q \ll P$ on $\mathcal{F}_T$, and let L be the induced likelihood process, given by

$$L_t = \frac{dQ}{dP}, \quad \text{on } \mathcal{F}_t, \quad 0 \le t \le T.$$ (5.6)

Under the new measure Q, the counting process N will have a predictable intensity $\lambda^Q \ne \lambda$ and our task is to derive an expression for the Q-intensity λ^Q.

In order to understand more clearly what is going on, we start with some heuristics, which will allow us to conjecture the formulation of the Girsanov theorem. Having made the conjecture it is then surprisingly easy to prove it. To this end we recall that the intuitive interpretation of λ_t^Q is given by the relation

$$\lambda_t^Q dt = E^Q [dN_t | \mathcal{F}_{t-}],$$

and in order to compute the expected value the obvious tool to use is the abstract Bayes formula. Using this and the fact that the likelihood process L is a P-martingale we have

$$\lambda_t^Q dt = E^Q [dN_t | \mathcal{F}_{t-}] = \frac{E^P [L_t dN_t | \mathcal{F}_{t-}]}{E^P [L_t | \mathcal{F}_{t-}]}$$

$$= \frac{E^P [(L_{t-} + dL_t) dN_t | \mathcal{F}_{t-}]}{L_{t-}}$$

$$= E^P [dN_t | \mathcal{F}_{t-}] + \frac{E^P [dL_t dN_t | \mathcal{F}_{t-}]}{L_{t-}}.$$

Recalling that $\lambda_t dt = E^P [dN_t | \mathcal{F}_{t-}]$ we thus obtain

$$\lambda_t^Q dt = \lambda_t dt + \frac{E^P [dL_t dN_t | \mathcal{F}_{t-}]}{L_{t-}}. \tag{5.7}$$

At this degree of generality we are not able to go further but, as in the Wiener case, the above expression will simplify considerably if we make some further assumptions about L. Since we know that L is a P-martingale and since also the compensated process,

$$dN_t - \lambda_t dt,$$

is a martingale increment under P, it natural to investigate what will happen if we assume that L has the particular structure

$$dL_t = g_t [dN_t - \lambda_t dt],$$

where g is a predictable process. Using the facts that $dN_t dt = 0$, $(dN_t)^2 = dN_t$ and that $g_t \in \mathcal{F}_{t-}$ (why?), we get

$$\frac{E^P [dL_t dN_t | \mathcal{F}_{t-}]}{L_{t-}} = \frac{E^P [g_t \{dN_t - \lambda_t dt\} dN_t | \mathcal{F}_{t-}]}{L_{t-}}$$

$$= g_t \frac{E^P [(dN_t)^2 | \mathcal{F}_{t-}]}{L_{t-}} = g_t \frac{E^P [dN_t | \mathcal{F}_{t-}]}{L_{t-}}$$

$$= g_t \frac{\lambda_t dt}{L_{t-}}.$$

From this we see that if we now define the predictable process φ by

$$\varphi_t = \frac{g_t}{L_{t-}},$$

the above expression simplifies to

$$\frac{E^P [dL_t dN_t | \mathcal{F}_{t-}]}{L_{t-}} = \varphi_t \lambda_t dt,$$

and if we plug this into (5.7) we obtain

$$\lambda_t^Q dt = \lambda_t dt + \varphi_t \lambda_t dt = \lambda_t (1 + \varphi_t) dt,$$

or

$$\lambda_t^Q = \lambda_t (1 + \varphi_t).$$

The point of these rather informal calculations is that we are able to guess what the Girsanov theorem will look like for a counting process. We can now state and prove the formal result.

Theorem 5.3 (The Girsanov Theorem) *Let N be an optional counting process on the filtered space* $(\Omega, \mathcal{F}, P, \mathbf{F})$ *and assume that N has the predictable intensity* λ. *Let* φ *be a predictable process such that*

$$\varphi_t \geq -1, \quad P - a.s. \tag{5.8}$$

and define the process L by

$$\begin{cases} dL_t &= L_{t-}\varphi_t \{dN_t - \lambda_t dt\}, \\ L_0 &= 1, \end{cases} \tag{5.9}$$

on the interval $[0, T]$. *Assume furthermore that*

$$E^P [L_T] = 1. \tag{5.10}$$

Now define a new probability measure $Q \ll P$ *on* $\mathcal{F}_T$ *by*

$$dQ = L_T dP. \tag{5.11}$$

Then N has the Q intensity λ^Q, *given by*

$$\lambda_t^Q = \lambda_t (1 + \varphi_t).$$

Proof We need to show that, for every non-negative predictable process g, we have

$$E^Q \left[\int_0^T g_t dN_t \right] = E^Q \left[\int_0^T g_t \lambda_t (1 + \varphi_t) dt \right]. \tag{5.12}$$

We start with the right-hand side to obtain

$$E^Q \left[\int_0^T g_t \lambda_t (1 + \varphi_t) dt \right] = \int_0^T E^Q [g_t \lambda_t (1 + \varphi_t)] dt$$

$$= \int_0^T E^P [L_t g_t \lambda_t (1 + \varphi_t)] dt = E^P \left[\int_0^T L_t g_t \lambda_t (1 + \varphi_t) dt \right]$$

$$= E^P \left[\int_0^T L_{t-} g_t \lambda_t (1 + \varphi_t) dt \right] \tag{5.13}$$

Turning to the left-hand side of (5.12) we obtain

$$E^Q \left[\int_0^T g_t dN_t \right] = E^P \left[L_T \int_0^T g_t dN_t \right],$$

and it is therefore natural to study the process Z, defined by

$$Z_t = L_t \int_0^t g_s \, dN_s,$$

which we write as $Z_t = L_t Y_t$ with

$$dY_t = g_t \, dN_t.$$

It is clear that Z is the product of two processes of bounded variation, so from the product rule (Proposition 2.20) we have, using $dN_t \, dt = 0$, $(dN_t)^2 = dN_t$,

$$
\begin{aligned}
dZ_t &= L_{t-} \, dY_t + Y_{t-} \, dL_t + \Delta L_t \Delta Y_t \\
&= L_{t-} g_t \, dN_t + Y_{t-} \, dL_t + L_{t-} g_t \varphi_t \, dN_t \\
&= L_{t-} g_t \left(1 + \varphi_t\right) dN_t + Y_{t-} \, dL_t.
\end{aligned}
$$

Integrating this, and recalling that λ is the P intensity of N, gives us

$$
\begin{aligned}
E^Q \left[\int_0^T g_t \, dN_t \right] = E^P \left[Z_T\right] &= E^P \left[\int_0^T dZ_t \right] \\
&= E^P \left[\int_0^T L_{t-} g_t \left(1 + \varphi_t\right) dN_t \right] + E^P \left[\int_0^T Y_{t-} \, dL_t \right] \\
&= E^P \left[\int_0^T L_{t-} g_t \left(1 + \varphi_t\right) \lambda_t \, dt \right],
\end{aligned}
\qquad (5.14)
$$

where we have also used the fact that, since L is a P-martingale and the process $t \to Y_{t-}$ is predictable, the dL integral has zero expected value. The equality (5.12) now follows from (5.13) and (5.14). $\qquad \square$

This result generalizes easily to the multidimensional case, and we can also include a finite number of driving Wiener processes in the obvious way.

Theorem 5.4 (Girsanov Theorem)   *Consider the filtered probability space $(\Omega, \mathcal{F}, P, \mathbf{F})$ and assume that $N^1, \ldots, N^k$ are optional counting processes with predictable intensities $\lambda^1, \ldots, \lambda^k$. Assume furthermore that $W^1, \ldots, W^d$ are standard independent $(\mathbf{F}, P)$-Wiener processes. Let $\varphi^1, \ldots, \varphi^k$ be predictable processes with*

$$\varphi_t^i \ge -1, \quad i = 1, \ldots, k, \quad P - a.s,$$

and let $\gamma^1, \ldots, \gamma^d$ be optional processes. Define the process L on $[0, T]$ by

$$
\begin{cases}
dL_t &= L_t \sum_{i=1}^d \gamma_t^i \, dW_t^i + L_{t-} \sum_{j=1}^k \varphi_t^i \left\{ dN_t^i - \lambda_t^i \, dt \right\}, \\
L_0 &= 1,
\end{cases}
\qquad (5.15)
$$

and assume that

$$E^P \left[L_T\right] = 1.$$

Define the measure Q on $\mathcal{F}_T$ by $dQ = L_T \, dP$. Then we have the following.

- *We can write*

$$dW_t^i = \gamma_t^i dt + dW_t^{Q,i}, \quad i = 1, \ldots, d,$$

  where $W^{Q,1}, \ldots, W^{Q,d}$ are Q-Wiener processes.
- *The Q intensities of $N^1, \ldots N^k$ are given by*

$$\lambda_t^{Q,i} = \lambda_t^i \left(1 + \varphi_t^i\right), \quad i = 1, \ldots, k.$$

5.3   The Converse of the Girsanov Theorem

If we start with a measure P and perform a Girsanov transformation according to (5.9)–(5.11) in order to define a new measure Q, then we know that $Q \ll P$. A natural question to ask is whether **all** measures $Q \ll P$ are obtained by the procedure (5.9)–(5.11).

In the general case it is obvious that the answer is no. Consider for example a case where the stochastic basis carries a Poisson process N with constant intensity λ as well as a $N[0,1]$-distributed random variable Z, which is independent of N. Suppose furthermore that we change P to Q by changing the distribution of Z from $N[0,1]$ to $N[5,1]$, while keeping the distribution fixed for N. It is then obvious that $Q \sim P$, but since the Girsanov transformation (5.9)–(5.11) is completely determined by N, and not in any way involving Z, it is intuitively obvious that the change from P to Q cannot be achieved by a Girsanov transformation of the type (5.9)–(5.11).

From this discussion it is reasonable to assume that if we restrict ourselves to the case when the filtration $\mathbf{F}$ is the *internal* one generated by the counting process N, then we may hope for a converse of the Girsanov theorem.

Proposition 5.5 (The Converse of the Girsanov Theorem)   *Let N be a counting process on $(\Omega, \mathcal{F}, P, \mathbf{F})$ with intensity process λ, and assume that the filtration $\mathbf{F}$ is the internal one, i.e. that*

$$\mathcal{F}_t = \mathcal{F}_t^N, \quad t \geq 0. \tag{5.16}$$

Assume furthermore that there exists a measure Q such that, for a fixed $T < \infty$, we have $Q \ll P$ on $\mathcal{F}_T$, and let L denote the corresponding likelihood process, i.e.

$$L_t = \frac{dQ}{dP}, \quad \text{on } \mathcal{F}_t, \quad 0 \leq t \leq T.$$

Then there exists a predictable process φ such that L has the dynamics

$$\begin{cases} dL_t &= L_{t-}\varphi_t \{dN_t - \lambda_t dt\}, \\ L_0 &= 1, \end{cases} \tag{5.17}$$

and the Q intensity is given by

$$\lambda_t^Q = \lambda_t(1 + \varphi_t).$$

Proof   Defining L as a above, we know from general theory that L is a P-martingale.

Since we have the internal filtration, the martingale representation theorem, 5.1, guarantees the existence of a predictable process g such that

$$dL_t = g_t \{dN_t - \lambda_t dt\},$$

and if we define φ by

$$\varphi_t = \frac{g_t}{L_{t-}},$$

we are done. There is a potential problem when $L_{t-} = 0$, but also this can be handled.   □

We can of course extend this result to the case of a multidimensional counting process and a multidimensional Wiener process. The proof is almost identical.

Proposition 5.6 (The Converse of the Girsanov Theorem) *Consider a filtered space* $(\Omega, \mathcal{F}, P, \mathbf{F})$ *carrying the counting processes* $N^1, \ldots, N^k$ *with predictable intensities* $\lambda^1, \ldots, \lambda^k$, *as well as the standard independent Wiener processes* $W^1, \ldots, W^d$. *We assume that the filtration is the* **internal** *one, generated by* N *and* W, *i.e.*

$$\mathcal{F}_t = \mathcal{F}_t^N \vee \mathcal{F}_t^W.$$

Assume furthermore that there exists a measure Q *such that for a fixed* T, *we have* $Q \ll P$ *on* $\mathcal{F}_T$, *and let* L *denote the corresponding likelihood process, i.e.*

$$L_t = \frac{dQ}{dP}, \quad \text{on } \mathcal{F}_t, \quad 0 \leq t \leq T.$$

Then there exists a k-dimensional predictable process φ, *and a d-dimensional predictable process* γ *such that* L *has the dynamics*

$$\begin{cases} dL_t &= L_{t-} \sum_{i=1}^{k} \varphi_s^i \left[dN_s^i - \lambda_s^i ds \right] + L_t \sum_{j=1}^{d} \int_0^t \gamma_s^j dW_s^j \\ L_0 &= 1. \end{cases} \tag{5.18}$$

5.4   Maximum Likelihood Estimation

In this section we give a brief introduction to maximum likelihood (ML) estimation for counting processes.

We need the concept of a statistical model.

Definition 5.7   A dynamic **statistical model** over a finite time interval $[0, T]$ consists of the following objects.

- A measurable space $(\Omega, \mathcal{F})$.
- A filtration $\mathbf{F}$.
- An indexed family of probability measures $\{P_\alpha; \ \alpha \in A\}$, defined on the space $(\Omega, \mathcal{F})$, where A is some index set and where all measures are assumed to be absolutely continuous on $\mathcal{F}_T$ with respect to some base measure P_{α_0}, i.e.

$$P_\alpha \ll P_{\alpha_0}, \quad \text{for all } \alpha \in A.$$

In most concrete applications (see examples below) the parameter α will be a real number or a finite-dimensional vector, i.e. A will be the real line or some finite-dimensional Euclidean space. The filtration will typically be generated by some observation process X.

The interpretation of all this is that the probability distribution is governed by some measure P_α, but we do not know which. We do have, however, access to a flow of information over time and this is formalized by the filtration above, so at time t we have the information contained in $\mathcal{F}_t$. Our problem is to try to estimate α given this flow of observations, or more precisely: for every t we want an estimate α_t of α, based upon the information contained in $\mathcal{F}_t$, i.e. based on the observations over the time interval $[0,t]$. The last requirement is formalized by requiring that the estimation process should be adapted to $\mathbf{F}$, i.e. that $\alpha_t \in \mathcal{F}_t$.

One of the most common techniques used in this context is that of finding, for each t, the **maximum likelihood** estimate of α. Formally the procedure works as follows.

- Compute, for each α, the corresponding likelihood process L^α (where α only has the role of being an index, and not a power) is defined by

$$L_t^\alpha = \frac{dP_\alpha}{dP_{\alpha_0}}, \quad \text{on } \mathcal{F}_t.$$

- For each fixed t (and ω), find the value of α which maximizes the likelihood ratio L_t^α.
- The optimal α is denoted by $\widehat{\alpha}_t$ and is called the **maximum likelihood estimate** of α based on the information gathered over $[0,t]$.

As the simplest possible example let us consider the problem of estimating the constant but unknown intensity of a scalar Poisson process.

In this example we do in fact have an obvious candidate for the intensity estimate. Indeed, if N is Poisson with intensity λ, then λ is the mean number of jumps per unit time, so the natural estimate is given by

$$\widehat{\lambda}_t = \frac{N_t}{t}, \quad t > 0.$$

To formalize our problem within the more abstract framework above, we need to build a statistical model. To this end we consider a filtered space $(\Omega, \mathcal{F}, P, \mathbf{F})$ carrying a Poisson process N with *unit intensity* under P. The filtration is assumed to be the internal one, i.e. $\mathcal{F}_t = \mathcal{F}_t^N$. For any non-negative real number λ we can now define the measure P_λ by the Girsanov transformation

$$\begin{cases} dL_t^\lambda &= L_{t-}^\lambda (\lambda - 1) \{dN_t - dt\}, \\ L_0^\lambda &= 1. \end{cases} \tag{5.19}$$

From the Girsanov theorem and the Watanabe characterization theorem it is clear that under P_λ, the process N will be Poisson with the constant intensity λ. The SDE (5.19) can easily be solved, for example by using Proposition 2.22, and we obtain

$$L_t^\lambda = e^{N_t \ln(\lambda) - t(\lambda - 1)}.$$

We thus have to maximize the expression

$$N_t \ln(\lambda) - t(\lambda - 1)$$

over $\lambda > 0$ and we immediately obtain the ML estimate as

$$\widehat{\lambda_t} = \frac{N_t}{t}. \tag{5.20}$$

We see that in this example the ML estimator actually coincides with our naive guess above. The point of using the ML technique is of course that in a more complicated situation (see the exercises) we may have no naive candidate, whereas the ML technique in principle is always applicable.

5.5 Cox Processes

In point process theory and its applications, such as for example in credit risk theory, a very important role is played by a particular class of counting processes known as "Cox processes", or "doubly-stochastic Poisson processes". In this section we will define the Cox process concept and then use the Girsanov theory developed above to prove the existence of Cox processes.

The intuitive idea of a Cox process is very simple and goes roughly as follows.

1. Consider a fixed random process λ on some probability space Ω.
2. Fix one particular trajectory of λ, say the one corresponding to the outcome $\omega \in \Omega$.
3. For this fixed ω, the mapping $t \to \lambda_t(\omega)$ is a deterministic function of time.
4. Construct, again for this fixed ω, a counting process N which is Poisson with intensity function $\lambda.(\omega)$.
5. Repeat this procedure for all choices of $\omega \in \Omega$.

If this informal procedure can be carried out (this is not at all clear), then it seems that it would produce a counting process N with the following property.

Conditional on the entire λ-trajectory, the process N is Poisson with that particular λ-trajectory as intensity function. This is, intuitively, the definition of a Cox process.

The construction above is of course not very precise from a mathematical point of view, and the same could be said for the statement about the properties of N. For example, what *exactly* do we mean by the sentence "Conditional on the entire λ-trajectory, the process N is Poisson with that particular λ-trajectory as intensity function"? We now have a small research program consisting of the following items.

- Define, in a mathematically precise way, the concept of a Cox process.
- Prove that, given an intensity process λ, the corresponding Cox processes exist.

The formal definition is as follows.

Definition 5.8 Consider a probability space $(\Omega, \mathcal{F}, P)$, carrying a counting process

N as well as a non-negative process λ. We say that N is a **Cox process** with intensity process λ if the relation

$$E\left[e^{iu(N_t - N_s)}\middle| \mathcal{F}_s^N \vee \mathcal{F}_\infty^\lambda\right] = e^{\Lambda_{s,t}(e^{iu}-1)} \tag{5.21}$$

holds for all $s < t$, where

$$\Lambda_{s,t} = \int_s^t \lambda_u du.$$

Equivalently we can write (5.21) as

$$P\left(N_t - N_s = n \middle| \mathcal{F}_s^N \vee \mathcal{F}_\infty^\lambda\right) = e^{-\Lambda_{s,t}}\frac{(\Lambda_{s,t})^n}{n!}, \quad n = 0, 1, 2, \dots \tag{5.22}$$

If we compare this with Definition 2.23 and Lemma 2.24 we see that the interpretation is indeed that, "conditional on the λ-trajectory the process N is Poisson with that particular λ-trajectory as intensity function".

We now go on to show that, for any given process λ, there actually exists a Cox process with λ as the intensity. The formal statement is given below, and the proof is a nice example of Girsanov technique.

Proposition 5.9   *Consider a probability space $(\Omega, \mathcal{F}, P_0)$, carrying a non-negative random process λ. Assume that the space also carries a Poisson process N, with unit intensity, and that N is independent of λ. Then there exists a probability measure $P \sim P_0$ with the following properties.*

- *The distribution of λ under P is the same as the distribution under P_0.*
- *Under P, the counting process N is a Cox process with intensity λ.*

Proof   We start by defining the filtration $\mathbf{F}$ by

$$\mathcal{F}_t = \mathcal{F}_t^N \vee \mathcal{F}_\infty^\lambda,$$

and note that this implies that $\mathcal{F}_\infty^\lambda \subseteq \mathcal{F}_0$. Next we define a likelihood process L in the obvious way by

$$\begin{cases} dL_t &= L_{t-}(\lambda_t - 1)\{dN_t - dt\}, \\ L_0 &= 1, \end{cases} \tag{5.23}$$

and define the new measure P by

$$L_t = \frac{dP}{dP_0} \quad \text{on } \mathcal{F}_t.$$

From the Girsanov theorem it is clear that N has the intensity λ under P but this is not enough. Denoting, by $\mathcal{L}^Q(X)$ the distribution of a random variable or process X under a measure Q, we have to show that

(a) $\mathcal{L}^P(\lambda) = \mathcal{L}^{P_0}(\lambda)$.

(b) Under P, the process N is Cox with intensity λ.

Item (a) is easy. Since $L_0 = 1$, the measures P_0 and P coincide on $\mathcal{F}_0$, and since $\sigma\{\lambda\} = \mathcal{F}_\infty^\lambda$ is included in $\mathcal{F}_0$, we conclude that $\mathcal{L}^P(\lambda) = \mathcal{L}^{P_0}(\lambda)$.

Item (b) requires a little bit of work. We need to show that

$$E^P\left[e^{iu(N_t - N_s)}\Big|\mathcal{F}_s\right] = e^{\Lambda_{s,t}(e^{iu}-1)}, \qquad (5.24)$$

and the obvious idea is of course to use the Bayes formula. We then have

$$E^P\left[e^{iu(N_t - N_s)}\Big|\mathcal{F}_s\right] = \frac{E^{P_0}\left[e^{iu(N_t - N_s)}L_t\big|\mathcal{F}_s\right]}{L_s}. \qquad (5.25)$$

where we have used the fact that $E^{P_0}[L_t|\mathcal{F}_s] = L_s$. In order to compute the last conditional expectation, we choose a fixed s, and consider the process Z on $[s, \infty)$ defined by

$$Z_t = e^{iu(N_t - N_s)}L_t.$$

Defining the process Y by

$$Y_t = e^{iu(N_t - N_s)},$$

we can write

$$Z_t = Y_t L_t,$$

and since both processes are of bounded variation we can use the product rule to obtain

$$dZ_t = Y_{t-}dL_t + L_{t-}dY_t + \Delta Y_t \Delta L_t.$$

The differential dL_t is given by (5.23), and for Y we easily obtain

$$dY_t = \left(e^{iu(N_{t-}+1)} - e^{iuN_{t-}}\right)e^{-iuN_s}\,dN_t = Y_{t-}\left(e^{iu}-1\right)dN_t.$$

From this expression and from (5.23) we have

$$\Delta Y_t \Delta L_t = Y_{t-}\left(e^{iu}-1\right)L_{t-}(\lambda_t - 1)\,dN_t = Z_{t-}\left(e^{iu}-1\right)(\lambda_t - 1)\,dN_t.$$

Denoting the P_0-martingale $N_t - t$ by M_t we thus obtain

$$dZ_t = Z_{t-}(\lambda_t - 1)\,dM_t + Z_{t-}\left(e^{iu}-1\right)dN_t + Z_{t-}\left(e^{iu}-1\right)(\lambda_t - 1)\,dN_t$$

$$= Z_{t-}\lambda_t\left(e^{iu}-1\right)dN_t + Z_{t-}(\lambda_t - 1)\,dM$$

$$= Z_{t-}\lambda_t\left(e^{iu}-1\right)dt + Z_{t-}\left(\lambda_t e^{iu}-1\right)dM_t.$$

Integrating this we obtain

$$Z_t = Z_s + \int_s^t Z_u \lambda_u\left(e^{iu}-1\right)du + \int_s^t Z_{u-}\left(\lambda_u e^{iu}-1\right)dM_u.$$

We now note that, since $\mathcal{F}_\infty^\lambda \subseteq \mathcal{F}_0$, we always have $\lambda_t \in \mathcal{F}_0 \subseteq \mathcal{F}_{t-}$, so the process λ is in fact **F**-predictable, implying that the dM integral is a P_0-martingale, Let $E_s^0[\cdot]$ denote the conditional expectation $E^{P_0}[\cdot|\mathcal{F}_s]$, and take E_s^0 expectations. Using the martingale property of the dM integral and the fact that $Z_s = L_s$ gives us

$$E_s^0[Z_t] = E_s^0[L_s] + \int_s^t E_s^0[Z_u\lambda_u]\left(e^{iu} - 1\right)du.$$

Since $\mathcal{F}_\infty^\lambda \subseteq \mathcal{F}_0$ we have $E_s^0[Z_u\lambda_u] = \lambda_u E_s^0[Z_u]$, yielding

$$E_s^0[Z_t] = L_s + \int_s^t \lambda_u\left(e^{iu} - 1\right)E_s^0[Z_u]\,du.$$

If we now denote $E_s^0[Z_t]$ by x_t (suppressing the fixed s) we have the integral equation

$$x_t = L_s + \int_s^t \lambda_u\left(e^{iu} - 1\right)x_u\,du,$$

which is the integral form of the ODE

$$\begin{cases} \dot{x}_t &= x_t\lambda_u\left(e^{iu} - 1\right), \\ x_s &= L_s, \end{cases}$$

with the solution

$$x_t = L_s e^{\Lambda_{s,t}\left(e^{iu} - 1\right)}.$$

We have thus shown that

$$E^0\left[e^{iu(N_t - N_s)}L_t\Big|\mathcal{F}_s\right] = L_s e^{\Lambda_{s,t}(e^{iu} - 1)},$$

and inserting this into (5.25) gives us (5.24).   □

The Watanabe theorem for Poisson processes is now easily extended to Cox processes.

Proposition 5.10 *Consider a filtered space $(\Omega, \mathcal{F}, P, \mathbf{F})$, and let N be an $\mathbf{F}$-adapted counting process. Assume that λ is a non-negative process satisfying the conditions*

$$\mathcal{F}_\infty^\lambda \subseteq \mathcal{F}_0,$$

$$\int_0^t \lambda_s\,ds < \infty, \quad 0 \le t < \infty.$$

Assume furthermore that the process M, defined by

$$dM_t = dN_t - \lambda_t dt, \quad M_0 = 0,$$

is an $\mathbf{F}$-martingale. Then N is a Cox process with intensity process λ.

Proof Copy the proof for the Poisson case.   □

5.6   Exercises

Exercise 5.1 Assume that the counting process N has the $\mathcal{F}_t^N$-predictable intensity

$$\lambda_t = \alpha g(N_{t-}).$$

where g is a non-negative known deterministic function, and α is an unknown parameter. Show that the ML estimate of α is given by

$$\widehat{\alpha}_t = \frac{N_t}{\int_0^t g\left(N_s\right) ds}.$$

Exercise 5.2   Assume that we can observe a process X with P-dynamics given by

$$dX_t = \mu dt + \sigma dW_t + dN_t.$$

where σ is a known constant, μ is an unknown parameter, N is Poisson with unknown intensity λ and W is standard Wiener. Our task is to estimate μ and λ, given observations of X.

(a) Define the process X under a base measure P_0 according to the dynamics

$$dX_t = \sigma dW_t^0 + dN_t,$$

where W^0 is Wiener and N is Poisson with unit intensity under P_0. Convince yourself that N and W^0 are observable, given the X observations, in the sense that

$$\mathcal{F}_t^N \subseteq \mathcal{F}_t^X, \quad \mathcal{F}_t^{W^0} \subseteq \mathcal{F}_t^X.$$

(b) Perform a Girsanov transformation from P_0 to P, such that X has the P-dynamics $dX_t = \mu dt + \sigma dW_t + dN_t$. Write down an explicit expression for the likelihood process L.

(c) Maximize L with respect to μ and λ and show that the ML estimates are given by

$$\widehat{\lambda}_t = \frac{N_t}{t}, \quad \widehat{\mu}_t = \frac{X_t - N_t}{t}.$$

Exercise 5.3   Prove proposition 5.10.

6 Connections between Stochastic Differential Equations and Partial Integro-Differential Equations

In this chapter we will study how stochastic differential equations (SDEs) driven by counting and Wiener processes are connected to certain types of partial integro-differential equations (PIDEs).

6.1 SDEs and Markov Processes

On a filtered space $(\Omega, \mathcal{F}, P, \mathbf{F})$ we consider a scalar SDE of the form

$$dX_t = \mu(t, X_t)dt + \sigma(t, X_t)dW_t + \beta(t-, X_{t-})dN_t, \tag{6.1}$$

$$X_0 = x_0. \tag{6.2}$$

Here we assume that $\mu(t, x)$, $\sigma(t, x)$ and $\beta(t, x)$ are given deterministic functions, and that x_0 is a given real number. The Wiener process W and the counting process N are allowed to be multidimensional, in which case σ and β are row vectors of the proper dimensions and the products σdW and βdN are interpreted as inner products. We need one more important assumption.

Assumption *We assume that the counting process N has a predictable intensity λ of the form*

$$\lambda_t = \lambda(t-, X_{t-}). \tag{6.3}$$

Here, with slight abuse of notation the λ in the right-hand side denotes a smooth deterministic function of (t, x).

It is reasonable to expect that under these assumptions the process X is Markov, and this is in fact true.

Proposition 6.1 *Under the assumptions above, the process X will be a Markov process.*

Proof The proof is rather technical and therefore omitted; the interested reader can find one in Øksendal & Sulem (2007). □

Remark The existence of a solution to the SDE above is not completely obvious. We discuss this in some detail in Section 6.4 below.

6.2   The Infinitesimal Generator

To every SDE of the form (6.1), and in fact to every Markov process, one can associate a natural operator, the "infinitesimal generator" $\mathcal{A}$ of the process. The infinitesimal generator is an operator on a function space, and in order to define it, let us consider a function $f : R_+ \times R \to R$. We now fix (t, x) and consider the difference quotient

$$\frac{1}{h} E_{t,x} \left[f(t + h, X_{t+h}) - f(t, x) \right].$$

The limit of this (if it exists) as $h \to 0$ would have the interpretation of a "mean derivative" of the composite process $t \longmapsto f(t, X_t)$, and this leads us to the following definition.

Definition 6.2   Let us by C_b denote the space of bounded continuous mappings $f : R_+ \times R \to R$. The **infinitesimal generator** $\mathcal{A} : \mathcal{D} \to C_b$ is defined by

$$[\mathcal{A}f](t, x) = \lim_{h \downarrow 0} \frac{1}{h} E_{t,x} \left[f(t + h, X_{t+h}) - f(t, x) \right], \tag{6.4}$$

where $\mathcal{D}$ is the subspace of C_b for which the limits exists for all (t, x).

The domain $\mathcal{D}$ is obviously important in the definition, but in what follows we will be rather imprecise about the exact description of $\mathcal{D}$, and we will also apply $\mathcal{A}$ to unbounded functions. The boundedness requirement above only serves to guarantee that the expected values are finite.

At first sight, the definition of $\mathcal{A}$ may look a bit forbidding. We have, however, a very natural intuitive interpretation. Multiplying by h and letting $h \to dt$ gives us the following important interpretation.

Note 6.3   We have

$$E_{t,x} \left[f(t + h, X_{t+h}) - f(t, x) \right] = \mathcal{A}f(t, x)h + o(h), \tag{6.5}$$

and informally

$$E_{t,x} \left[f(t + dt, X_{t+dt}) - f(t, x) \right] = \mathcal{A}f(t, x)dt, \tag{6.6}$$

or alternatively

$$E_{t,x} \left[df(t, X_t) \right] dt = \mathcal{A}f(t, x). \tag{6.7}$$

We can thus interpret $\mathcal{A}f$ as the conditionally expected mean differential per unit time.

It turns out that the infinitesimal generator provides a huge amount of information about the underlying process X. We have for example the following central result.

Proposition 6.4   *The distribution of a Markov process X is uniquely determined by the infinitesimal generator $\mathcal{A}$.*

We will not be able to prove this result for a general Markov process, but we will at least make it believable for the case of a SDE like (6.1).

We thus consider that SDE and go on to determine the shape of the infinitesimal generator. For simplicity we consider only the case when W and N are scalar. If $f \in C^{1,2}$ we have, from the Itô formula,

$$df(t, X_t) = \left\{ \frac{\partial f}{\partial t}(t, X_t) + \mu(t, X_t)\frac{\partial f}{\partial x}(t, X_t) + \frac{1}{2}\sigma^2(t, X_t)\frac{\partial^2 f}{\partial x^2}(t, X_t) \right\} dt$$

$$+ \sigma(t, X_t)\frac{\partial f}{\partial x}(t, X_t)dW_t + f_\beta(t-, X_{t-})dN_t,$$

where f_β is defined by

$$f_\beta(t, x) = f(t, x + \beta(t, x)) - f(t, x). \tag{6.8}$$

We compensate the counting process by adding and subtracting the term $\lambda(T-, X_{t-})dt$ to dN_t. We then have

$$df(t, X_t) = \mathcal{A}f(t, X_t)dt + \sigma(t, X_t)\frac{\partial f}{\partial x}(t, X_t)dW_t + f_\beta(t-, X_{t-})d\tilde{N}_t, \tag{6.9}$$

where the operator $\mathcal{A}$ is defined by

$$\mathcal{A}f(t, x) = \frac{\partial f}{\partial t}(t, x) + \mu(t, x)\frac{\partial f}{\partial x}(t, x) + \frac{1}{2}\sigma^2(t, x)\frac{\partial^2 f}{\partial x^2}(t, x) + f_\beta(t, x)\lambda(t, x),$$

and where the compensated increment $d\tilde{N}$ is defined by

$$d\tilde{N}_t = dN_t - \lambda(t-, X_{t-})dt.$$

Remark   We note that the last term of $\mathcal{A}$ in (6.9) should formally be written as

$$f_\beta(t-, X_{t-})\lambda(t-, X_{t-}).$$

However, because of the assumed continuity of f, β and λ, and the fact that we are integrating with respect to Lebesgue measure dt, we are allowed to evaluate this term at (t, X_t).

The point of writing df as in (6.9) is that we have decomposed df into a *drift*, given by the dt term, and a *martingale*, given by the sum of the dW and $d\tilde{N}$ terms. Let us now fix (t, x) as initial conditions for X. We may then integrate (6.9) to obtain

$$f(t + h, X_{t+h}) = f(t, x) + \int_t^{t+h} \mathcal{A}f(s, X_s)ds$$

$$+ \int_t^{t+h} \sigma(s, X_s)\frac{\partial f}{\partial x}(s, X_s)dW_s$$

$$+ \int_t^{t+h} f_\beta(s-, X_{s-})d\tilde{N}_s.$$

Here the dW integral is obviously a martingale and since $\tilde{N}$ is a martingale and the f_β term is predictable (why?) we see that also the $d\tilde{N}$ integral is a martingale. Taking expectations we thus obtain

$$\frac{1}{h}E_{t,x}\left[f(t + h, X_{t+h}) - f(t, x)\right] = E_{t,x}\left[\frac{1}{h}\int_t^{t+h} \mathcal{A}f(s, X_s)ds\right].$$

Letting $h \to 0$ and using the fundamental theorem of calculus, we obtain

$$\frac{1}{h} E_{t,x} [f(t + h, X_{t+h}) - f(t, x)] = \mathcal{A} f(t, x),$$

with $\mathcal{A}$ defined as above. We have thus proved the following main result.

Proposition 6.5 *Assume that X has the dynamics*

$$dX_t = \mu(t, X_t)dt + \sigma(t, X_t)dW_t + \beta(t-, X_{t-})dN_t, \tag{6.10}$$

and that the intensity of N given by $\lambda(t-, X_{t-})$. Then the following hold.

- *The infinitesimal generator of X is given by*

$$\mathcal{A} f = \frac{\partial f}{\partial t}(t, x) + \mu(t, x)\frac{\partial f}{\partial t}(t, x) + \frac{1}{2}\sigma^2(t, x)\frac{\partial^2 f}{\partial x^2} + f_\beta(t, x)\lambda(t, x), \tag{6.11}$$

  where

$$f_\beta(t, x) = f(t, x + \beta(t, x)) - f(t, x). \tag{6.12}$$

- *The process $f(t, X_t)$ is a (local) martingale if and only if it satisfies the equation*

$$\mathcal{A} f(t, x) = 0, \quad (t, x) \in R_+ \times R. \tag{6.13}$$

We note that the equation (6.13) contains a number of partial derivatives and a (degenerate) integral term (the f_β term). It is thus a partial integro-differential equation (PIDE).

Remark It is common in the literature to split the generator above as

$$\mathcal{A} = \frac{\partial}{\partial t} + \mathcal{G} \tag{6.14}$$

and define the generator as just $\mathcal{G}$. It is in general clear from the context which terminology one uses.

Remark The result extends, in the obvious way, to the case of a multidimensional Wiener process and a k-variate counting process.

We end this section by noting that the second item above can be generalized to any Markov process. We have in fact the following general result, which in fact holds for a Markov processes on a very general state space.

Proposition 6.6 (Dynkin's formula) *Assume that X is a Markov process with infinitesimal generator $\mathcal{A}$. Then, for every f in the domain of $\mathcal{A}$, the process*

$$f(t, X_t) - \int_0^t \mathcal{A} f(s, X_s)ds$$

is a martingale. Furthermore, the process $f(t, X_t)$ is a martingale if and only if

$$\mathcal{A} f(t, x) = 0, \quad (t, x) \in R_+ \times R.$$

Proof The proof in the general case is a quite technical so we omit it. From an intuitive point of view the result is, however, more or less obvious, and we give an informal argument. Letting $h \to dt$ in the definition of $\mathcal{A}$ we recall from (6.7) that

$$E\left[df(t, X_t)|\, \mathcal{F}_t\right] = \mathcal{A}f(t, X_t)dt.$$

with the interpretation $df(t, X_t) = f(t + dt, X_{t+dt}) - f(t, X_t)$. From this it is clear that the "conditionally detrended" difference

$$df(t, X_t) - \mathcal{A}f(t, X_t)dt$$

should be a martingale increment, and this is precisely the content of the Dynkin formula. The second statement follows directly from the Dynkin formula. □

6.3   The Feynman–Kac and the Kolmogorov Backward Equations

We continue to study the SDE

$$dX_t = \mu(t, X_t)dt + \sigma(t, X_t)dW_t + \beta(t-, X_{t-})dN_t, \tag{6.15}$$

where N has the intensity function $\lambda(t, x)$. Let us now consider a fixed point in time T and a real-valued function Φ. The object of this section is to understand how one can compute the expectation $E\left[\Phi(X_T)\right]$. In order to do this we consider the process Z defined by

$$Z_t = E\left[\Phi(X_T)|\, \mathcal{F}_t^X\right]. \tag{6.16}$$

We first note that, since X is Markov, we can write Z as

$$Z_t = E\left[\Phi(X_T)|\, X_t\right],$$

so in fact we have

$$Z_t = f(t, X_t),$$

where the deterministic function f is defined by $f(t, x) = E\left[\Phi(X_T)|\, X_t = x\right]$ or, equivalently by

$$f(t, x) = E_{t,x}\left[\Phi(X_T)\right].$$

Second, we note that since Z is given by the conditional expectation (6.16), the process Z is a martingale. From Proposition 6.5 we thus see that f must satisfy

$$\mathcal{A}f(t, x) = 0,$$

with the obvious boundary condition $f(T, x) = \Phi(x)$. We have thus proved the following result.

Proposition 6.7 (The Kolmogorov Backward Equation)   *Let X be the solution of (6.10), T a fixed point in time and Φ any function such that $\Phi(X_T) \in L^1$. Define the function f by*

$$f(t, x) = E_{t,x}\left[\Phi(X_T)\right].$$

Then f satisfies the Kolmogorov backward equation

$$\begin{cases} \mathcal{A}f(t,x) &= 0, & (t,x) \in R_+ \times R \\ f(T,x) &= \Phi(x), & x \in R \end{cases} \tag{6.17}$$

where $\mathcal{A}$ is given by (6.11).

Remark It is clear from the discussion around the Dynkin formula that the Kolmogorov backward equation is valid, not only for the solution to an SDE of the form (6.10), but in fact for a general Markov process.

In particular we may choose $\Phi(x) = I_A(x)$ where $A \subseteq R$ is a Borel set in R, and I denotes the indicator function. In this case we see that

$$f(t,x) = P\left(X_T \in A \,|\, X_t = x\right).$$

and we see that these **transition probabilities** must satisfy the backward equation with the boundary condition

$$f(T,x) = I_A(x).$$

Even more in particularly, if we assume that X has transition **densities** $p(t,x;T,z)$ with the interpretation

$$p(t,x;T,z)dz = P\left(X_T \in dz \,|\, X_t = x\right)$$

then also these transition densities must satisfy the Kolmogorov equation in the (t,x) variables, with boundary condition

$$p(T,x;T,z) = \delta_z(x),$$

where δ_z is the Dirac measure at z. We thus see that the transition probabilities, and thus the entire distribution, of the process X are completely determined by the infinitesimal generator $\mathcal{A}$.

We can also turn the Kolmogorov equation around. Instead of starting with an expected value and deriving a PIDE, we may start with the PIDE and derive an expected value.

Proposition 6.8 (Feynman–Kac) *Assume that X is as above, and suppose that a function f solves the PIDE*

$$\begin{cases} \mathcal{A}f(t,x) &= 0, & (t,x) \in R_+ \times R \\ f(T,x) &= \Phi(x), & x \in R \end{cases},$$

where $\mathcal{A}$ is given by (6.11). Then we have

$$f(t,x) = E_{t,x}\left[\Phi(X_T)\right].$$

Proof Assume that f satisfies the backward equation. If we consider the process $f(t,X_t)$ then, since $\mathcal{A}f = 0$, it is clear from the Dynkin formula that $f(t,X_t)$ is a martingale. Using the Markov property and the boundary condition we then obtain

$$f(t,X_t) = E\left[f(T,X_T) | \mathcal{F}_t\right] = E\left[\Phi(X_T) | X_t\right]. \qquad \square$$

In many finance applications it is natural to consider problems with discounting. A small variation of the arguments above gives us the following result.

Proposition 6.9 (Feynman–Kac) *Assume that X is as above, and suppose that a function f solves the PIDE*

$$\begin{cases} \mathcal{A}f(t,x) - rf(t,x) &= 0, \quad (t,x) \in R_+ \times R \\ f(T,x) &= \Phi(x), \quad x \in R \end{cases},$$

where $\mathcal{A}$ is given by (6.11) and r is a real number Then we have

$$f(t,x) = e^{-r(T-t)} E_{t,x}[\Phi(X_T)].$$

6.4   Existence of a Solution to the SDE

In this section we discuss the existence of a solution to the SDE

$$dX_t = \mu(t,X_t)dt + \sigma(t,X_t)dW_t + \beta(t-,X_{t-})dN_t, \tag{6.18}$$
$$X_0 = x_0, \tag{6.19}$$

where the counting process N is assumed to have a predictable intensity λ of the form

$$\lambda_t = \lambda(t-,X_{t-}). \tag{6.20}$$

The problem with this SDE is that existence of a solution is not completely obvious. We have seen earlier, that if N is a counting process with an *exogenously given* intensity process λ_t, then a solution will exist under mild assumptions. The problem with the SDE above is that the intensity process depends on the solution of the SDE, so λ is not exogenously given.

This problem can, however, be resolved using Girsanov theory and we have the following result.

Proposition 6.10 *Let $\mu(t,x)$, $\sigma(t,x)$, $\beta(t,x)$ and $\lambda(t,x)$ be given functions. Assume furthermore that μ, σ and β are regular enough to guarantee a strong solution to the SDE (6.18)–(6.19), under the assumption that W is Wiener and N is Poisson with unit intensity. Then there exists a filtered probability space $(\Omega, \mathcal{F}, P^\lambda, \mathbf{F})$, and a process X such that the following hold.*

- *The process X solves the SDE*

$$dX_t = \mu(t,X_t)dt + \sigma(t,X_t)dW_t + \beta(t-,X_{t-})dN_t, \tag{6.21}$$
$$X_0 = x_0. \tag{6.22}$$

- *The counting process N has a $(P^\lambda, \mathbf{F})$-predictable intensity given by*

$$\lambda_t = \lambda(t-,X_{t-}).$$

- *The process W is $(P^\lambda, \mathbf{F})$-Wiener.*

Proof Consider a filtered space $(\Omega, \mathcal{F}, P, \mathbf{F})$ carrying a Wiener process W and a Poisson process N with unit intensity (which then is independent of W) and assume that $\mathcal{F}_t = \mathcal{F}_t^W \vee \mathcal{F}_t^N$. Let X be the solution to the SDE (6.18)–(6.19), Now perform a Girsanov transformation from P to P^λ, using the likelihood process L defined by

$$dL_t = L_{t-}\left[\lambda(t-, X_{t-}) - 1\right]\{dN_t - dt\},$$
$$L_0 = 1.$$

The Girsanov theorem then implies that N has the $(P^\lambda, \mathbf{G})$-intensity $\lambda_t = \lambda(t-, X_{t-})$. It remains to show that W is still Wiener under P^λ and this is left as an exercise. □

6.5   Exercises

Exercise 6.1   Complete the proof of Proposition 6.10 by showing that W is a $(P^\lambda, \mathbf{F})$-Wiener process.

Hint: Use the Lévy characterization and the fact that M is a P^λ-martingale if and only if $L_t M_t$ is a P-martingale.

7   Marked Point Processes

In this chapter we will generalize the concept of a k-variate point process to that of a market point process.

7.1   Introduction

Consider a k-variate point process $(N^1, \ldots, N^k)$ on some filtered space $(\Omega, \mathcal{F}, P, \mathbf{F})$, and let us denote by $\lambda^1, \ldots \lambda^k$ the corresponding $\mathbf{F}$-predictable intensities. Let us now define the **mark space** E by

$$E = \{1, 2, \ldots, k\},$$

and on E we define the σ-algebra $\mathcal{E}$ as the power set of E, so $\mathcal{E} = 2^E$.

We recall from Remark 2.17 that the process $(N^1, \ldots, N^k)$ can equivalently be described in terms of the points (T_n, Z_n). We will now use this representation to define a random measure on $R_+ \times E$.

Definition 7.1   On the measurable space $(R_+ \times E, \mathcal{B} \otimes \mathcal{E})$ we define the random measure Ψ by

$$\Psi(D) = \sum_{n=1}^{\infty} I_D \{(T_n, Z_n)\}, \tag{7.1}$$

for all $\mathcal{B} \otimes \mathcal{E}$-measurable sets $D \subseteq R_+ \times E$.

In everyday language this means that, for every ω, the entity $\Psi(D)(\omega)$ is the number of points $(T_n(\omega), Z_n(\omega))$ which belong to D. We note the following properties of Ψ.

- The measure Ψ is a random **counting measure** on $R_+ \times E$, in the sense that it only takes integer values.
- For every fixed t we have (because of no common jumps).

$$\Psi(\{t\} \times E) = 0 \text{ or } 1.$$

- For every t we have

$$\Psi([0, t] \times E) < \infty, \quad P - a.s.$$

- We can recover the counting process N^i from Ψ by the formula

$$N_t^i = \Psi([0, t] \times \{i\}).$$

- For every $A \subseteq E$ we may define the counting process N^A by

$$N_t^A = \Psi([0,t] \times A)$$

  and the corresponding intensity is given by

$$\lambda_t(A) = \sum_{i=1}^{k} \lambda_t^i I_A(i).$$

- For every fixed t, the intensity $\lambda_t(\cdot)$ is a (random) measure on E.

7.2   Defining a Marked Point Process

We now generalize the simple idea in the previous section to a more general situation.

Definition 7.2   Consider a filtered probability space $(\Omega, \mathcal{F}, P, \mathbf{F})$ and a measurable **mark space** $(E, \mathcal{E})$. A **marked point process** Ψ is a random measure on $(R_+ \times E, \mathcal{B} \otimes \mathcal{E})$ with the following properties.

1. The measure Ψ is a counting measure in the sense that it only takes integer values.
2. For every fixed t it holds that

$$\Psi(\{t\} \times E) = 0 \text{ or } 1.$$

3. For every $A \in \mathcal{E}$ the counting process N^A, defined by

$$N_t^A = \Psi([0,t] \times A),$$

   is $\mathbf{F}$-optional.
4. For every t we have

$$\Psi([0,t] \times E) < \infty, \quad P-a.s.$$

Note that for any $D \in \mathcal{B} \otimes \mathcal{E}$ we should really write $\Psi(D, \omega)$ to indicate the dependence on ω, but most often we will suppress ω. It is clear that since $\Psi(\cdot, \omega)$ is a counting measure for every ω, it will be completely determined by its atoms. Since, by definition, two atoms cannot have the same t-coordinate, we can equally well describe the measure Ψ by a double sequence $\{(T_n, Z_n)\}_{n=1}^{\infty}$, where $(T_n, Z_n) \in R_+ \times E$ and $T_1 < T_2, \ldots$ The time sequence $\{T_n\}_n$ is of course the sequence of jump times for the counting process N_t^E, so

$$T_n = \inf \{t \geq 0 : \Psi([0,t] \times E) = n\}.$$

We can thus identify a marked point process either with the random measure Ψ or with the random point sequence $\{(T_n, Z_n)\}_{n=1}^{\infty}$. We will often use the acronym MPP to denote a marked point process, and we will sometimes refer to the jump times $\{T_n\}_n$ as **event times**.

We finish this section by giving the definition of the internal filtration generated by an MPP.

Definition 7.3   Given a marked point process Ψ, the internal filtration $\mathbf{F}^{\Psi}$ is defined by

$$\mathcal{F}_t^{\Psi} = \sigma\left\{\Psi([0,s] \times A) : A \in \mathcal{E}, \ s \le t\right\}.$$

so

$$\mathcal{F}_t^{\Psi} = \bigvee_{A \in \mathcal{E}} \mathcal{F}_t^{N^A}.$$

7.3   Point-Process Integrals and Differentials

Given a marked point process Ψ and a random process β, where β is a mapping $\beta : R_+ \times E \times \Omega \to R$, we can form the integral $X = \beta \star \Psi$ defined by

$$X_t = \int_0^t \int_E \beta_s(z) \Psi(ds, dz). \tag{7.2}$$

The interpretation of this is clear, and we will use it over and over again.

Interpretation 7.4   If the MPP has a point at time t with mark z then X has a jump of size $\beta_t(z)$.

In more concrete terms we can write X as

$$X_t = \sum_{T_n \le t} \beta_{T_n}(Z_n) \tag{7.3}$$

where $\{(T_n, Z_n)\}_{n=1}^{\infty}$ are the event times and marks of Ψ. If X is as above we will write the **stochastic differential**

$$dX_t = \int_E \beta_t(z) \Psi(dt, dz) \tag{7.4}$$

as a shorthand notation for (7.2)–(7.3).

7.4   The Doob–Meyer Decomposition

In this section we will discuss an important result, the so called called Doob–Meyer decomposition, which will be very useful in the next section when we discuss point processes with stochastic intensities. We start with the discrete-time theory and then turn to the continuous-time case.

7.4.1   Discrete Time

Consider a filtered space $(\Omega, \mathcal{F}, P, \mathbf{F})$ in discrete time. We then have the following, rather trivial, result.

Proposition 7.5 (Doob–Meyer in Discrete Time)   *Suppose that X is a submartingale. Then there exists a unique process A with the following properties*

- *A is predictable and increasing, and $A_0 = 0$.*
- *The process M defined by*

$$M_n = X_n - A_n, \quad n = 0, 1, \ldots$$

  is a martingale.

Proof   Define the process A, by setting $A_0 = 0$ and then defining the increments by

$$\Delta A_n = E[\Delta X_n | \mathcal{F}_{n-1}], \quad n = 1, 2, \ldots, \tag{7.5}$$

where we recall that, by definition, $\Delta A_n = A_n - A_{n-1}$. We then have

$$A_n = \sum_{k=1}^{n} \Delta A_n, \quad n = 1, 2, \ldots$$

By construction it is clear that $A_n \in \mathcal{F}_{n-1}$ so A is clearly predictable. Since X is a submartingale we have $\Delta A_n = E[\Delta X_n | \mathcal{F}_{n-1}] \geq 0$, so A is increasing. We also have

$$E[\Delta M_n | \mathcal{F}_{n-1}] = E[\Delta X_n | \mathcal{F}_{n-1}] - E[\Delta A_n | \mathcal{F}_{n-1}] = E[\Delta X_n | \mathcal{F}_{n-1}] - \Delta A_n = 0,$$

which shows that M is a martingale. To prove uniqueness is easy and left to the reader.   □

7.4.2   Continuous Time

We now consider a filtered space $(\Omega, \mathcal{F}, P, \mathbf{F})$ in continuous time, and assume that X is a submartingale. It is then natural to expect that there should be a continuous-time version of the Doob–Meyer decomposition. As we saw above, proving the existence of the decomposition in discrete time was more or less trivial, but in continuous time it is much harder. We do, however, have a result but it requires an integrability condition which we now present.

Definition 7.6   A submartingale X is said to be of **class** $\mathcal{D}$ if the family of random variables $\{X_T\}$ where T varies over the class of finite stopping times, is uniformly integrable.

We can now state our main result.

Theorem 7.7 (Doob–Meyer Decomposition)   *Assume that X is a submartingale of class $\mathcal{D}$. Then there exists a unique process A with the following properties.*

- *A is predictable and increasing, and $A_0 = 0$.*
- *The process M defined by*

$$M_t = X_t - A_t, \quad t \geq 0,$$

  is a uniformly integrable martingale.

Proof   The full proof is difficult and we refer the reader to Cohen & Elliott (2015) or Jacod & Shiryaev (1987).   □

Although the full proof is rather technical, the intuitive idea is quite simple. The obvious idea is to define the process A by setting $A_0 = 0$ and then define the differential of A as a continuous-time version of (7.5). We would thus define the differential dA_t by the expression

$$dA_t = E[dX_t | \mathcal{F}_{t-}]. \tag{7.6}$$

This looks nice, but from a formal point of view the right-hand side of the equation has a syntax error, which is why the full proof is a bit hard. Nevertheless, from an *informal* and intuitive point of view, the formula (7.6) is of great value, and we will refer to it at several occasions below, so we highlight it as follows.

Intuition 7.8 *Informally we have the interpretation*

$$dA_t = E[dX_t | \mathcal{F}_{t-}]. \tag{7.7}$$

We finish by stating the local version of the Doob–Meyer decomposition.

Proposition 7.9 *Assume that X is a local submartingale. Then there exists a unique predictable, increasing process A with $A_0 = 0$, such that the process M, defined by $M_t = X_t - A_t$, is a local martingale.*

7.5   The Intensity of an MPP

Consider a given MPP Ψ with mark space $(E, \mathcal{E})$, and assume that the process N_t^A has an **F**-predictable intensity $\lambda_t(A)$ for every $A \in \mathcal{E}$. It is then easy to see that $\lambda_t(\cdot)$ is a random measure on $(E, \mathcal{E})$. This leads us to the definition of the intensity measure of an MPP.

Definition 7.10 We say that the marked point process Ψ has the **predictable intensity** λ, if λ is a mapping $\lambda : R_+ \times \mathcal{E} \times \Omega \to R$ with the following properties.

1. For each (t, ω) it holds that $\lambda_t(\cdot, \omega)$ is a non-negative measure on $(E, \mathcal{E})$.
2. For each $A \in \mathcal{E}$ the process $t \longmapsto \lambda_t(A)$ is **F**-predictable.
3. For each $A \in \mathcal{E}$, the counting process

$$N_t^A = \Psi([0, t] \times A)$$

has the predictable **F**-intensity process $\lambda(A)$.

For questions concerning the existence of a predictable intensity, see Last & Brandt (1995). Loosely speaking, the program is the following.

- For every $A \in \mathcal{E}$ the Doob–Meyer decomposition from Theorem 7.7 will guarantee the existence of an increasing predictable process $\nu_t(A)$ such that the process

$$N_t^A - \nu_t(A)$$

  is a martingale. The process ν is called the **compensator** of the point process, and the martingale above is called the compensated martingale.

- One can then show that, for fixed t, $v_t(\cdot)$ is a (random) measure on $(E, \mathcal{E})$.
- If the mapping $t \longmapsto v_t(A)$ is absolutely continuous with respect to Lebesgue measure P-a.s. for every fixed $A \in \mathcal{E}$, then we can write

$$v_t(A) = \int_0^t \lambda_s(A) ds.$$

- Assuming the existence of an intensity is thus equivalent to assuming that the compensator v is absolutely continuous. The intuitive content of this is roughly that we exclude deterministic jump times. The assumption does in fact imply that for every jump time T_n and every fixed t we have $P(T_n = t) = 0$.

For the rest of the text we will boldly (and shamelessly) assume that λ does exist. The intuitive interpretation of λ is just as before:

$$\lambda_t(A) dt = E\left[dN_t^A \big| \mathcal{F}_{t-} \right].$$

and we have the following extension of Proposition 4.5.

Proposition 7.11

$$P\left(Z_n \in A \big| \mathcal{F}_{T_n-}^{\Psi} \right) = \frac{\lambda_{T_n}(A)}{\lambda_{T_n}(E)}, \quad on \ \{T_n < \infty\}.$$

Proof The proof of this requires some knowledge of the detailed structure of the internal filtration $\mathbf{F}^{\Psi}$ and is omitted. See Last & Brandt (1995). □

We thus see that the intensity λ will generate a measure-valued process

$$\Gamma_t(dz) = \frac{\lambda_t(dz)}{\lambda_t(E)}$$

and that $\Gamma_t(dz)$ is a probability measure. The pair $(\lambda_t(E), \Gamma_t)$ are often referred to as the local characteristics of Ψ.

Definition 7.12 If Ψ is such that $\lambda_t(E) < \infty$ we define the **local characteristics** λ^E and Γ by

$$\lambda_t^E = \lambda_t(E), \tag{7.8}$$

$$\Gamma(dz) = \frac{\lambda_t(dz)}{\lambda_t(E)}. \tag{7.9}$$

The structure of the process Ψ can thus schematically be described as follows.

- The intensity process λ_t^E governs the distribution of the event times $\{T_n\}_n$.
- Given the fact that we have an event at time t, the distribution over marks is governed by the probability measure $\Gamma_t(dz)$.

We finish this section by studying the case when the intensity is deterministic.

Proposition 7.13 *Consider a marked point process Ψ with corresponding intensity process λ. Assume that $\lambda_t(dz)$ is deterministic. Then the following hold.*

- *The counting process N_t^E is Poisson with intensity function λ_t^E.*

- *Suppose that $A_1, \ldots, A_k$ are $\mathcal{E}$-measurable disjoint subsets of E. Then the counting processes $N^{A_1}, \ldots, N^{A_k}$ are independent Poisson processes with intensity functions $\lambda_t(A_1), \ldots, \lambda_t(A_k)$ respectively.*

Proof   Since the intensities $\lambda_t(A_1), \ldots, \lambda_t(A_k)$ are deterministic, the Watanabe theorem implies that $N^{A_1}, \ldots, N^{A_k}$ are Poisson. Since, by construction, they have no common jumps, Proposition 3.2 implies that they are independent.   □

   A particular case of this situation occurs when $\lambda_t(dz)$ is constant over time, so the intensity process is a fixed deterministic measure $\lambda(dz)$. In this case the interpretation is as follows.

- Events occur according to a Poisson process with constant intensity λ^E.
- When an event occurs, the mark $z \in E$ is (independently of the past history) determined by the probability measure

$$\Gamma(dz) = \frac{\lambda(dz)}{\lambda(E)}.$$

In such a case we say that the underlying marked point process Ψ is **compound Poisson**.

7.6   Sums of Marked Point Processes

Suppose that we have a filtered space $(\Omega, \mathcal{F}, P, \mathbf{F})$ carrying two marked point processes Ψ_1 and Ψ_2, having the same mark space E, and admitting the intensity measures $\lambda_t^1(dz)$ and $\lambda_t^2(dz)$ respectively. We then have the following easy result.

Proposition 7.14   *Assume that Ψ_1 and Ψ_2 have no common event times. Then the stochastic measure Ψ, defined by*

$$\Psi(dt, dz) = \Psi_1(dt, dz) + \Psi_2(dt, dz),$$

is a marked point process with intensity measure

$$\lambda_t(dz) = \lambda_t^1(dz) + \lambda_t^2(dz).$$

Proof   The absence of common event times guarantees that Ψ is indeed a point process. The intensity part is obvious from the definition of intensity.   □

7.7 $\mathcal{P}$-Predictability and Martingales

We note that the process M_t^A, defined by

$$dM_t^A = dN_t^A - \lambda_t(A)dt,$$

is a martingale for each $A \in \mathcal{E}$, and this implies that an integral of the form

$$\int_0^t h_s(A) \left\{ dN_s^A - \lambda_t(A)ds \right\}$$

will be a martingale provided that the process $t \longmapsto h_t(A)$ (where A only serves as an index) is predictable. Infinitesimally we should also be allowed to think that

$$dN_t^{dz} - \lambda_t(dz)dt = \Psi(dt, dz) - \lambda_t(dz)dt$$

is a martingale increment (in the time variable), where dz is the integration element in $\mathcal{E}$. Since the sum of martingales is a new martingale, we furthermore expect that an integral of the form

$$\int_0^t \int_E H_s(z) \{\Psi(ds, dz) - \lambda_s(dz)ds\}$$

should be a martingale if the process $H_t(z)$ is predictable in some reasonable sense. The correct definition of predictability turns out to be the following, where we recall that, given a filtered space $(\Omega, \mathcal{F}, P, \mathbf{F})$, we denote the predictable σ-algebra on $R_+ \times \Omega$ by Σ_p.

Definition 7.15   Given a filtered space $(\Omega, \mathcal{F}, P, \mathbf{F})$ and a mark space $(E, \mathcal{E})$, the **predictable** σ-algebra $\mathcal{P}$ on $(R_+ \times \Omega) \times E$ is defined by

$$\mathcal{P} = \Sigma_p \otimes \mathcal{E}.$$

A mapping $H : R_+ \times E \times \Omega \to R$ is $\mathcal{P}$-**predictable** if it is measurable with respect to $\mathcal{P}$.

   In more concrete terms $\mathcal{P}$-predictable means that the process $H_t(z)$ is predictable in the standard sense for each $z \in E$. We then have the following rather expected result.

Proposition 7.16   *Let* $\Psi(dt, dz)$ *be a marked point process with predictable intensity measure* $\lambda_t(dz)$. *Assume that* $H_t(z)$ *is predictable and that*

$$E\left[\int_0^\infty \int_E |H_t(z)| \lambda_t(dz)dt\right] < \infty. \tag{7.10}$$

Then the process

$$M_t = \int_0^t \int_E H_t(z) \{\Psi(ds, dz) - \lambda_s(dz)ds\} \tag{7.11}$$

is a martingale. If H satisfies

$$\int_0^\infty \int_E |H_t(z)| \lambda_t(dz)dt < \infty, \quad P - a.s. \tag{7.12}$$

then M is a local martingale.

Proof   Prove the result for a generator system of $\mathcal{P}$ and then use a monotone class argument.   □

8   The Itô Formula

In this chapter we will derive the relevant Itô formula when we, apart from a Wiener process, also have a driving marked point process.

8.1   Deriving the Itô Formula

Consider a filtered probability space $(\Omega, \mathcal{F}, P, \mathbf{F})$ carrying a Wiener process W and a marked point process Ψ with intensity process λ. If a cadlag process X can be represented as

$$X_t = x_0 + \int_0^t \mu_s \, ds + \int_0^t \sigma_s \, dW_s + \int_0^t \int_E \beta_s(z) \Psi(dz, ds) \tag{8.1}$$

where μ and σ are optional and β is $\mathcal{P}$-predictable, we say that X has a **stochastic differential**, and we write

$$dX_t = \mu_t \, dt + \sigma_t \, dW_t + \int_E \beta_t(z) \Psi(dz, dt), \tag{8.2}$$

$$X_0 = x_0. \tag{8.3}$$

We note that the differential formulation (8.2)–(8.3) is merely a shorthand expression for the integral equation (8.1).

The interpretation of (8.2) is very simple and goes as follows.

- Between the event times of q, we have a standard stochastic differential

$$dX_t = \mu_t \, dt + \sigma_t \, dW_t.$$

- If we have an event at time t with mark $z \in E$, then X will have a jump with size $\beta_t(z)$, so

$$X_t = X_{t-} + \beta_t(z).$$

Suppose now that X has the stochastic differential (8.2) above, and that $F(t, x)$ is a $C^{1,2}$ function. The question is now to determine the differential $dF(t, X_t)$ and this is quite easy.

- Between the event times of Ψ, we have the standard expression

$$dF(t, X_t) = \left\{ F_t(t, X_t) \, dt + \mu_t F_x(t, X_t) + \frac{1}{2} \sigma_t^2 F_{xx}(t, X_t) \right\} dt$$
$$+ \sigma_t F_x(t, X_T) \, dW_t.$$

- If Ψ has an event at time t with mark z, the process X will have a jump of size $\beta_t(z)$.
- Consequently, the jump of F, defined by

$$\Delta F(t, X_t) = F(t, X_t) - F(t-, X_{t-}),$$

can be written as

$$\Delta F(t, X_t) = F(t-, X_{t-} + \beta_t(z)) - F(t-, X_{t-}).$$

- We can thus write

$$\Delta F(t, X_t) = \int_E [F(t-, X_{t-} + \beta_t(z)) - F(t-, X_{t-})] \Psi(dt, dz)$$

and we note (with some satisfaction) that the integrand is predictable. Since F is smooth we may write this as

$$\Delta F(t, X_t) = \int_E [F(t, X_{t-} + \beta_t(z)) - F(t, X_{t-})] \Psi(dt, dz).$$

By this argument we have in fact proved the following Itô formula. The multidimensional extension is obvious.

Proposition 8.1 (The Itô formula)   *Assume that X has the differential (8.2) and that F is smooth. Then we have*

$$dF(t, X_t) = \left\{ F_t(t, X_t) + \mu_t F_x(t, X_t) + \frac{1}{2}\sigma_t^2 F_{xx}(t, X_t) \right\} dt$$

$$+ \sigma_t F_x(t, X_T) dW_t$$

$$+ \int_E [F(t, X_{t-} + \beta_t(z)) - F(t, X_{t-})] \Psi(dt, dz). \qquad (8.4)$$

8.2   The Semimartingale Decomposition

The Itô formula above gives us the dynamics of $F(t, X_t)$ but in many applications we would also like to have the "special semimartingale decomposition" of X, i.e. we would like to write dX_t in the form

$$dX_t = \alpha_t dt + dM_t,$$

where M is a (local) martingale. In other words, we would like to decompose dX_t in a drift part and a martingale part. The formal definition of a semimartingale is as follows.

Definition 8.2   A process X is a **semimartingale** if it can be decomposed as

$$X_t = A_t + M_t,$$

where A is an optional process of bounded variation and M is a local martingale. If A is also predictable, then we say that X is a **special semimartingale**

The decomposition of a semimartingale is not unique. If, for example X is a Poisson process N with intensity λ, we can write

$$X_t = N_t,$$

with $M = 0$ and $A = N$ (and A is of course not predictable), but we can also decompose X as a special semimartingale in the form

$$X_t = \lambda t + [N_t - \lambda t],$$

with predictable part $A_t = \lambda t$ and martingale part $M_t = N_t - \lambda t$. The second decomposition obviously seems to be more canonical than the first, and one can in fact prove the following result.

Proposition 8.3   *If X is a special semimartingale, then the decomposition*

$$X_t = A_t + M_t$$

is unique up to an additive constant, so if we require that $M_0 = 0$ then it is unique.

In all applications in this book, we can always find a decomposition $X = A + M$ such that

$$A_t = \int_0^t a_s ds$$

for some optional process a. This implies that A is continuous and hence predictable. For ease of notation we therfore introduce the following notational convention.

Note   In this book all semimartingales are special semimartingales, so in the text the term "semimartingale" henceforth means "special semimartingale".

Going back to (8.4) we can easily find the semimartingale decomposition by compensating the point process Ψ with the intensity λ. In order to have more compact notation we introduce the notation

$$F_\beta(t, x, z) = F(t, x + \beta_t(z)) - F(t, x). \tag{8.5}$$

We then have

$$dF(t, X_t) = \left\{ F_t(t, X_t) + \mu_t F_x(t, X_t) + \frac{1}{2}\sigma_t^2 F_{xx}(t, X_t) + \int_E F_\beta(t, X_t, z)\lambda_t(dz) \right\} dt$$
$$+ \sigma_t F_x(t, X_T) dW_t$$
$$+ \int_E F_\beta(t, X_{t-}, z) \left\{ \Psi(dt, dz) - \lambda_t(dz)dt \right\}, \tag{8.6}$$

where the first term is predictable and of bounded variation. The dW-term is of course a local martingale differential and, by predictability and Proposition 7.16, the last term is also a (local) martingale increment.

8.3 The Quadratic Variation $\langle M, N \rangle$

In this section we introduce a new concept, namely the "predictable quadratic variation process" (or "angular bracket process") $\langle M, N \rangle$ for square-integrable martingales. This is a very important concept in stochastic analysis and in applications. We will for example use it in Section 19.1 to define the total volatility of a jump-diffusion price process.

8.3.1 Construction and Basic Properties

The construction of $\langle M, N \rangle$ goes as follows. Let M be a square-integrable martingale. Then, by the Jensen inequality, the process M_t^2 will be a submartingale, and by Doob's inequality it follows that it will belong to the class $\mathcal{D}$. The Doob–Meyer decomposition theorem then implies that there exists a unique predictable increasing process A with $A_0 = 0$ such that $M_t^2 - A_t$ is a martingale. We now simply define $\langle M, M \rangle$ by $\langle M, M \rangle_t = A_t$. By localization we can in fact prove the following result.

Proposition 8.4 *Let M and N be locally square-integrable martingales. Then there exist a unique predictable process $\langle M, N \rangle$, with $\langle M, N \rangle_0 = 0$, of finite variation such that the process*

$$M_t N_t - \langle M, N \rangle_t \tag{8.7}$$

is a local martingale. Moreover, we have the polarization formula

$$\langle M, N \rangle_t = \frac{1}{4} \{ \langle M + N, M + N \rangle_t - \langle M - N, M - N \rangle_t \} .$$

Proof By localization we can assume that M and N are square-integrable, which allows us to define $\langle M + N, M + N \rangle$ and $\langle M - N, M - N \rangle$ using Doob–Meyer. We then simply define $\langle M, N \rangle$ by the polarization formula above and check that the process in (8.7) is a martingale. $\square$

We now collect some properties of the quadratic variation process.

Proposition 8.5 *The angular bracket process has the following properties.*

- *It is symmetric, i.e*

$$\langle M, N \rangle = \langle N, M \rangle.$$

- *It is bilinear: if M, N and K are local martingales, and α and β are real numbers, then*

$$\langle \alpha M + \beta K, N \rangle = \alpha \langle M, N \rangle + \beta \langle K, N \rangle.$$

- *If h is predictable then*

$$\langle h \star M, N \rangle = h \star \langle M, N \rangle,$$

i.e.

$$d\langle h \star M, N \rangle_t = h_t d\langle M, N \rangle_t.$$

Proof The first two point are obvious. The last point is expected given the bilinearity but the precise proof is omitted. See Protter (2004).   □

8.3.2   Intuitive Interpretation of $\langle M, N \rangle$

The construction above is very elegant, but the intuitive interpretation of $\langle M, N \rangle$ is perhaps not so clear. To get a more intuitive feeling for the angular bracket, let us start by going back to the process $\langle M, M \rangle$. By definition, and Doob–Meyer, the process $\langle M, M \rangle$ is the unique predictable increasing process A such that the process $M_t^2 - A_t$ is a martingale. In Section 7.4 we saw that the intuitive interpretation of $A = \langle M, M \rangle$, was to write it in differential form as

$$d\langle M, M \rangle_t = E\left[dM_t^2 \middle| \mathcal{F}_{t-} \right],$$

and this is our first interpretation of the angular bracket. There is, however, a slightly different but more interesting interpretation, and to see this let us go back to discrete time. If M is a discrete-time martingale then we can of course define the corresponding discrete-time process $\langle M, M \rangle$ by

$$\Delta \langle M, M \rangle_n = E\left[\Delta (M^2)_n \middle| \mathcal{F}_{n-1} \right] = E\left[M_n^2 - M_{n-1}^2 \middle| \mathcal{F}_{n-1} \right].$$

By iterated expectations it is easy to see that

$$E\left[M_n^2 - M_{n-1}^2 \middle| \mathcal{F}_{n-1} \right] = E\left[(M_n - M_{n-1})^2 \middle| \mathcal{F}_{n-1} \right],$$

so we do in fact have

$$\Delta \langle M, M \rangle_n = E\left[(\Delta M_n)^2 \middle| \mathcal{F}_{n-1} \right].$$

We can easily extend this to the bracket $\langle M, N \rangle$ and, going to continuous time, we have the following informal result.

Intuition 8.6 *The angular bracket process $\langle M, N \rangle$ has the following informal interpretation*

$$d\langle M, N \rangle_t = E\left[d(M_t \cdot N_t) \middle| \mathcal{F}_{t-} \right], \tag{8.8}$$

or

$$d\langle M, N \rangle_t = E\left[dM_t \cdot dN_t \middle| \mathcal{F}_{t-} \right], \tag{8.9}$$

or alternatively

$$d\langle M, N \rangle_t = \mathrm{Cov}\left[(dM_t, dN_t) \middle| \mathcal{F}_{t-} \right]. \tag{8.10}$$

Given this interpretation, the results in Proposition 8.5 are more or less obvious.

8.3.3   The Bracket $\langle X, Y \rangle$ for Special Semimartingales

The extension of the angular bracket to semimartingales is an easy one.

Definition 8.7 If X and Y are (special) semimartingales with decompositions

$$X_t = A_t + M_t,$$
$$Y_t = B_t + N_t,$$

where A, B are predictable processes of bounded variations, and M, N are locally square-integrable martingales, then the angular bracket is defined as

$$\langle X, Y \rangle_t = \langle M, N \rangle_t. \tag{8.11}$$

With a little bit of effort one can show that the previously derived intuitive interpretation is still valid.

Intuition 8.8 *The angular bracket process $\langle X, Y \rangle$ has the following informal interpretation*

$$d\langle X, Y \rangle_t = E[d(X_t \cdot Y_t)| \mathcal{F}_{t-}], \tag{8.12}$$

or

$$d\langle X, Y \rangle_t = E[dX_t \cdot dY_t| \mathcal{F}_{t-}], \tag{8.13}$$

or alternatively

$$d\langle X, Y \rangle_t = \text{Cov}[(dX_t, dY_t)| \mathcal{F}_{t-}]. \tag{8.14}$$

8.3.4 The Bracket $\langle M, M \rangle$ for a Jump Diffusion

We now move on to actually compute the angular bracket for the case of a jump diffusion. We thus consider a filtered space $(\Omega, \mathcal{F}, P, \mathbf{F})$ carrying a d-dimensional Wiener process W and a marked point process $\Psi(dt, dz)$ with mark space $(E, \mathcal{E})$, and intensity measure $\lambda_t(dz)$. In this setting we study the process M with dynamics

$$dM_t = \sigma_t dW_t + \int_E \beta_t(z)[\Psi(dt, dz) - \lambda_t(dz)dt], \tag{8.15}$$

where σ is an optional d-dimensional row vector process, and β is $\mathcal{P}$-predictable. It is clear that M is a local martingale and we now go on to compute $\langle M, M \rangle$. From the Itô formula we have

$$dM_t^2 = 2M_t \sigma_t dW_t + \|\sigma_t\|^2 dt - 2M_t \left(\int_E \beta_t(z)\lambda_t(dz) \right) dt + \Delta(M_t^2),$$

where $\|\cdot\|$ denotes the Euclidian norm in R^d. We have

$$\Delta(M_t^2) = M_t^2 - M_{t-}^2 = (M_{t-} + \Delta M_t)^2 - M_{t-}^2 = 2M_{t-}\Delta M_t + (\Delta M_t)^2$$
$$= 2M_{t-} \int_E \beta_t(z)\Psi(dt, dz) + \int_E \beta_t^2(z)\Psi(dt, dz),$$

where we have used the fact that if we have an event with mark z at time t then $\Delta M_t = \beta_t(z)$ and $(\Delta M_t)^2 = \beta_t^2(z)$. We thus have

$$
dM_t^2 = 2M_t\sigma_t dW_t + 2M_{t-}\int_E \beta_t(z)\left[\Psi(dt,dz) - \lambda_t(dz)dt\right]
$$

$$
+ \int_E \beta_t^2(z)\left[\Psi(dt,dz) - \lambda_t(dz)dt\right]
$$

$$
+\|\sigma_t\|^2 dt + \int_E \beta_t^2(z)\lambda_t(dz)dt.
$$

The first three terms are martingale increments, and the last two terms are obviously predictable (why?), so we have proved the following result.

Proposition 8.9 *For the jump-diffusion local martingale M defined by (8.15) we have*

$$
\langle M, M\rangle_t = \int_0^t \left\{\|\sigma_s\|^2 + \int_E \beta_s^2(z)\lambda_s(dz)\right\} ds \tag{8.16}
$$

or, equivalently,

$$
d\langle M, M\rangle_t = \left[\|\sigma_t\|^2 + \int_E \beta_t^2(z)\lambda_t(dz)\right] dt. \tag{8.17}
$$

In connection with stock price dynamics we will later use the following informal interpretation.

Intuition *On the informal level we have*

$$
\mathrm{Var}\left[dM_t|\mathcal{F}_{t-}\right] = \left[\|\sigma_t\|^2 + \int_E \beta_t^2(z)\lambda_t(dz)\right] dt. \tag{8.18}
$$

8.4   The Compound Poisson Process

We consider a point process Ψ with the real line as mark space, so $E = R$, and we assume that the intensity measure process $\lambda(dz)$ is deterministic and constant over time. We then define the process X by

$$
dX_t = \int_R z\Psi(dz,dt), \quad x_0 = 0. \tag{8.19}
$$

We can also write this as

$$
X_t = \sum_{n \leq N_t} Z_n,
$$

where $N_t = \Psi([0,t], R)$ is the counting process which counts the number of events regardless of marks, whereas $\{Z_n\}$ are the i.i.d. marks. In this case event N is Poisson, with intensity $\lambda = \lambda(R)$, and the distribution for the generic mark Z, which is also the jump size of X, is given by $\Gamma(dz) = \lambda(dz)/\lambda(R)$. A process X of this kind is referred to as a **compound Poisson process** and we also say that Ψ is compound Poisson.

We recall that two Poisson processes are independent if and only if they have no common jumps. The same holds for compound Poisson processes.

Proposition 8.10   *Assume that the space $(\Omega, \mathcal{F}, P, \mathbf{F})$ carries two compound Poisson processes X^1 and X^2 corresponding to the point process Ψ_1 and Ψ_2. Then the following hold.*

1. *The processes X^1 and X^2 are independent if and only if Ψ_1 and Ψ_2 have no common event times.*
2. *If Ψ_1 and Ψ_2 have no common event times, then the process X, defined by*

$$X_t = X_t^1 + X_t^2,$$

is compound Poisson.

Proof   The proof is left to the reader.   □

8.5   A Simple Lévy Process

As an application of the Itô formula let us consider a Wiener process W and a point process Ψ with the real line as mark space, so $E = R$. We assume that the intensity measure process is deterministic and constant over time. The intensity measure is denoted by $\lambda(dz)$ and we now study the process X defined by

$$dX_t = \mu dt + \sigma dW_t + \int_R z \Psi(dz, dt). \tag{8.20}$$

We see that in this case, the mark z equals the jump size of the process X, and it is obvious that the increments of X are independent and stationary. The X-process is in fact a simple example of a Lévy process, and our task is to compute the characteristic function

$$\varphi_t^X(u) = E\left[e^{iuX_t}\right].$$

To this end we define the process Y by

$$Y_t = e^{iuX_t}$$

and apply the Itô formula. Given a mark z at time t, the jump in X is z and the induced jump in Y is given by

$$\Delta Y_t = e^{iu(X_{t-} + z)} - e^{iuX_{t-}} = Y_{t-}\left(e^{iuz} - 1\right).$$

Using the Itô formula we thus have the dynamics

$$dY_t = Y_t \left\{ iu\mu - \frac{1}{2}\sigma^2 u^2 \right\} dt + iu\sigma Y_t dW_t + Y_{t-} \int_R \left(e^{iuz} - 1\right) \Psi(dt, dz).$$

Compensating the point process we obtain

$$dY_t = Y_t \left\{ iu\mu - \frac{1}{2}\sigma^2 u^2 + \int_R \left(e^{iuz} - 1\right) \lambda(dz) \right\} dt + iu\sigma Y_t dW_t$$

$$+ Y_{t-} \int_R \left(e^{iuz} - 1\right) [\Psi(dt, dz) - \lambda(dz)dt],$$

where the last term is a martingale increment. We now integrate and take expected values to obtain

$$y_t = 1 + \int_0^t y_s \left\{ iu\mu - \frac{1}{2}\sigma^2 u^2 + \int_R \left(e^{iuz} - 1 \right) \lambda(dz) \right\} ds,$$

where $y_t = E[Y_t]$. Taking the derivative we obtain the ODE

$$\dot{y}_t = \left\{ iu\mu - \frac{1}{2}\sigma^2 u^2 + \int_R \left(e^{iuz} - 1 \right) \lambda(dz) \right\} y_t,$$

$$y_0 = 1,$$

so we have our final result.

Proposition 8.11 *Let X be defined by (8.20) above. Then the characteristic function is given by*

$$\varphi_t^X(u) = e^{iu\mu t - \frac{1}{2}\sigma^2 u^2 t + t \int_R (e^{iuz} - 1)\lambda(dz)}.$$

This is a special case of the Lévy–Khinchine formula. For (much) more on this see Protter (2004).

8.6   Linear and Geometric Stochastic Differential Equations

In this section we consider a Wiener process W and a point process Ψ with mark space E, and we present two very common SDEs.

8.6.1   The Linear SDE

The **linear SDE** is defined by

$$\begin{cases} dX_t &= (aX_t + b)dt + \sigma dW_t + \int_E \beta(z)\Psi(dz, dt), \\ X_0 &= x_0, \end{cases} \tag{8.21}$$

where a, b, σ are real numbers, and β is a deterministic function.

Multiplying by the integrating factor e^{-at} gives us

$$e^{-at}dX_t - ae^{-at}X_t = e^{-at}bdt + \sigma e^{-at}dW_t + e^{-at}\int_E \beta(z)\Psi(dz, dt),$$

so, By Itô, we obtain

$$d\left(e^{-at}X_t\right) = e^{-at}bdt + \sigma e^{-at}dW_t + e^{-at}\int_E \beta(z)\Psi(dz, dt).$$

Integrating this gives us

$$e^{-at}X_t - x_0 = \int_0^t e^{-as}bds + \int_0^t \sigma e^{-as}dW_s + \int_0^t e^{-as}\int_E \beta(z)\Psi(dz, ds)$$

and we have our result.

Proposition 8.12 *The SDE (8.21) has the solution*

$$X_t = e^{at}x_0 + \int_0^t e^{a(t-s)}bds + \int_0^t \sigma e^{-a(t-s)}dW_s + \int_0^t \int_E e^{a(t-s)}\beta(z)\Psi(dz,ds). \quad (8.22)$$

8.6.2 The Geometric Lévy Process

The **geometric SDE** is defined by

$$\begin{cases} dX_t & = & \alpha X_t dt + X_t \sigma dW_t + X_{t-} \int_E \beta(z)\Psi(dz,dt), \\ X_0 & = & x_0 \end{cases} \quad (8.23)$$

with a, σ and β as in the previous section. Using the methodology of Proposition 2.22 we easily obtain the following result.

Proposition 8.13 *The solution to (8.23) is given by the formula*

$$X_t = x_0 e^{(\alpha - \frac{1}{2}\sigma^2)t + \sigma W_t} \cdot \prod_{T_n \le t} [1 + \beta(Z_n)], \quad (8.24)$$

where (T_n, Z_n) denotes the event times and marks of Ψ. If $x_0 > 0$ and $\beta > -1$ we can write the solution as

$$X_t = x_0 \exp\left\{ (\alpha - \frac{1}{2}\sigma^2)t + \sigma W_t + \int_0^t \int_E \ln[1 + \beta(z)]\,\Psi(ds,dz) \right\}. \quad (8.25)$$

If, in addition, the point process Ψ has a deterministic and constant intensity measure $\lambda(dz)$, we say that the process is (a simple version of) a **Geometric Lévy Process**.

8.7 Exercises

Exercise 8.1 Assume that the point process Ψ in Sections 8.6.1–8.6.2 has a deterministic and constant intensity measure $\lambda(dz)$.

(a) Compute $E[X_t]$ when X solves the linear SDE (8.21).
(b) Compute $E[X_t]$ when X solves the geometric SDE (8.23).

Exercise 8.2 Suppose that

$$dX_t = \alpha_X X_t dt + X_t \sigma_X dW_t + X_{t-} \int_E \beta_X(z)\Psi(dz,dt),$$

$$dY_t = \alpha_Y Y_t dt + X_t \sigma_Y dW_t + Y_{t-} \int_E \beta_Y(z)\Psi(dz,dt),$$

where W is a two-dimensional Wiener process. Define $Z_t = X_t/Y_t$ and compute dZ_t.

9 Martingale Representation, Girsanov and Kolmogorov

In this chapter we will extend our earlier results on martingale representations, measure transformations and the Kolmogorov equation to the marked point-process case.

9.1 Martingale Representation

We first study the possibility of representing a point-process martingale by a stochastic integral. Given our experience from the k-variate case, the following result should come as no surprise.

Theorem 9.1 (The Martingale Representation Theorem) *Consider a filtered space $(\Omega, \mathcal{F}, P, \mathbf{F})$ carrying a marked point process Ψ with mark space E and predictable $\mathbf{F}$-intensity $\lambda_t(dz)$. Assume that the filtration is the internal one, so that $\mathbf{F} = \mathbf{F}^\Psi$. Assume furthermore that X is a $(P, \mathbf{F})$-martingale (or local martingale). Then there exists a predictable process $H_t(z)$ and a real number x such that*

$$X_t = x + \int_0^t \int_E H_s(z) \{\Psi(ds, dz) - \lambda_s(dz)ds\}, \qquad (9.1)$$

or, on differential form,

$$dX_t = \int_E H_t(z) \{\Psi(dt, dz) - \lambda_t(dz)dt\}. \qquad (9.2)$$

The process H satisfies (7.10) in the martingale case and (7.12) in the local martingale case.

Proof   See Last & Brandt (1995).   □

9.2 The Girsanov Theorem

We now turn to the marked point-process version of the Girsanov theorem. With the k-variate theory in mind, the result is quite expected.

Theorem 9.2 (The Girsanov Theorem) *Suppose we have a filtered probability space $(\Omega, \mathcal{F}, P, \mathbf{F})$ carrying a marked point process $\Psi(dt, dz)$ with predictable intensity $\lambda_t(dz)$. Let $\varphi_t(z)$ be a predictable process satisfying*

$$\varphi_t(z) \geq 0, \quad P\text{-a.s.} \quad \text{for all } (t, z)$$

and define the likelihood process L by

$$\begin{cases} dL_t & = & L_{t-} \int_E [\varphi_s(z) - 1] \{ \Psi(dt, dz) - \lambda_t(dz)dt \}, \\ L_0 & = & 1. \end{cases} \tag{9.3}$$

Assume that

$$E^P [L_T] = 1,$$

and define the probability measure Q by

$$L_T = \frac{dQ}{dP}, \quad on \ \mathcal{F}_T.$$

Then Ψ has the predictable $(Q, \mathbf{F})$-intensity λ^Q given by

$$\lambda_t^Q(dz) = \varphi_t(z)\lambda_t(dz). \tag{9.4}$$

Proof   The proof is very similar to the proof for the counting process case. $\square$

With a small effort one can derive the following representation for the likelihood process L.

Lemma 9.3   *The likelihood process L above can be expressed as*

$$L_t = \prod_{k=1}^{N_t} \varphi_{T_k}(Z_k) \cdot e^{\int_0^t \int_E [1 - \varphi_s(z)]\lambda_s(dz)ds} \tag{9.5}$$

where $\{T_n\}$ and $\{Z_n\}$ denotes the event times and marks of Ψ. The formula can also be written as

$$L_t = e^{\int_0^t \int_E \ln \varphi_s(z)\Psi(ds,dz) + \int_0^t \int_E [1 - \varphi_s(z)]\lambda_s(dz)ds} \tag{9.6}$$

Proof   Exercise for the reader. $\square$

The Girsanov theorem above is easily extended to include a Wiener process.

Theorem 9.4 (The Girsanov Theorem)   *Consider a filtered space $(\Omega, \mathcal{F}, P, \mathbf{F})$ carrying a d-dimensional Wiener process W and marked point process $\Psi(dt, dz)$ with predictable intensity $\lambda_t(dz)$. Let γ_t be a d-dimensional column vector process and let $\varphi_t(z)$ be a predictable process satisfying*

$$\varphi_t(z) \geq 0, \quad P - a.s. \quad for \ all \ (t, z)$$

and define the likelihood process L by

$$\begin{cases} dL_t & = & L_t \gamma_t^* dW_t + L_{t-} \int_E [\varphi_s(z) - 1] \{ \Psi(ds, dz) - \lambda_s(dz)ds \}, \\ L_0 & = & 1. \end{cases} \tag{9.7}$$

*where * denotes transpose. Assume that*

$$E^P [L_T] = 1,$$

and define the probability measure Q by

$$L_T = \frac{dQ}{dP}, \quad on \ \mathcal{F}_T.$$

Then Ψ has the predictable $(Q, \mathbf{F})$-intensity λ^Q given by

$$\lambda_t^Q(dz) = \varphi_t(z)\lambda_t(dz). \tag{9.8}$$

and we can write

$$dW_t = \gamma_t dt + dW_t^Q, \tag{9.9}$$

where W^Q is Q-Wiener.

9.3   Stochastic Differential Equations and Partial Integro-Differential Equations

On a filtered space $(\Omega, \mathcal{F}, P, \mathbf{F})$ we consider a scalar SDE of the form

$$dX_t = \mu(t, X_t)dt + \sigma(t, X_t)dW_t + \int_E \beta(t-, X_{t-}, z)\Psi(dt, dz), \tag{9.10}$$

$$X_0 = x_0. \tag{9.11}$$

Here we assume that $\mu(t, x)$, $\sigma(t, x)$ and $\beta(t, x, z)$ are given deterministic functions, and that x_0 is a given real number. The Wiener process W is allowed to be multidimensional, in which case σ is a row vector of the proper dimension and the product σdW is interpreted as an inner product. We need one more important assumption.

Assumption   We assume that the point process Ψ has a predictable intensity λ of the form

$$\lambda_t(dz) = \lambda_t(X_{t-}, dz), \tag{9.12}$$

where $\lambda_t(x, dz)$ in the right-hand side denotes a deterministic measure indexed by (t, x).

As one expects, the solution of the SDE above is a Markov process and we now want to compute the corresponding infinitesimal generator.

Proposition 9.5   *The infinitesimal generator of the process X above is given by*

$$\mathcal{A}F(t, x) = F_t(t, x)dt + \mu_t F_x(t, x) + \frac{1}{2}\sigma_t^2 F_{xx}(t, x) + \int_E F_\beta(t, x, z)\lambda_t(dz), \tag{9.13}$$

where

$$F_\beta(t, x, z) = F(t, x + \beta(t, x, z)) - F(t, x). \tag{9.14}$$

Proof   This follows immediately from formula (8.6).   □

We will of course have the usual Kolmogorov equation.

Proposition 9.6 (The Kolmogorov Backward Equation)   *Let X be the solution of (9.10), T a fixed point in time and K any function such that $K(X_T) \in L^1$. Define the function f by*

$$f(t, x) = E_{t,x}[K(X_T)].$$

Then f satisfies the Kolmogorov backward equation

$$\begin{cases} \mathcal{A}f(t,x) & = & 0, & (t,x) \in R_+ \times R \\ f(T,x) & = & K(x), & x \in R \end{cases} \tag{9.15}$$

where $\mathcal{A}$ is given by (9.13).

9.4   Exercises

Exercise 9.1   Prove the formula (9.5).

Exercise 9.2   Suppose that Ψ has intensity $\lambda_t(dz)$. We can then decompose it as

$$\lambda_t(dz) = \lambda_t^E \Gamma_t(dz),$$

where $\lambda_t^E = \lambda_t(E)$ is the intensity process for the counting process N^E, and Γ_t is a probability measure for each t. Now consider a predictable process λ_t^0 and a measure-valued process $\Gamma_t^0(dz)$ such that Γ_t^0 is a probability measure for each t. How should you determine φ in the Girsanov theorem in order to transform (λ^E, Γ) into (λ^0, Γ^0)?

Exercise 9.3   Consider a point process Ψ with mark space R, where $\lambda_t(dz) = f(z)dz$ and f is a density function for a probability distribution. This means that events occur according to a Poisson process with unit intensity and that marks are i.i.d. with density f, so we have compound Poisson. Now consider a family of density functions $\{f(z,a)\}_{a \in R}$ all of which are equivalent to f.

1. Find, for each $\gamma > 0$ and $a \in R$, the Girsanov transformation which transforms Ψ into a process where events occur according to a Poisson process with intensity γ and i.i.d. marks with density $f(z,a)$.
2. Study the ML estimate for γ and a based on the filtration $\mathbf{F}^N$. Show that the ML estimates decouple to an estimate of γ based on N^E and an ML estimate of a based on $Z_1, \ldots, Z_{N_t^E}$.

Exercise 9.4   Consider the SDE (9.10) and two smooth functions $F(t,x)$ and $k(t,x)$. Now define the process Z_t by

$$Z_t = F(t,X_t) + \int_0^t k(s,X_s)ds, \quad t \geq t.$$

Prove that, under suitable integrability conditions, the following hold.

- If $k(t,x) + \mathcal{G}F(t,x) = 0$ for all (t,x) then Z is a martingale.
- If $k(t,x) + \mathcal{G}F(t,x) \geq 0$ for all (t,x) then Z is a submartingale.
- If $k(t,x) + \mathcal{G}F(t,x) \leq 0$ for all (t,x) then Z is a supermartingale.

Exercise 9.5   Consider the same setup as in the previous exercise. Assume that F solves the boundary value problem

$$\begin{cases} k(t,x) + \mathcal{A}F(t,x) & = & 0, & 0 \leq t \leq T, \; x \in R^n \\ F(T,x) & = & K(x) \end{cases}$$

Prove that, under suitable integrability conditions, we have the representation formula

$$F(t, x) = E_{t,x}\left[\int_t^T k(s, X_s)ds + K(X_T)\right].$$

Part II

Optimal Control in Discrete Time

10 Dynamic Programming for Markov Processes

In this chapter we give a brief introduction to standard discrete-time dynamic programming. We will give the main arguments but we go lightly on some more technical details, so measurability and integrability issues are often swept under the carpet.

10.1 Setup

As our basic setup we will consider the case of an optimal control of a recursive process with finite-time horizon. The basic ingredients are as follows.

- A probability space $(\Omega, \mathcal{F}, P)$.
- A measurable space $(\mathbf{X}, \sigma_X)$, called the **state space**.
- A measurable space $(\mathbf{U}, \sigma_U)$, called the **control space**.
- A measurable space $(\mathbf{D}, \sigma_D)$, called the **noise space**.
- A sequence $\{f_n\}_{n=1}^{\infty}$ of mappings where $f_n : \mathbf{X} \times \mathbf{U} \times \mathbf{D} \to \mathbf{X}$.
- A sequence $\{\xi_n\}_{n=1}^{\infty}$ of $\mathbf{D}$-valued independent random variables. These variables are referred to as the **noise variables**.

Given the objects above, we can define a **controlled recursive process** as follows

$$X_{n+1} = f(X_n, u_n, \xi_{n+1}), \quad n = 0, 1, 2, \ldots \tag{10.1}$$

and we now have to clarify more precisely what this means.

If, for a moment we assume that f_n does not depend on u, we would have the recursion

$$X_{n+1} = f(X_n, \xi_{n+1}), \quad n = 0, 1, 2, \ldots \tag{10.2}$$

so the value of the **state** X at time $n + 1$ would be determined by X_n and the outcome of the **noise** ξ_{n+1}. In this case it is rather clear that, given an initial point $X_0 = x_0$, the process X would be Markov. In other words: X_{n+1} is determined by X_n and the noise ξ_{n+1}. In this way the process X will generate the filtration $\mathbf{F} = \mathbf{F}^X$.

The difference with (10.1) from (10.2) is that in (10.1) we can influence the outcome of X_{n+1} by choosing a **control** $u_n \in \mathbf{U}$, so X_{n+1} will be determined by the value X_n, by the noise ξ_{+1} and by our action u_n.

In the most general case the control sequence $\{u_n\}_{n=1}^{\infty}$ is allowed to be any $\mathbf{F}^X$-adapted random process, but since the X process is partly determined by the control process u, the filtration is also partly determined by u, so when we speak of an adapted process we have to be careful. After a moment of reflection it is, however, clear that the natural restriction of u should be that u_n is of the form

$$u_n = u_n(X_0, X_1, \ldots, X_n) \tag{10.3}$$

where u_n in the right-hand side, with a slight abuse of notation, is a deterministic measurable mapping $u_n : \mathbf{X}^{n+1} \to \mathbf{X}$

10.2   The Control Problem

In order to have an optimization problem we now add the final three ingredients.

- A sequence $\{H_n\}_{n=1}^{\infty}$ of mappings where $H_n : \mathbf{X} \times \mathbf{U} \to \mathbf{R}$. The mapping H_n is referred to as the **local utility function**.
- A mapping $F : \mathbf{X} \to \mathbf{R}$, referred to as the **terminal utility function**.
- An indexed family $\{U_n(x) : x \in \mathbf{X}, n \in \mathbf{N}\}$ of subsets of $\mathbf{U}$, so $U_n(x) \subseteq \mathbf{U}$ for all $x \in \mathbf{X}$ and all $n = 0, 1, 2, \ldots, T$.

The family $\{U_n(x)\}$ provides us with **control constraints** in the sense that if $X_n = x$ then we must choose the control u_n such that $u_n \in U_n(x)$.

We can now state our main problem.

Problem 10.1   *The problem to be solved is to choose a control process u of the form (10.3) which maximizes*

$$E\left[\sum_{n=0}^{T-1} H_n(X_n, u_n) + F(X_T)\right]$$

given the dynamics

$$X_{n+1} = f_n(X_n, u_n, \xi_{n+1}),$$
$$X_0 = x_0,$$

subject to the constraints

$$u_n \in U_n(X_n), \quad n = 0, 1, 2, \ldots$$

In principle the control process u is allowed to be any process of the form

$$u_n = u_n(X_0, X_1, \ldots, X_n)$$

satisfying the constraints above, but we will restrict ourselves to the case of so-called *feedback control laws*.

Definition 10.2   A **feedback control law** is a mapping $\mathbf{u} : \mathbf{N} \times \mathbf{X} \to \mathbf{U}$.

The interpretation of this is that, given the control law $\mathbf{u}$, the control process u will be of the form

$$u_n = \mathbf{u}_n(X_n),$$

so the action taken today depends only on current time n and on the current state X_n.

Note the important difference between the objects u and $\mathbf{u}$: The object u is a control *value*, so u is a point in the control space $\mathbf{U}$, i.e. $u \in \mathbf{U}$. The object $\mathbf{u}$ is a control *law*, so $\mathbf{u}$ is a mapping $\mathbf{u} : \mathbf{N} \times \mathbf{X} \to \mathbf{U}$.

Our final problem can now be stated as follows.

Problem 10.3

$$\text{maximize} \quad E\left[\sum_{n=0}^{T-1} H_n\left(X_n, \mathbf{u}_n(X_n)\right) + F(X_T)\right]$$

given the dynamics

$$X_{n+1} = f_n(X_n, \mathbf{u}_n(X_n), \xi_{n+1}),$$
$$X_0 = x_0,$$

subject to the constraints

$$u_n \in U_n(X_n), \quad n = 0, 1, 2, \ldots$$

over the class of feedback control laws $\mathbf{u}$.

The class of feedback control laws is of course smaller than the class of adapted controls. It is however possible to prove that the optimal control is always realized by a feedback law, so from an optimality point of view there is no restriction to limit ourselves to feedback laws.

Definition 10.4 The class $\mathcal{U}$ of **admissible feedback laws** is defined as the class of feedback laws $\mathbf{u}$ satisfying the constraints

$$u_n \in U_n(X_n), \quad n = 0, 1, 2, \ldots$$

10.3 Embedding the Problem

The way to attack our optimization problem is to embed the problem in a family of problems indexed by time and space, and then to connect all these problem by a recursive equation known as the **Bellman equation**. We will then see that solving the Bellman equation is equivalent to solving the optimal control problem. The embedding is as follows.

Definition 10.5 For each fixed initial point (n, x) we define $\mathcal{P}_{n,x}$ to be the problem of maximizing

$$E_{n,x}\left[\sum_{k=n}^{T-1} H_k\left(X_k, \mathbf{u}_k(X_k)\right) + F(X_T)\right]$$

given the dynamics

$$X_{k+1} = f_k(X_k, \mathbf{u}_k(X_k), \xi_{k+1}), \quad n \le k \le T-1$$
$$X_n = x,$$

over the class of feedback laws $\mathbf{u}$ satisfying the constraints

$$\mathbf{u}_k(x) \in U_k(x), \quad \text{for all } k \geq n, \ x \in \mathbf{X}.$$

We now define the *value function* and the *optimal value function*. Recall that $\mathcal{U}$ is the class of admissible feedback laws.

Definition 10.6

- The **value function**

$$J : \mathbf{N} \times \mathbf{X} \times \mathcal{U} \to \mathbf{R}$$

  is defined by

$$J_n(x, \mathbf{u}) = E_{n,x} \left[\sum_{k=n}^{T-1} H_k \left(X_k, \mathbf{u}_k(X_k) \right) + F(X_T) \right]$$

- The **optimal value function**

$$V : \mathbf{N} \times \mathbf{X} \to \mathbf{R}$$

  is defined by

$$V_n(x) = \sup_{\mathbf{u} \in \mathcal{U}} J_n(x, \mathbf{u})$$

Thus $J_n(x, \mathbf{u})$ is the expected utility of using the law $\mathbf{u}$ of the time interval $[n, T]$ if you start in state x at time n. The optimal value function $V_n(x)$ gives you the optimal utility over $[n, T]$ if you start in state x at time n.

10.4   Time-Consistency and the Bellman Optimality Principle

Let us now go back to Definition 10.5 and the problem $\mathcal{P}_{n,x}$. We make the following simplifying assumption.

Assumption   *We assume that for every initial point (n, x) there exists an optimal control law for problem $\mathcal{P}_{n,x}$. This control law is denoted by $\widehat{\mathbf{u}}^{nx}$.*

Note that the upper index (n, x) denotes the fixed initial point for problem $\mathcal{P}_{n,x}$. The object $\widehat{\mathbf{u}}^{nx}$ is thus a mapping $\widehat{\mathbf{u}}^{nx} : [n, T] \times \mathbf{X} \to \mathbf{R}$, so the control applied at some time $k \geq n$ will be given by the expression

$$\widehat{\mathbf{u}}_k^{nx}(X_k).$$

It is important to realize that *a priori* the optimal law for the problem $\mathcal{P}_{n,x}$ could very well depend on the choice of the starting point (n, x). It turns out, however, that the optimal law does *not* depend on the choice of the initial point. The formalization and proof of this statement is as follows.

Theorem 10.7 (The Bellman Optimality Principle)   *Fix an initial point (n, x) and consider the corresponding optimal law $\widehat{\mathbf{u}}^{nx}$. Then the law $\widehat{\mathbf{u}}^{nx}$ is also optimal for any subinterval of the form $[m, T]$ where $m \geq n$. In other words*

$$\widehat{\mathbf{u}}_k^{nx}(y) = \widehat{\mathbf{u}}_k^{m, X_m}(y)$$

for all $k \geq m$ and all $y \in \mathbf{X}$. In particular, the optimal law for the initial point $n = 0$ will be optimal for all subintervals. This law will be denoted by $\widehat{\mathbf{u}}$.

In more pedestrian terms this means the following. Suppose that you optimize at time $n = 0$ and follow $\widehat{\mathbf{u}}$ up to time n, where you now have reached the state X_n. At time n you reconsider and decide to forget your original problem and instead solve problem $\mathcal{P}_{n, X_n}$. What the Bellman principle tells you is that the law $\widehat{\mathbf{u}}$ (restricted to the time interval $[n, T]$ is optimal, not only for your original problem, but also for your new problem. In decision-theoretic jargon we could say that our family of problems are **time consistent** and in particular this implies that the expression "the optimal law" has a well-defined meaning – it does not depend on your choice of starting point.

Proof   The proof is by contradiction. Let us thus assume that for some $n > 0$ there exists a law $\bar{\mathbf{u}}$ on the interval $[n, T]$ such that

$$E_{n,x}\left[\sum_{k=n}^{T-1} H_k\left(X_k, \bar{\mathbf{u}}_k(X_k)\right) + F(X_T)\right] \geq E_{n,x}\left[\sum_{k=n}^{T-1} H_k\left(X_k, \widehat{\mathbf{u}}_k(X_k)\right) + F(X_T)\right]$$

for all $x \in \mathbf{X}$ and strict inequality for some $x \in \mathbf{X}$. We can then construct a new law $\mathbf{u}^\star$ on $[0, T]$ by the following formula (note that here, and through to the end of the next section, $\star$ does *not* indicate transpose):

$$\mathbf{u}_k^\star(y) = \begin{cases} \widehat{\mathbf{u}}_k(y) & \text{for} \quad 0 \leq k < n - 1 \\ \bar{\mathbf{u}}_k(y) & \text{for} \quad n \leq k < T - 1. \end{cases}$$

We then have

$$J_0(x_0, \mathbf{u}^\star) = E_{0, x_0}\left[\sum_{k=0}^{T-1} H_k\left(X_k, \mathbf{u}_k^\star\right) + F(X_T)\right]$$

$$= E_{0, x_0}\left[\sum_{k=0}^{n-1} H_k\left(X_k, \widehat{\mathbf{u}}_k\right)\right] + E_{0, x_0}\left[\sum_{k=n}^{T-1} H_k\left(X_k, \bar{\mathbf{u}}_k\right) + F(X_T)\right]$$

$$= E_{0, x_0}\left[\sum_{k=0}^{n-1} H_k\left(X_k, \widehat{\mathbf{u}}_k\right)\right] + E_{0, x_0}\left[E_{n, X_n}\left[\sum_{k=n}^{T-1} H_k\left(X_k, \bar{\mathbf{u}}_k\right) + F(X_T)\right]\right],$$

where we have used iterated expectations and the Markov property to obtain the last term. It now follows from the assumption concerning $\bar{\mathbf{u}}$ that we have

$$E_{n, X_n}\left[\sum_{k=n}^{T-1} H_k\left(X_k, \bar{\mathbf{u}}_k\right) + F(X_T)\right] \geq E_{n, X_n}\left[\sum_{k=n}^{T-1} H_k\left(X_k, \widehat{\mathbf{u}}_k\right) + F(X_T)\right]$$

with strict inequality with positive probability so, again using iterated expectations and the Markov property, we obtain

$$
J_0(x_0, \mathbf{u}^\star) > E_{0,x_0} \left[\sum_{k=0}^{n-1} H_k\left(X_k, \widehat{\mathbf{u}}_k\right) \right] + E_{0,x_0} \left[E_{n,X_n} \left[\sum_{k=n}^{T-1} H_k\left(X_k, \widehat{\mathbf{u}}_k\right) \right] + F(X_T) \right]
$$

$$
= E_{0,x_0} \left[\sum_{k=0}^{n-1} H_k\left(X_k, \widehat{\mathbf{u}}_k\right) \right] + E_{0,x_0} \left[\sum_{k=n}^{T} H_k\left(X_k, \widehat{\mathbf{u}}_k\right) \right]
$$

$$
= E_{0,x_0} \left[\sum_{k=0}^{T} H_k\left(X_k, \widehat{\mathbf{u}}_k\right) + F(X_T) \right] = J_0(x_0, \widehat{\mathbf{u}}).
$$

We have thus obtained the inequality

$$
J_0(x_0, \mathbf{u}^\star) > J_0(x_0, \widehat{\mathbf{u}}),
$$

which contradicts the optimality of $\widehat{\mathbf{u}}$ on the interval $[0, T]$. $\square$

10.5   The Bellman Equation

We now go on to derive the Bellman equation, which is a recursion scheme for the optimal value function. To this end we fix an arbitrarily chosen initial point (n, x) and an arbitrary control value $u \in U_n(x)$. We now define the control law $\mathbf{u}^\star$ on $[n, T]$ by

$$
\mathbf{u}_n^\star(x) = u,
$$
$$
\mathbf{u}_k^\star(y) = \widehat{\mathbf{u}}_k(y), \quad \text{for all } k \in [n+1, T] \text{ and for all } y \in \mathbf{X}.
$$

To use the law $\mathbf{u}^\star$ simply means that at time n we use the arbitrary control value u, and from time $n+1$ and onwards we use the optimal control $\widehat{\mathbf{u}}$. We can now describe the basic idea of dynamic programming, which simply consists of the following scheme.

- Given the initial point (n, x) as above we consider the following two control laws on $[n, T]$:
  - the optimal law $\widehat{\mathbf{u}}$;
  - the law $\mathbf{u}^\star$ defined above.
- We then compute the expected utilities generated by the two laws.
- Using the obvious fact that the utility from $\widehat{\mathbf{u}}$ must be greater than or equal to the utility from $\mathbf{u}^\star$ we obtain our recursion scheme.

We now carry out this program.

1. Expected utility for $\widehat{\mathbf{u}}$: This is trivial. Since $\widehat{\mathbf{u}}$ is the optimal law we have, by definition, the result

$$
J_n(x, \widehat{\mathbf{u}}) = V_n(x).
$$

2. Expected utility for $\mathbf{u}^\star$: By definition we have

$$J_n(x, \mathbf{u}^\star) = E_{n,x} \left[\sum_{k=n}^{T-1} H_k \left(X_k^{\mathbf{u}^\star}, \mathbf{u}_k^\star(X_k) \right) + F(X_T) \right]$$

$$= H_n(x, u) + E_{n,x} \left[\sum_{k=n+1}^{T-1} H_k \left(X_k^{\mathbf{u}^\star}, \mathbf{u}_k^\star(X_k) \right) + F(X_T) \right],$$

where we have used the notation $X_k^{\mathbf{u}^\star}$ to emphasize that the distribution of the X-process depends on the control law $\mathbf{u}^\star$. Using iterated expectations and the Markov property we obtain

$$E_{n,x} \left[\sum_{k=n+1}^{T-1} H_k \left(X_k^{\mathbf{u}^\star}, \mathbf{u}_k^\star(X_k) \right) + F(X_T) \right]$$

$$= E_{n,x} \left[E_{n+1,X_{n+1}^u} \left[\sum_{k=n+1}^{T-1} H_k \left(X_k^{\mathbf{u}^\star}, \mathbf{u}_k^\star(X_k) \right) + F(X_T) \right] \right],$$

where it is important to notice that the distribution of X_{n+1} only depends on the chosen control value u at time n, hence the notation X_{n+1}^u. We now recall that $\mathbf{u}^\star = \widehat{\mathbf{u}}$ on the time interval $[n+1, T]$, so we have

$$E_{n+1,X_{n+1}^u} \left[\sum_{k=n+1}^{T} H_k \left(X_k^{\mathbf{u}^\star}, \mathbf{u}_k^\star(X_k) \right) + F(X_T) \right] = V_{n+1} \left(X_{n+1}^u \right).$$

This gives us the result

$$J_n(x, \mathbf{u}^\star) = H_n(x, u) + E_{n,x} \left[V_{n+1} \left(X_{n+1}^u \right) \right].$$

3. Comparing the control laws: We obviously have the relation

$$J_n(x, \widehat{\mathbf{u}}) \geq J_n(x, \mathbf{u}^\star), \quad \text{for all } u \in U_n(x),$$

with equality when $u = \widehat{u}_n(x)$. Using our results from above we thus have

$$V_n(x) \geq H_n(x, u) + E_{n,x} \left[V_{n+1} \left(X_{n+1}^u \right) \right],$$

and since this holds for all $u \in U_n(x)$, with equality for $u = \widehat{u}_n(x)$, we have in fact proved our main result.

Theorem 10.8 (The Bellman Equation)   *The optimal value function satisfies the recursive equation*

$$V_n(x) = \sup_{u \in U_n(x)} \left\{ H_n(x, u) + E_{n,x} \left[V_{n+1} \left(X_{n+1}^u \right) \right] \right\},$$

$$V_T(x) = F(x).$$

Furthermore, the supremum in the equation is realized by the optimal control law $\widehat{u}_n(x)$.

10.6   Handling the Bellman Equation

In the general case one cannot hope for an analytic solution to the Bellman equation. There are indeed cases when you can solve the equation explicitly (see Chapter 10.7 for an example), but these are very rare. There are, however, at least three standard ways of attacking the Bellman equation.

1. You try to solve it numerically by implementing the recursion on a computer. This may be feasible when the state variable is low dimensional, but in higher dimensions you may run into hard numerical problems.
2. You completely give up hope for an analytic or numerical solution. Instead you use the equation in order to derive qualitative properties of the solution. If you have a qualitative conjecture about the solution, you can try to prove it by using the recursive structure of the equation in an induction proof.
3. If you have reasonable hopes for an analytic solution, then you make an *Ansatz*, i.e. you make an educated guess about the structure of V. Typically the Ansatz is a conjecture that V belongs to some family of functions parameterized by a finite number of time-dependent parameters. You then plug the Ansatz into the Bellman equation and if you are lucky you will get a finite-dimensional system of difference equations for the parameters. The making of a good Ansatz is really an art form, but a standard procedure is to look at the local utility functions H_n and F and hope that the structure of these is inherited by the optimal value function V. We will see how this works in Chapter 10.7.

10.7   The Linear Quadratic Regulator: I

In order to illustrate the use of dynamic programming and the Bellman equation we now consider a classical engineering problem: The linear quadratic regulator. For simplicity we only study a scalar version of the problem. The reward functional, which in this case is to be minimized, is given by

$$J_n(x, \mathbf{u}) = E_{n,x} \left[\sum_{t=n}^{T-1} \{QX_t^2 + Ru_t^2\} \right] + HE_{n,x} \left[X_T^2 \right],$$

with scalar dynamics

$$X_{t+1} = AX_t + Bu_t + C\xi_{t+1}.$$

The scalars Q, R, H, A, B and C are assumed to be known, and we suppose also that $R > 0$. The random variables $\{\xi_t; \ t = 0, 1, \ldots, T\}$ are i.i.d. with distribution $N[0, 1]$. There is no restriction on the control.

This is a time-consistent problem and the Bellman equation reads as follows:

$$V_n(x) = \inf_{u \in \mathbf{R}} \{Qx^2 + Ru^2 + E_{n,x} \left[V_{n+1} \left(X_{n+1}^u \right) \right] \}, \tag{10.4}$$

$$V_T(x) = Hx^2. \tag{10.5}$$

By inspecting the boundary condition Hx^2 and the quadratic term Qx^2 in the sum, we are led to the following Ansatz:

$$V_n(x) = P_n x^2 + q_n.$$

It would of course be natural to also include a linear term $K_n x$ in the Ansatz. This can be done, but it turns out that $K = 0$, so the Ansatz has been proposed with a certain amount of hindsight.

Given the Ansatz and the condition $X_n = x$ we use the X-dynamics to obtain

$$V_{n+1}\left(X_{n+1}^u\right) = P_{n+1}\left(Ax + Bu + CW_{n+1}\right)^2 + q_n$$

which gives us

$$E_{n,x}\left[V_{n+1}\left(X_{n+1}^u\right)\right] = P_{n+1}A^2 x^2 + P_{n+1}B^2 u^2 + 2P_{n+1}ABxu + P_{n+1}C^2 + q_n.$$

The minimization problem in (10.4) is thus to minimize

$$Ru^2 + P_{n+1}B^2 u^2 + 2P_{n+1}ABxu$$

for $u \in R$. The candidate optimal control is easily computed as

$$\hat{u}_n = -\frac{P_{n+1}AB}{R + P_{n+1}B^2}x.$$

Plugging this into (10.4), using the Ansatz and simplifying, gives us

$$P_n x^2 + q_n = \left\{Q + P_{n+1}A + \frac{P_{n+1}^2 A^2 B^2}{R + P_{n+1}B}\right\}x^2 + P_{n+1}C^2 + q_{n+1}.$$

Since this holds for all x, we can use separation of variables to prove the following result.

Proposition 10.9 *For the linear quadratic regulator, the optimal value function has the form*

$$V_n(x) = P_n x^2 + q_n,$$

and the optimal control is given by

$$\hat{u}_n = -\frac{P_{n+1}AB}{R + P_{n+1}B^2}x,$$

where P_n satisfies the recursive system

$$\begin{cases} P_n &= Q + P_{n+1}A + \frac{A^2 B^2 P_{n+1}^2}{R + B P_{n+1}}, \\ P_T &= H. \end{cases}$$

There is also an obvious recursion for q_n but that recursion is of little interest since it does not affect the optimal control.

Part III

Optimal Control in Continuous Time

11 Continuous-Time Dynamic Programming

We now move on to dynamic programming theory in continuous time. This can be done within the framework of a general controlled Markov process, but in order to keep the theory reasonably concrete we restrict ourselves to the case of a controlled SDE, driven by a finite-dimensional Wiener process W and a marked point process Ψ. As the reader will see, the arguments in the discrete-time case will be almost exactly the same as the arguments in the discrete-time case.

11.1 Setup

The basic setup is that we consider a filtered probability space $(\Omega, \mathcal{F}, P, \mathbf{F})$. On this space we have the following driving processes.

- A d-dimensional Wiener process W.
- A marked point process Ψ with mark space $(\mathbf{E}, \mathcal{E})$.

We make the following assumption concerning the point process.

Assumption 11.1 *We assume that Ψ has a deterministic $\mathbf{F}$-intensity measure process $\lambda_t(dz)$.*

We also have the following exogenously given objects:

- The **drift function** $\mu(t, x, u)$
$$\mu : \mathbf{R}_+ \times \mathbf{R}^n \times \mathbf{R}^k \to \mathbf{R}^n$$

- The **diffusion matrix function** $\sigma(t, x, u)$
$$\sigma : \mathbf{R}_+ \times \mathbf{R}^n \times \mathbf{R}^k \to \mathbf{R}^{n \times d}$$

- The **jump function** $\beta(t, x, u, z)$
$$\beta : \mathbf{R}_+ \times \mathbf{R}^n \times \mathbf{R}^k \times \mathbf{E} \to \mathbf{R}^n$$

- The **local utility function** $H(t, x, u)$
$$H : \mathbf{R}_+ \times \mathbf{R}^n \times \mathbf{R}^k \to \mathbf{R}$$

- The **terminal utility function** $F(x)$
$$F : \mathbf{R}^n \to \mathbf{R}$$

- For each (t, x) we have a given **constraint set** $U_t(x) \subseteq \mathbf{R}^k$.

We can now state our main problem.

Problem 11.2   *Find a predictable $\mathbf{R}^k$-valued control process u which maximizes*

$$\mathcal{J}_0(u) = E\left[\int_0^T H(t, X_t, u_t)dt + F(X_T)\right] \tag{11.1}$$

given the state dynamics

$$dX_t = \mu(t, X_t, u_t)\, dt + \sigma(t, X_t, u_t)\, dW_t + \int_E \beta(t-, X_{t-}, u_t, z)\Psi(dt, dz),$$

$$X_0 = x_0$$

and the constraints

$$u_t \in U_t(X_{t-}), \quad t \geq 0.$$

We view the n-dimensional process X as a **state process**, which we are trying to "control" (or "steer"). We can (partly) control the state process X by choosing the k-dimensional **control process** u in a suitable way, and we must now try to give a precise mathematical meaning to the formal SDE above.

Our first modeling problem concerns the class of admissible control processes and, just as we did in the discrete-time case, we will restrict ourselves to **feedback control laws**.

Definition 11.3   A **feedback control law** is a deterministic mapping

$$\mathbf{u} : \mathbf{R}_+ \times \mathbf{R}^n \rightarrow \mathbf{R}^k.$$

The interpretation of this is that, given the control law $\mathbf{u}$, the control process u will be of the form

$$u_t = \mathbf{u}(t, X_{t-}),$$

where we evaluate X at $t-$ in order to obtain a predictable control process.

We use the notation $\mathbf{u}$ (boldface) in order to indicate that $\mathbf{u}$ is a **function** $\mathbf{u} : \mathbf{R}_+ \times \mathbf{R}^n \rightarrow \mathbf{R}^k$. In contrast to this we use the notation u (italics) to denote the **value** of a control at a certain time. Thus $\mathbf{u}$ denotes a mapping, whereas u denotes a point in $\mathbf{R}^k$.

Suppose now that we have chosen a fixed control law $\mathbf{u}(t, x)$. Then we can insert $\mathbf{u}$ into the SDE above to obtain the standard SDE

$$dX_t = \mu(t, X_t, \mathbf{u}(t, X_t))\, dt + \sigma(t, X_t, \mathbf{u}(t, X_t))\, dW_t$$

$$+ \int_E \beta(t-, X_{t-}, \mathbf{u}(t, X_{t-}), z)\Psi(dt, dz). \tag{11.2}$$

In most concrete cases we also have to satisfy some **control constraints**, and we model this by the condition $\mathbf{u}(t, x) \in U_t(x)$ for each (t, x). We can now define the class of **admissible control laws**.

Definition 11.4   A control law $\mathbf{u}$ is called **admissible** if

- $\mathbf{u}(t, x) \in U_t(x)$ for all $t \in \mathbf{R}_+$ and all $x \in \mathbf{R}^n$.

- For any given initial point (t, x) the SDE

$$dX_s = \mu\,(s, X_s, \mathbf{u}(s, X_s))\,ds + \sigma\,(s, X_s, \mathbf{u}(s, X_s))\,dW_s$$
$$+ \int_E \beta(s-, X_{s-}, \mathbf{u}(s, X_{s-}), z)\Psi(ds, dz)$$
$$X_t = x$$

has a unique solution.
- The reward functional (11.1) is well defined and finite.

The class of admissible control laws is denoted by $\mathcal{U}$.

For a given control law $\mathbf{u}$, the solution process X will of course depend on the initial value x, as well as on the chosen control law $\mathbf{u}$. To be precise we should therefore denote the process X by $X^{x,\mathbf{u}}$, but sometimes we will suppress x or $\mathbf{u}$. We note that equation (11.2) looks rather messy and, since we will also have to deal with the Itô formula in connection with (11.2), we need some more streamlined notation. We thus introduce the *controlled infinitesimal operator* as follows.

Definition 11.5   Consider equation (11.2), and let $\star$ denote matrix transpose.

- For any fixed vector $u \in R^k$, the functions μ^u, σ^u, β^u and C^u are defined by

$$\mu^u(t, x) = \mu(t, x, u),$$
$$\sigma^u(t, x) = \sigma(t, x, u),$$
$$\beta^u(t, x, z) = \beta(t, x, u, z),$$
$$C^u(t, x) = \sigma(t, x, u)\sigma(t, x, u)^\star.$$

- For any control law $\mathbf{u}$, the functions $\mu^{\mathbf{u}}$, $\sigma^{\mathbf{u}}$, $\beta^{\mathbf{u}}$, $C^{\mathbf{u}}$ and $F^{\mathbf{u}}$ are defined by

$$\mu^{\mathbf{u}}(t, x) = \mu(t, x, \mathbf{u}(t, x)),$$
$$\sigma^{\mathbf{u}}(t, x) = \sigma(t, x, \mathbf{u}(t, x)),$$
$$C^{\mathbf{u}}(t, x) = \sigma(t, x, \mathbf{u}(t, x))\sigma(t, x, \mathbf{u}(t, x))^\star,$$
$$H^{\mathbf{u}}(t, x) = H(t, x, \mathbf{u}(t, x)).$$

- For any fixed vector $u \in \mathbf{R}^k$, the partial integral-differential operator $\mathcal{G}^u$ is defined by

$$\mathcal{G}^u f(t, x) = \sum_{i=1}^{n} \mu_i^u(t, x)\frac{\partial f}{\partial x_i}(t, x) + \frac{1}{2}\sum_{i,j=1}^{n} C_{ij}^u(t, x)\frac{\partial^2 f}{\partial x_i \partial x_j}(t, x)$$
$$+ \int_E \{f(t, x + \beta^u(t, x, z)) - f(t, x)\}\,\lambda_t(dz).$$

- For any control law $\mathbf{u}$, the partial integral-differential operator $\mathcal{G}^{\mathbf{u}}$ is defined by

$$\mathcal{G}^{\mathbf{u}} f(t, x) = \sum_{i=1}^{n} \mu_i^{\mathbf{u}}(t, x)\frac{\partial f}{\partial x_i}(t, x) + \frac{1}{2}\sum_{i,j=1}^{n} C_{ij}^{\mathbf{u}}(t, x)\frac{\partial^2 f}{\partial x_i \partial x_j}(t, x)$$
$$+ \int_E \{f(t, x + \beta^{\mathbf{u}}(t, x, z)) - f(t, x)\}\,\lambda_t(dz).$$

Comparing with Proposition 9.5 we see that, for a given control law **u**, the infinitesimal operator of the controlled X-process is given by

$$\mathcal{A}^{\mathbf{u}} = \frac{\partial}{\partial t} + \mathcal{G}^{\mathbf{u}}.$$

Given a control law **u** we will sometimes write equation (11.2) in a convenient shorthand notation as

$$dX_t = \mu^{\mathbf{u}}(t, X_t)dt + \sigma^{\mathbf{u}}(t, X_t)dW_t + \int_E \beta^{\mathbf{u}}(t, X_{t-}, z)\Psi(dt, dz).$$

For a given control law **u** with a corresponding controlled process $X^{\mathbf{u}}$ we will also often use the shorthand notation $\mathbf{u}_t$ instead of the clumsier expression $\mathbf{u}\left(t, X_t^{\mathbf{u}}\right)$.

Our formal problem can thus be written as that of maximizing $\mathcal{J}_0(\mathbf{u})$ over all $\mathbf{u} \in U$, and we define the **optimal value** $\hat{\mathcal{J}}_0$ by

$$\hat{\mathcal{J}}_0 = \sup_{\mathbf{u} \in \mathcal{U}} \mathcal{J}_0(\mathbf{u}).$$

If there exists an admissible control law $\hat{\mathbf{u}}$ with the property that

$$\mathcal{J}_0(\hat{\mathbf{u}}) = \hat{\mathcal{J}}_0,$$

then we say that $\hat{\mathbf{u}}$ is an **optimal control law** for the given problem. Note that, as for any optimization problem, the optimal law may not exist. For a given concrete control problem our main objective is of course to find the optimal control law (if it exists), or at least to learn something about the qualitative behavior of the optimal law.

11.2   Embedding the Problem

The way to attack our optimization problem is, exactly as in the discrete-time case, to embed the problem in a family of problems indexed by time and space. We will then connect all these problem by a non-linear partial differential equation equation known as the **Hamilton–Jacobi–Bellman Equation**. We will then see that solving the HJB equation is equivalent to solving the optimal control problem.

We will now describe the embedding procedure, and for that purpose we choose a fixed point t in time, with $0 \le t \le T$. We also choose a fixed point x in the state space, i.e. $x \in R^n$. For this fixed pair (t, x) we now define the following control problem.

Definition 11.6   The control problem $\mathcal{P}_{t,x}$ is defined as the problem of maximizing

$$E_{t,x}\left[\int_t^T H(s, X_s^{\mathbf{u}}, \mathbf{u}_s)ds + F\left(X_T^{\mathbf{u}}\right)\right]$$

given the dynamics

$$dX_s^{\mathbf{u}} = \mu\left(s, X_s^{\mathbf{u}}, \mathbf{u}_s\right)ds + \sigma\left(s, X_s^{\mathbf{u}}, \mathbf{u}_s\right)dW_s, \tag{11.3}$$

$$+ \int_E \beta(s-, X_{s-}^{\mathbf{u}}, \mathbf{u}_{s-}, z)\Psi(ds, dz) \tag{11.4}$$

$$X_t = x, \tag{11.5}$$

and the constraints

$$\mathbf{u}(s, y) \in U_s(y), \quad \text{for all } (s, y) \in [t, T] \times R^n.$$

Observe that we use the notation s and y above because the letters t and x are already used to denote the fixed chosen point (t, x). We note that in terms of the definition above, our original problem is the problem $\mathcal{P}_{0, x_0}$.

We now define the **value function** and the **optimal value function**.

Definition 11.7

- The **value function**

$$\mathcal{J} : R_+ \times R^n \times U \to R$$

   is defined by

$$\mathcal{J}(t, x, \mathbf{u}) = E\left[\int_t^T H(s, X_s^{\mathbf{u}}, \mathbf{u}_s)ds + F\left(X_T^{\mathbf{u}}\right)\right]$$

   given the dynamics (11.4)–(11.5).
- The **optimal value function**

$$V : R_+ \times R^n \to R$$

   is defined by

$$V(t, x) = \sup_{\mathbf{u} \in \mathcal{U}} \mathcal{J}(t, x, \mathbf{u}).$$

Thus $\mathcal{J}(t, x, \mathbf{u})$ is the expected utility of using the control law $\mathbf{u}$ over the time interval $[t, T]$, given the fact that you start in state x at time t. The optimal value function gives you the optimal expected utility over $[t, T]$ under the same initial conditions.

11.3 Time-Consistency and the Bellman Principle

Let us now go back to Definition 11.6, the problem $\mathcal{P}_{t, x}$. We make the following simplifying assumption.

Assumption *We assume that for every initial point (t, x) there exists an optimal control law for problem $\mathcal{P}_{t, x}$. This control law is denoted by $\widehat{\mathbf{u}}^{tx}$.*

Note that the upper index (t, x) denotes the fixed initial point for problem $\mathcal{P}_{t, x}$. The object $\widehat{\mathbf{u}}^{tx}$ is thus a mapping $\widehat{\mathbf{u}}^{tx} : [t, T] \times R^n \to R^k$, so the control applied at some time $s \geq t$ and state $y \in R^n$ will be given by the expression

$$\widehat{\mathbf{u}}^{nx}(s, y).$$

It is important to realize that *a priori* the optimal law for the problem $\mathcal{P}_{t, x}$ could very well depend on the choice of the starting point (n, x). It turns out, however, that the optimal law does *not* depend on the choice of the initial point. The formalization and proof of this statement is as follows.

Theorem 11.8 (The Bellman Optimality Principle)   *Fix an initial point (t,x) and consider the corresponding optimal law $\widehat{\mathbf{u}}^{tx}$. Then the law $\widehat{\mathbf{u}}^{tx}$ is also optimal for any subinterval of the form $[r,T]$ where $t \leq r \leq T$. In other words*

$$\widehat{\mathbf{u}}^{tx}(s,y) = \widehat{\mathbf{u}}^{r,X_r}(s,y)$$

for all $s \geq r$ and all $y \in \mathbf{R}^n$. In particular, the optimal law for the initial point $n = 0$ will be optimal for all subintervals. This law will be denoted by $\widehat{\mathbf{u}}$.

Proof   The proof is identical to the proof for the discrete-time case.   □

Recalling our argument from Chapter 10.4 this means the following. Suppose that you optimize at time $t = 0$ and follow $\widehat{\mathbf{u}}$ up to time t, where you now have reached the state X_t. At time t you reconsider and decide to forget your original problem and instead solve problem $\mathcal{P}_{t,X_t}$. What the Bellman principle tells you is that the law $\widehat{\mathbf{u}}$ (restricted to the time interval $[t,T]$ is optimal, not only for your original problem, but also for your new problem. In decision-theoretic jargon we could say that our family of problems are **time consistent**, and in particular this implies that the expression "the optimal law" has a well-defined meaning – it does not depend on your choice of starting point.

11.4   The Hamilton–Jacobi–Bellman Equation

The main object of interest for us is the optimal value function V, and we now go on to derive a PDE for V. It should be noted that this derivation is largely heuristic. We make some rather strong regularity assumptions, and we disregard a number of technical problems. We will comment on these problems later, but to see exactly which problems we are ignoring we now make some basic assumptions.

Assumption 11.9   *We assume the following.*

1. *There exists an optimal control law $\widehat{\mathbf{u}}$.*
2. *The optimal value function V is regular in the sense that $V \in C^{1,2}$.*
3. *A number of limiting procedures in the following arguments can be justified.*

We now go on to derive the PDE, and to this end we fix $(t,x) \in (0,T) \times R^n$. Furthermore, we choose a real number h (interpreted as a "small" time increment) such that $t + h < T$. We choose a fixed but arbitrary control law $\mathbf{u}$, and define the control law $\mathbf{u}^\star$ (where $\star$ is just a marker, and does not denote transpose) by

$$\mathbf{u}^\star(s,y) = \begin{cases} \mathbf{u}(s,y), & (s,y) \in [t,t+h] \times R^n \\ \widehat{\mathbf{u}}(s,y), & (s,y) \in (t+h,T] \times R^n. \end{cases}$$

In other words, if we use $\mathbf{u}^\star$ then we use the arbitrary control $\mathbf{u}$ during the time interval $[t,t+h]$, and then we switch to the optimal control law during the rest of the time period.

The whole idea of dynamic programming actually boils down to the following procedure, and the reader will recognize that the philosophy is exactly the same as the one we used in discrete time.

- First, given the point (t, x) as above, we consider the following two strategies over the time interval $[t, T]$:
  1: Use the optimal law $\hat{\mathbf{u}}$.
  2: Use the control law $\mathbf{u}^\star$ defined above.
- We then compute the expected utilities obtained by the respective strategies.
- Finally, using the obvious fact that $\hat{\mathbf{u}}$ by definition has to be at least as good as the strategy $\mathbf{u}^\star$, and letting h tend to zero, we obtain our fundamental PDE.

We now carry out this program.

1. Expected utility for $\hat{\mathbf{u}}$:
This is trivial. Since by definition the control is the optimal one we obtain

$$\mathcal{J}(t, x, \hat{\mathbf{u}}) = V(t, x).$$

2. Expected utility for strategy $\mathbf{u}^\star$:
We divide the time interval $[t, T]$ into two parts, the intervals $[t, t + h]$ and $(t + h, T]$ respectively.

$$\mathcal{J}(t, x, \mathbf{u}^\star) = E_{t,x} \left[\int_t^{t+h} H(s, X_s^{\mathbf{u}^\star}, \mathbf{u}_s^\star) ds + \int_{t+h}^T H(s, X_s^{\mathbf{u}^\star}, \mathbf{u}_s^\star) ds + F(X_T^{\mathbf{u}^\star}) \right].$$

Recalling that $\mathbf{u}^\star = \mathbf{u}$ on $[t, t+h]$, and that $\mathbf{u}^\star = \hat{\mathbf{u}}$ on $(t+h, T]$, using iterated expectations and the Markov structure, we can write this as

$$E_{t,x} \left[\int_t^{t+h} H\left(s, X_s^{\mathbf{u}}, \mathbf{u}_s\right) ds + E_{t+h, X_{t+h}^{\mathbf{u}}} \left[\int_{t+h}^T H\left(s, X_s^{\hat{\mathbf{u}}}, \hat{\mathbf{u}}_s\right) ds + F(X_T^{\hat{\mathbf{u}}}) \right] \right].$$

It now follows from the Bellman optimality principle that we have

$$E_{t+h, X_{t+h}^{\mathbf{u}}} \left[\int_{t+h}^T H\left(s, X_s^{\hat{\mathbf{u}}}, \hat{\mathbf{u}}_s\right) ds + F(X_T^{\hat{\mathbf{u}}}) \right] = V(t + h, X_{t+h}^{\mathbf{u}})$$

The total expected utility for $\mathbf{u}^\star$ is thus given by

$$\mathcal{J}(t, x, \mathbf{u}^\star) = E_{t,x} \left[\int_t^{t+h} H\left(s, X_s^{\mathbf{u}}, \mathbf{u}_s\right) ds + V(t + h, X_{t+h}^{\mathbf{u}}) \right].$$

3. Comparing the strategies:
We now go on to compare the two strategies and since, by definition, $\hat{\mathbf{u}}$ is the optimal control, we must have the inequality

$$\mathcal{J}(t, x, \mathbf{u}^\star) \le \mathcal{J}(t, x, \hat{\mathbf{u}}) \tag{11.6}$$

We also note that the inequality sign is due to the fact that the arbitrarily chosen control law $\mathbf{u}$ which we use on the interval $[t, t + h]$ need not be the optimal one. In particular we have the following obvious fact.

Remark We have equality in (11.6) if and only if the control law $\mathbf{u}$ is an optimal law $\hat{\mathbf{u}}$. (Note that the optimal law does not have to be unique.)

Given our previous calculations, we can write the inequality (11.6) as

$$V(t,x) \geq E_{t,x}\left[\int_t^{t+h} H\left(s, X_s^u, \mathbf{u}_s\right) ds\right] + E_{t,x}\left[V(t+h, X_{t+h}^u)\right],$$

or, equivalently as

$$E_{t,x}\left[\int_t^{t+h} H\left(s, X_s^u, \mathbf{u}_s\right) ds\right] + \mathcal{A}_h^u V(t,x) \leq 0,$$

where $\mathcal{A}_h V(t,x)$ is given by by

$$\mathcal{A}_h V(t,x) = E_{t,x}\left[V(t+h, X_{t+h}^u)\right] - V(t,x).$$

We now divide by h and let h go to zero. We then have

$$\lim_{h \to 0} \frac{1}{h} E_{t,x}\left[\int_t^{t+h} H\left(s, X_s^u, \mathbf{u}_s\right) ds\right] = H(t,x,u)$$

$$\lim_{h \to 0} \frac{1}{h} \mathcal{A}_h^u V(t,x) = \mathcal{A}^u V(t,x),$$

where we use the notation $u = \mathbf{u}(t,x)$. We have thus derived the inequality

$$H(t,x,u) + \mathcal{A}^u V(t,x) \leq 0.$$

Since the control law $\mathbf{u}$ was arbitrary, this inequality will hold for all choices of $u \in U$, and we will have equality if and only if $u = \hat{\mathbf{u}}(t,x)$. We thus have the following equation

$$\sup_{u \in U_t(x)} \{H(t,x,u) + \mathcal{A}^u V(t,x)\} = 0.$$

Using the notation

$$\mathcal{A}^u = \frac{\partial}{\partial t} + \mathcal{G}^u$$

we can also write this as

$$\frac{\partial V}{\partial t}(t,x) + \sup_{u \in U_t(x)} \{H(t,x,u) + \mathcal{G}^u V(t,x)\} = 0.$$

During the discussion the point (t,x) was fixed, but since it was chosen as an arbitrary point we see that the equation holds in fact for all $(t,x) \in (0,T) \times R^n$. Thus we have a non-standard type of PDE, and we obviously need some boundary conditions. One such condition is easily obtained, since we obviously (why?) have $V(T,x) = F(x)$ for all $x \in R^n$. We have now arrived at our goal, namely the Hamilton–Jacobi–Bellman equation, often referred to simply as the HJB equation.

Theorem 11.10 (Hamilton–Jacobi–Bellman equation)   *Under Assumption 11.9, the following hold.*

1. V satisfies the Hamilton–Jacobi–Bellman equation

$$\begin{cases} \dfrac{\partial V}{\partial t}(t,x) + \sup_{u \in U_t(x)} \{H(t,x,u) + \mathcal{G}^u V(t,x)\} &= 0, \quad (t,x) \in (0,T) \times R^n \\[2mm] V(T,x) &= F(x), \quad x \in R^n. \end{cases}$$

2. *For each $(t, x) \in [0, T] \times R^n$ the supremum in the HJB equation above is attained by*
   $u = \hat{u}(t, x)$.

It is important to note that this theorem has the form of a *necessary* condition. It says that *if* V is the optimal value function, and *if* $\hat{u}$ is the optimal control, *then* V satisfies the HJB equation, and $\hat{u}(t, x)$ realizes the supremum in the equation. We also note that Assumption 11.9 is *ad hoc*. One would prefer to have conditions in terms of the initial data μ, σ, F and F which would guarantee that Assumption 11.9 is satisfied. This can in fact be done, but at a fairly high price in terms of technical complexity. The reader is referred to the specialist literature; see, for example, Øksendal (2004).

A gratifying, and perhaps surprising, fact is that the HJB equation also acts as a *sufficient* condition for the optimal control problem. This result is known as the **verification theorem** for dynamic programming, and we will use it repeatedly below. Note that, as opposed to the necessary conditions above, the verification theorem is very easy to prove rigorously.

Theorem 11.11 (Verification theorem) *Suppose that we have two functions $K(t, x)$ and $g(t, x)$, such that*

- *K is sufficiently integrable (see Remark 11.12 below) and solves the HJB equation*

$$\begin{cases} \dfrac{\partial K}{\partial t}(t, x) + \sup_{u \in U} \{H(t, x, u) + \mathcal{G}^u K(t, x)\} &= 0, \quad (t, x) \in (0, T) \times R^n \\ K(T, x) &= F(x), \quad x \in R^n. \end{cases}$$

- *The function g is an admissible control law.*
- *For each fixed (t, x), the supremum in the expression*

$$\sup_{u \in U} \{H(t, x, u) + \mathcal{G}^u K(t, x)\}$$

is attained by the choice $u = g(t, x)$.

Then the following hold.

1. *The optimal value function V to the control problem is given by*

$$V(t, x) = K(t, x).$$

2. *There exists an optimal control law $\hat{u}$, and in fact $\hat{u}(t, x) = g(t, x)$.*

Remark Note that we have used the letter K (instead of V) in the HJB equation above. This is because the letter V by definition denotes the optimal value function.

Proof Assume that K and g are given as above. Now choose an arbitrary control law, $\mathbf{u} \in \mathcal{U}$, and fix a point (t, x). We define the process $X^{\mathbf{u}}$ on the time interval $[t, T]$ as the solution to the equation

$$dX_s^{\mathbf{u}} = \mu^{\mathbf{u}}\left(s, X_s^{\mathbf{u}}\right) ds + \sigma^{\mathbf{u}}\left(s, X_s^{\mathbf{u}}\right) dW_s,$$
$$X_t = x.$$

Since K solves the HJB equation we see that

$$\frac{\partial K}{\partial t}(s, y) + H(s, y, u) + G^u K(s, y) \le 0$$

for all $(s, y) \in R_+ \times R^n$ and all $u \in U$. In particular we have

$$\frac{\partial K}{\partial t}(s, y) + H(s, y, \mathbf{u}(s, y)) + G^\mathbf{u} K(s, y) \le 0.$$

This implies that the process

$$K(s, X_s^\mathbf{u}) + \int_t^s H(r, X_r^\mathbf{u}, \mathbf{u}(r, X_r^\mathbf{u}))dr$$

is a supermartingale. Using this fact and the boundary condition of the HJB equation, we obtain

$$K(t, x) \ge E_{t,x} \left[\int_t^T H^\mathbf{u}(s, X_s^\mathbf{u})ds + F(X_T^\mathbf{u}) \right] = \mathcal{J}(t, x, \mathbf{u}).$$

Since the control law $\mathbf{u}$ was arbitrarily chosen this gives us

$$K(t, x) \ge \sup_{\mathbf{u} \in \mathcal{U}} \mathcal{J}(t, x, \mathbf{u}) = V(t, x). \tag{11.7}$$

To obtain the reverse inequality we choose the specific control law $\mathbf{u}(t, x) = \mathbf{g}(t, x)$. Going through the same calculations as above, and using the fact that by assumption we have

$$\frac{\partial K}{\partial t}(t, x) + H^\mathbf{g}(t, x) + G^\mathbf{g} K(t, x) = 0, \quad \text{for all } (t, x) \in R_+ \times R^n,$$

we obtain the equality

$$K(t, x) = E_{t,x} \left[\int_t^T H^\mathbf{g}(s, X_s^\mathbf{g})ds + F(X_T^\mathbf{g}) \right] = \mathcal{J}(t, x, \mathbf{g}). \tag{11.8}$$

On the other hand we have the trivial inequality

$$V(t, x) \ge \mathcal{J}(t, x, \mathbf{g}), \tag{11.9}$$

so, using (11.7)–(11.9), we obtain

$$K(t, x) \ge V(t, x) \ge \mathcal{J}(t, x, \mathbf{g}) = K(t, x).$$

This shows that in fact

$$K(t, x) = V(t, x) = \mathcal{J}(t, x, \mathbf{g}),$$

which proves that $K = V$, and that $\mathbf{g}$ is the optimal control law. $\square$

Remark 11.12   The assumption that K is "sufficiently integrable" in the theorem above is needed to guarantee that the stochastic processes

$$\int_0^t \nabla_x K(s, X_s^\mathbf{u})\sigma^\mathbf{u}(s, X_s^\mathbf{u})dW_s$$

and

$$\int_0^t \int_E \{K(s, X^{\mathbf{u}}_{s-} + \beta^{\mathbf{u}}(s, X^{\mathbf{u}}_{s-}, z)) - K(s, X^{\mathbf{u}}_{s-})\} [\Psi(ds, dz) - \lambda_s(dz)]$$

are true (and not merely local) martingales. This will be the case if, for example, K satisfies the conditions

$$E\left[\int_0^T \|\nabla_x K(s, X^{\mathbf{u}}_s)\sigma^{\mathbf{u}}(s, X^{\mathbf{u}}_s)\|^2)ds\right] < \infty$$

and

$$E\left[\int_0^T \int_E |K(s, X^{\mathbf{u}}_{s-} + \beta^{\mathbf{u}}(s, X^{\mathbf{u}}_{s-}, z)) - K(s, X^{\mathbf{u}}_{s-})|\lambda_t(dz)ds\right] < \infty$$

for all admissible control laws **u**.

Remark Sometimes, instead of a maximization problem, we consider a minimization problem. Of course we now make the obvious definitions for the value function and the optimal value function. It is then easily seen that all the results above still hold if the expression

$$\sup_{u \in U} \{H(t, x, u) + \mathcal{G}A^u V(t, x)\}$$

in the HJB equation is replaced by the expression

$$\inf_{u \in U} \{H(t, x, u) + \mathcal{G}A^u V(t, x)\}.$$

Remark In the verification theorem we may allow the control constraint set U to be state- and time-dependent, i.e. of the form $U(t, x)$.

11.5   Handling the HJB Equation

In this section we will describe the actual handling of the HJB equation, and in the next section we will study a classical example – the linear quadratic regulator. We thus consider our standard optimal control problem with the corresponding HJB equation:

$$\begin{cases} \dfrac{\partial V}{\partial t}(t, x) + \sup_{u \in U} \{H(t, x, u) + \mathcal{G}^u V(t, x)\} & = & 0, \\ V(T, x) & = & F(x). \end{cases} \qquad (11.10)$$

Schematically we now proceed as follows.

1. Consider the HJB equation as a PDE for an unknown function V.
2. Fix an arbitrary point $(t, x) \in [0, T] \times R^n$ and solve, for this fixed choice of (t, x), the static optimization problem

$$\max_{u \in U} [H(t, x, u) + \mathcal{G}^u V(t, x)].$$

Note that in this problem u is the only variable, whereas t and x are considered to be fixed parameters. The functions F, μ, σ and V are considered as given.

3. The optimal choice of u, denoted by $\hat{u}$, will of course depend on our choice of t and x, but it will also depend on the function V and its various partial derivatives (which are hiding under the sign $G^u V$). To highlight these dependencies we write $\hat{u}$ as

$$\hat{u} = \hat{u}(t, x; V). \tag{11.11}$$

4. The function $\hat{u}(t, x; V)$ is our candidate for the optimal control law, but since we do not know V this description is incomplete. Therefore we substitute the expression for $\hat{u}$ in (11.11) into the PDE (11.10), giving us the PDE

$$\frac{\partial V}{\partial t}(t, x) + H^{\hat{u}}(t, x) + G^{\hat{u}} V(t, x) = 0,$$

$$V(T, x) = F(x).$$

5. Now we solve the PDE above! (See the remark below.) Then we put the solution V into expression (11.11). Using the verification theorem 11.11 we can now identify V as the optimal value function and $\hat{u}$ as the optimal control law.

The hard work of dynamic programming consists in solving the highly non-linear PDE in step 5 above. There are of course no general analytic methods available for this, so the number of known optimal control problems with an analytic solution is very small indeed. In an actual case one usually tries to *guess* a solution, i.e. we typically make an *Ansatz* for V, parameterized by a finite number of parameters, and then we use the PDE in order to identify the parameters. The making of an Ansatz is often helped by the intuitive observation that if there is an analytical solution to the problem, then it seems likely that V inherits some structural properties from the boundary function F as well as from the instantaneous utility function H.

For a general problem there is thus very little hope of obtaining an analytic solution and it is worth pointing out that many of the known solved control problems have, to some extent, been "rigged" in order to be analytically solvable.

11.6   The Linear Quadratic Regulator: II

We now want to put the ideas from the previous section into action, and for this purpose we study the best known of all control problems, namely the linear quadratic regulator, which was studied in discrete time in Section 10.7. We will start by studying the purely Wiener-driven model, and in an exercise we then extend it to a jump-diffusion model. In this classical engineering example we wish to *minimize*

$$E\left[\int_0^T \{X_t^\star Q X_t + u_t^\star R u_t\} \, dt + X_T^\star H X_T \right],$$

(where $\star$ now denotes transpose) given the dynamics

$$dX_t = \{A X_t + B u_t\} \, dt + C \, dW_t.$$

One interpretation of this problem is about controlling a vehicle in such a way that it stays close to the origin (the terms $x^\star Qx$ and $x^\star Hx$) while at the same time keeping the "energy" $u^\star Ru$ small.

As usual $X_t \in \mathbf{R}^n$ and $\mathbf{u}_t \in \mathbf{R}^k$, and we impose no control constraints on u. The matrices Q, R, H, A, B and C are assumed to be known. Without loss of generality we may assume that Q, R and H are symmetric, and we assume that R is positive definite (and thus invertible).

The HJB equation now becomes

$$
\begin{cases}
\dfrac{\partial V}{\partial t}(t,x) \;+\; \displaystyle\inf_{u\in\mathbf{R}^k} \left\{ x^\star Qx + u^\star Ru + [\nabla_x V](t,x)\,[Ax + Bu] \right\} \\[2mm]
\qquad\qquad +\; \dfrac{1}{2} \displaystyle\sum_{i,j} \dfrac{\partial^2 V}{\partial x_i \partial x_j}(t,x)\,[CC^\star]_{i,j} = 0, \\[2mm]
V(T,x) \;=\; x^\star Hx.
\end{cases}
$$

For each fixed choice of (t,x) we now have to solve the static unconstrained optimization problem to minimize

$$
u^\star Ru + [\nabla_x V](t,x)\,[Ax + Bu].
$$

Since by assumption $R > 0$, we get the solution by setting the gradient equal to zero, thus giving us the equation

$$
2u^\star R = -(\nabla_x V)B,
$$

which gives us the optimal u as

$$
\hat{u} = -\frac{1}{2} R^{-1} B^\star (\nabla_x V)^\star.
$$

Here we see clearly (compare point 2 in the scheme above) that in order to use this formula we need to know V, and we thus try to make an educated guess about the structure of V. From the boundary-value function $x^\star Hx$ and the quadratic term $x^\star Qx$ in the instantaneous cost function it seems reasonable to assume that V is a quadratic function. Consequently we make the following Ansatz:

$$
V(t,x) = x^\star P(t)x + q(t),
$$

where we assume that $P(t)$ is a deterministic symmetric matrix function of time, whereas $q(t)$ is a scalar deterministic function. It would of course also be natural to include a linear term of the form $L(t)x$, but it turns out that this is not necessary.

With this trial solution we have, suppressing the t-variable and denoting time derivatives by a dot,

$$
\frac{\partial V}{\partial t}(t,x) = x^\star \dot{P}x + \dot{q},
$$

$$
\nabla_x V(t,x) = 2x^\star P,
$$

$$
\nabla_{xx} V(t,x) = 2P
$$

$$
\hat{u} = -R^{-1} B^\star Px.
$$

Inserting these expressions into the HJB equation we get

$$x^\star \dot{P} x + \dot{q} + x^\star Q x + x^\star P B R^{-1} R R^{-1} B^\star P x + 2 x^\star P A x$$
$$-2 x^\star P B R^{-1} B^\star P x + \sum_{i,j} P_{ij} [CC^\star]_{ij} = 0.$$

We note that the last term above equals $\mathrm{Tr}[C^\star P C]$, where Tr denotes the trace of a matrix, and furthermore we see that $2 x^\star P A x = x^\star A^\star P x + x^\star P A x$ (this is just cosmetic). Collecting terms gives us

$$x^\star \left\{ \dot{P} + Q - P B R^{-1} B^\star P + A^\star P + P A \right\} x + \dot{q} + \mathrm{Tr}[C^\star P C] = 0.$$

If this equation is to hold for all x and all t then firstly the bracket must vanish, leaving us with the matrix ODE

$$\dot{P} = P B R^{-1} B^\star P - A^\star P - P A - Q.$$

We are then left with the scalar equation

$$\dot{q} = - \mathrm{Tr}[C^\star P C].$$

We now need some boundary values for P and q, but these follow immediately from the boundary conditions of the HJB equation. We thus end up with the following pairs of equations

$$\begin{cases} \dot{P} &= P B R^{-1} B^\star P - A^\star P - P A - Q, \\ P(T) &= H. \end{cases} \tag{11.12}$$

$$\begin{cases} \dot{q} &= - \mathrm{Tr}[C^\star P C], \\ q(T) &= 0. \end{cases} \tag{11.13}$$

The matrix equation (11.12) is known as a **Riccati equation**, and there are powerful algorithms available for solving it numerically. The equation for q can then be integrated directly.

Summing up, we see that the optimal value function and the optimal control law are given by the following formulas (note that the optimal control is linear in the state variable):

$$V(t, x) = x^\star P(t) x + \int_t^T \mathrm{Tr}[C^\star P(s) C] ds,$$
$$\hat{u}(t, x) = -R^{-1} B^\star P(t) x.$$

11.7   A Generalized HJB Equation

In many concrete applications, in particular in economics, it is natural to consider an optimal control problem, where the state variable is constrained to stay within a pre-specified domain. For example, it may be reasonable to demand that the wealth of an investor is never allowed to become negative. We will now generalize our class of optimal control problems to allow for such considerations.

Let us therefore consider the following controlled SDE

$$dX_t = \mu(t, X_t, u_t)\, dt + \sigma(t, X_t, u_t)\, dW_t,$$
$$X_0 = x_0,$$

where as before we impose the control constraint $u_t \in U$. We also consider as given a fixed time interval $[0, T]$ and a fixed domain $D \subseteq [0, T] \times \mathbf{R}^n$. The basic idea is that when the state process hits the boundary ∂D of D, then the activity is at an end. It is thus natural to define the **stopping time** τ by

$$\tau = \inf\{t \geq 0 \,|(t, X_t) \in \partial D\} \wedge T,$$

where $x \wedge y = \min[x, y]$. We consider as given an instantaneous utility function $H(t, x, u)$ and a "bequest function" $F(t, x)$, i.e. a mapping $F : \partial D \to \mathbf{R}$. The control problem to be considered is that of maximizing

$$E\left[\int_0^\tau H(s, X_s^{\mathbf{u}}, \mathbf{u}_s)ds + F\left(\tau, X_\tau^{\mathbf{u}}\right)\right].$$

In order for this problem to be interesting we have to demand that $X_0 \in D$, and the interpretation is that when we hit the boundary ∂D, the game is over and we obtain the bequest $F(\tau, X_\tau)$. We see immediately that our earlier situation corresponds to the case when $D = [0, T] \times \mathbf{R}^n$ and when F is constant in the t-variable.

In order to analyze our present problem we may proceed as in the previous sections, introducing the value function and the optimal value function exactly as before. The only new technical problem encountered is that of considering a stochastic integral with a stochastic limit of integration. Since this will take us outside the scope of the present text we will confine ourselves to giving the results. The proofs are (modulo the technicalities mentioned above) exactly as before.

Theorem 11.13 (HJB equation) *Assume that*

- *The optimal value function V is in $C^{1,2}$.*
- *An optimal law $\hat{u}$ exists.*

Then the following hold.

1. V satisfies the HJB equation

$$\begin{cases} \dfrac{\partial V}{\partial t}(t, x) + \sup_{u \in U}\{H(t, x, u) + \mathcal{G}^u V(t, x)\} &= 0, \quad \text{for all } (t, x) \in D \\[2mm] V(t, x) &= F(t, x), \quad \text{for all } (t, x) \in \partial D. \end{cases}$$

2. For each $(t, x) \in D$ the supremum in the HJB equation above is attained by $u = \hat{u}(t, x)$.

Theorem 11.14 (Verification theorem) *Suppose that we have two functions $H(t, x)$ and $g(t, x)$, such that*

- *K is sufficiently integrable and solves the HJB equation*

$$\begin{cases} \dfrac{\partial K}{\partial t}(t, x) + \sup_{u \in U}\{H(t, x, u) + \mathcal{G}^u K(t, x)\} &= 0, \quad \text{for all } (t, x) \in D \\[2mm] K(t, x) &= F(t, x), \quad \text{for all } (t, x) \in \partial D. \end{cases}$$

- *The function g is an admissible control law.*
- *For each fixed (t, x), the supremum in the expression*

$$\sup_{u \in U} \{H(t, x, u) + \mathcal{A}^u K(t, x)\}$$

is attained by the choice $u = g(t, x)$.

Then the following hold.

1. *The optimal value function V to the control problem is given by*

$$V(t, x) = K(t, x).$$

2. *There exists an optimal control law $\hat{u}$, and in fact $\hat{u}(t, x) = g(t, x)$.*

11.8   Optimal Consumption and Investment

11.8.1   A Diffusion Model with Power Utility

We will now turn to a classical optimal consumption/investment problem; the reader unfamiliar with financial economics is referred to Björk (2020) for more detailed information concerning the economic background.

We consider an a economic agent over a fixed time interval $[0, T]$. The agent wants to maximize her expected utility over the time period, and she has the following investment opportunities.

- Investment can be made in a risk-free asset (bank account) B with price process

$$dB_t = rB_t dt,$$

where the *short rate r* is assumed to be constant.
- Investment can be made in a risky asset (stock) S with price dynamics given by

$$dS_t = \alpha S_t dt + \sigma S_t dW_t.$$

The constant μ is the *mean rate of return* per annum, and the constant σ is the *volatility* per annum. The time unit is one year.

We denote the portfolio weight on the risky asset by u_t^S, the portfolio weight on the bank account by u_t^B and the consumption rate by c_t. The wealth process X_t of a self-financing portfolio will then have the following dynamics:

$$dX_t = X_t \left\{\alpha u_t^S + r u_t^B\right\} dt - c_t dt + X_t u_t^S \sigma dW_t.$$

The objective of the agent is to maximize the expected utility of consumption according to the utility functional

$$E\left[\int_0^T U(t, c_t) dt\right],$$

given the wealth dynamics

$$dX_t = X_t \left[u_t^B r + u_t^B \alpha \right] dt - c_t dt + u_t^S \sigma X_t dW_t,$$

and the constraints

$$c_t \geq 0, \quad \text{for all } t \geq 0,$$
$$u_t^B + u_t^S = 1, \quad \text{for all } t \geq 0.$$

In a control problem of this kind it is important to be aware of the fact that one may quite easily formulate a nonsensical problem. To take a simple example, suppose that the utility function U is increasing and unbounded in the c-variable. Then the problem above degenerates completely. It does not possess an optimal solution at all the reason being of course that the consumer can increase his/her utility to any given level by simply consuming an arbitrarily large amount at every t. The consequence of this hedonistic behavior is of course the fact that the wealth process will, with very high probability, become negative, but this is neither prohibited by the control constraints, nor punished by any bequest function.

An elegant way out of this dilemma is to choose the domain D of Section 11.7 as $D = [0,T] \times \{x \in \mathbf{R} : x > 0\}$. With τ defined as above this means, in concrete terms, that

$$\tau = \inf \{t > 0 \,|\, X_t = 0\} \wedge T.$$

A natural objective function in this case is thus given by

$$E \left[\int_0^\tau U(t, c_t) dt \right],$$

which automatically ensures that when the consumer has no wealth, then all activity is terminated.

We will now analyze this problem in some detail. First we notice that we can get rid of the constraint $u_t^0 + u_t^1 = 1$ by defining a new control variable w as $w = u^1$, and then substituting $1 - w$ for u^0. This gives us our final formulation of the problem as follows.

Problem

$$\underset{w,\,c}{\text{maximize}} \quad E \left[\int_0^\tau U(t, c_t) dt \right],$$

given the state dynamics

$$dX_t = w_t \left[\alpha - r \right] X_t dt + (r X_t - c_t) dt + w_t \sigma X_t dW_t,$$

and the constraint

$$w_t \geq 0, \quad t \in [0,T].$$

The corresponding HJB equation is

$$\begin{cases} \dfrac{\partial V}{\partial t} + \underset{c \geq 0, w \in R}{\sup} \left\{ U(t,c) + wx(\alpha - r)\dfrac{\partial V}{\partial x} + (rx - c)\dfrac{\partial V}{\partial x} + \dfrac{1}{2}x^2 w^2 \sigma^2 \dfrac{\partial^2 V}{\partial x^2} \right\} & = & 0, \\[2mm] V(T,x) & = & 0, \\ V(t,0) & = & 0. \end{cases}$$

The static optimization problem to be solved with respect to c and w is thus that of maximizing

$$U(t,c) + wx(\alpha - r)\frac{\partial V}{\partial x} + (rx - c)\frac{\partial V}{\partial x} + \frac{1}{2}x^2 w^2 \sigma^2 \frac{\partial^2 V}{\partial x^2},$$

and, assuming an interior solution, the first-order conditions are

$$U_c(t,c) = V_x, \tag{11.14}$$

$$w = \frac{-V_x}{x \cdot V_{xx}} \cdot \frac{\alpha - r}{\sigma^2}, \tag{11.15}$$

where we have used subscripts to denote partial derivatives. For an economist, equation (11.14) is a well-known relation. It says that the marginal (direct) utility of consumption must equal the marginal (indirect) utility of wealth.

We now specialize our example to the case when U is of the form

$$U(t,c) = e^{-\delta t} c^{\gamma},$$

where $0 < \gamma < 1$. The economic reasoning behind this is that we now have an infinite marginal utility at $c = 0$. This will force the optimal consumption to be positive throughout the planning period, a fact which will facilitate the analytical treatment of the problem. We are thus, in some sense, "rigging" the problem.

We again see that in order to implement the optimal consumption–investment plan (11.14)–(11.15) we need to know the optimal value function V. We therefore suggest an Ansatz and, in view of the shape of the instantaneous utility function, it is natural to try one of the form

$$V(t,x) = e^{-\delta t} h(t) x^{\gamma}, \tag{11.16}$$

where, because of the boundary conditions, we must demand that

$$h(T) = 0. \tag{11.17}$$

Given a V of this form we have (using $\cdot$ to denote the time derivative)

$$\frac{\partial V}{\partial t} = e^{-\delta t} \dot{h} x^{\gamma} - \delta e^{-\delta t} h x^{\gamma}, \tag{11.18}$$

$$\frac{\partial V}{\partial x} = \gamma e^{-\delta t} h x^{\gamma - 1}, \tag{11.19}$$

$$\frac{\partial^2 V}{\partial x^2} = \gamma(\gamma - 1) e^{-\delta t} h x^{\gamma - 2}. \tag{11.20}$$

Inserting these expressions into (11.14)–(11.15) we get

$$\hat{w}(t,x) = \frac{\alpha - r}{\sigma^2 (1 - \gamma)}, \tag{11.21}$$

$$\hat{c}(t,x) = x h(t)^{-1/(1-\gamma)}. \tag{11.22}$$

This looks very promising: we see that the candidate optimal portfolio is constant and that the candidate optimal consumption rule is linear in the wealth variable. In order to use the verification theorem we now want to show that a V-function of the form (11.16) actually solves the HJB equation. We therefore substitute the expressions (11.18)–(11.22) into the HJB equation. This gives us the equation

$$x^\gamma \left\{ h(t) + Ah(t) + Bh(t)^{-\gamma/(1-\gamma)} \right\} = 0,$$

where the constants A and B are given by

$$A = \frac{\gamma(\alpha - r)^2}{\sigma^2(1 - \gamma)} + r\gamma - \frac{1}{2}\frac{\gamma(\alpha - r)^2}{\sigma^2(1 - \gamma)} - \delta$$

$$B = 1 - \gamma.$$

If this equation is to hold for all x and all t, then we see that h must solve the ODE

$$\dot{h}(t) + Ah(t) + Bh(t)^{-\gamma/(1-\gamma)} = 0, \tag{11.23}$$

$$h(T) = 0. \tag{11.24}$$

An equation of this kind is known as a **Bernoulli equation**, and it can be solved explicitly.

Summing up, we have shown that if we define V as in (11.16) with h defined as the solution to (11.23)–(11.24), and if we further define $\hat{w}$ and $\hat{c}$ by (11.21)–(11.22), then V satisfies the HJB equation, with $\hat{w}$ and $\hat{c}$ attaining the supremum in the equation. The verification theorem then tells us that we have indeed found the optimal solution.

11.8.2   A Jump-Diffusion Model with Log Utility

In this section we consider a model which is very similar to the one above, but with two differences.

- We assume that the stock price dynamics are given by

$$dS_t = \alpha S_t dt + \sigma S_t dW_t + S_{t-} \int_E z \Psi(dt, dz),$$

  where Ψ is a marked point process with deterministic and constant (over time) intensity measure $\lambda(dz)$. The economic interpretation is that if we have a jump at t with mark z, then the relative change in S, given by $\Delta S_t / S_{t-}$, will be z. If, for example, $z = -0.2$ this means that the stock price decreases by 20%. In order to guarantee non-negative stock prices we must therefore assume that mark space E is (a subset of) the interval $E = [-1, \infty)$.
- We assume log utility, i.e

$$U(t, c) = e^{-\delta t} \ln(c),$$

  and we also add a utility function of terminal wealth, given by

$$e^{-\delta T} K \ln(x).$$

We can of course also write the S-dynamics on semimartingale form as

$$dS_t = \left(\alpha + \int_E z\lambda(dz) \right) S_t dt + \sigma S_t dW_t + S_{t-} \int_E z \left\{ \Psi(dt, dz) - \lambda(dz)dt \right\}.$$

so we see that α is the mean rate of return *between jumps*, whereas the mean rate of return *including jumps* is given by $\alpha + \int_E z\lambda(dz)$. Given the new setup our problem is the following.

Problem

$$\underset{w,c}{\text{maximize}} \quad E\left[\int_0^T e^{-\delta t}\ln(c_t)dt + e^{-\delta T}K\ln(X_T)\right],$$

given the state dynamics

$$dX_t = w_t[\alpha - r]X_t dt + (rX_t - c_t)dt + w_t\sigma X_t dW_t + X_{t-}w_t\int_E z\Psi(dt,dz),$$

and the constraints

$$c_t \geq 0, \quad 0 \leq w_t \leq 1, \quad t \in [0,T].$$

It turns out that we need the following restriction for our model.

Assumption 11.15 *We assume that the following conditions hold*

$$\alpha + \int_E z\lambda(dz) \geq r$$

$$(\alpha - r) - \sigma^2 + \int_E \frac{z}{1+z}\lambda(dz) \leq 0.$$

The corresponding HJB equation is

$$\begin{cases} \dfrac{\partial V}{\partial t} + \underset{c \geq 0, w \in R}{\sup} \ \{U(t,c) + GV(t,x)\} &= 0, \\ \\ \hspace{3.5cm} V(T,x) &= e^{-\delta T}K\ln(x), \end{cases}$$

where

$$GV(t,x) = wx(\alpha - r)\frac{\partial V}{\partial x} + (rx - c)\frac{\partial V}{\partial x} + \frac{1}{2}x^2 w^2 \sigma^2 \frac{\partial^2 V}{\partial x^2}$$
$$+ \int_E \{V(t,x(1+wz)) - V(t,x)\}\lambda(dz).$$

Assuming an interior optimum, the first-order conditions are

$$e^{-\delta t}\frac{1}{c} = V_x$$

$$x(\alpha - r)V_x + wx^2\sigma^2 V_{xx} + x\int_E zV_x(t,x(1+wz))\lambda(dz) = 0.$$

At this point we make the Ansatz

$$V(t,x) = e^{-\delta t}g(t)\ln(x) + e^{-\delta t}h(t),$$

where g and h are deterministic functions of time. Given the Ansatz, we obtain

$$V_t = -\delta e^{-\delta t}g\ln(x) + e^{-\delta t}\dot{g}\ln(x) + \dot{h},$$

$$V_x = e^{-\delta t}g\frac{1}{x},$$

$$V_{xx} = -e^{-\delta t}g(t)\frac{1}{x^2}.$$

After some manipulations, this gives us

$$\hat{c} = \frac{x}{g},$$

$$(\alpha - r) + \int_E \frac{z}{1 + wz} \lambda(dz) = w\sigma^2,$$

and it follows from Assumption 11.15 that there is a unique solution to the equation for w. This implies that $\hat{w}$ is a constant (not depending on t or x). Plugging everything into the HJB equation results in the following equation

$$[\dot{g} - \delta g]\ln(x) + \dot{h} - \delta h + g\left\{\hat{w}(\alpha - r) + r - 1 - \frac{1}{2}\sigma^2\hat{w}^2 + D\right\} = 0,$$

where the constant D is given by

$$D = \int_E \ln(1 + \hat{w}z)\lambda(dz).$$

This gives us the following ODE for g:

$$\begin{cases} \dot{g} &= \delta g, \\ g(T) &= e^{-\delta T}K, \end{cases}$$

with solution

$$g(t) = Ke^{-\delta(T-t)}.$$

We also get an ODE for h, but since the optimal controls are completely determined by g, the ODE for h is not so interesting.

Proposition 11.16   *Under Assumption 11.15, the following hold*

- *The optimal consumption plan is given by*

$$\hat{c}(t, x) = Ke^{\delta(T-t)}x.$$

- *The optimal portfolio weight $\hat{w}$ on the risky asset is constant, and given as the unique solution to*

$$(\alpha - r) + \int_E \frac{z}{1 + wz} \lambda(dz) = w\sigma^2.$$

11.9   Intensity Control Problems

In this section we will study some problems where the control affects the state process through the intensity of the driving point process.

11.9.1   A Quadratic Objective Functional

In this example we consider a controlled counting process N with $\lambda_t = u_t$, where $u \geq 0$ is the control variable. The objective is to maximize

$$E \left[A N_T^2 - \int_0^T u_t^2 \, dt \right].$$

where $0 \leq A < 1/T$, given the control constraint $u_t \geq 0$. The upper bound on A is needed in order to avoid explosion (see below).

In this case we have an optimal value function of the form $V(t, n)$, where $n = 0, 1, 2, \ldots$ and the HJB equation is as follows.

$$
\begin{cases}
\dfrac{dV}{dt}(t, n) + \sup_{u \geq 0} \left\{ -u^2 + u \left[V(t, n+1) - V(t, n) \right] \right\} &= 0 \\[2mm]
V(T, n) &= An^2
\end{cases}
$$

The first-order condition for u gives us

$$\hat{u}_t = \frac{1}{2} \left[V(t, n+1) - V(t, n) \right]$$

and since V obviously is increasing in n (why?) we see that the constraint $u \geq 0$ is satisfied. The HJB equation now reduces to

$$\frac{dV}{dt}(t, n) + \frac{1}{4} \left[V(t, n+1) - V(t, n) \right]^2 = 0.$$

We now need to make an Ansatz, and a natural choice is a quadratic form, so our Ansatz is

$$V(t, n) = a_t n^2 + b_t n + c_t,$$

where a, b and c are deterministic functions of time. Given this, and again denoting time-derivatives by a dot, we obtain

$$\frac{dV}{dt}(t, n) = \dot{a} n^2 + \dot{b} n + \dot{c}$$

$$V(t, n+1) - V(t, n) = 2an + a + b,$$

$$\hat{u}(t, n) = an + \frac{a+b}{2}.$$

Plugging this back into HJB gives us

$$\dot{a} n^2 + \dot{b} n + \dot{c} + \frac{1}{4} (2an + a + b)^2 = 0.$$

so we have

$$\dot{a} n^2 + \dot{b} n + \dot{c} + a^2 n^2 + an(a + b) + \frac{a+b}{4} = 0.$$

Identifying coefficients we thus obtain the ODE system

$$\dot{a} = -a^2,$$

$$\dot{b} = -a(a + b),$$

$$\dot{c} = -\frac{a+b}{4},$$

with boundary conditions

$$a_T = A, \quad b_T = 0, \quad c_T = 0.$$

The equation for a is easily solved as

$$a_t = \frac{1}{\frac{1}{A} + t - T}$$

and we see that we need the assumption that $1/A > T$ in order to avoid explosion. The equation for b is linear and c can then be obtained by direct integration.

11.9.2   Maximizing a Target-Hitting Probability

The problem for this section is taken from Brémaud (1981). We consider a scalar counting process N and make the following assumption.

Assumption   *The intensity of N is of the form*

$$\lambda_t = u_t,$$

where u is any non-negative $\mathbf{F}^N$-adapted control process with $a \leq u \leq b$.

The problem under study is to maximize the probability of hitting a predetermined target K, given constraints on the control. We thus want to maximize

$$P(N_T^u = K)$$

subject to the constraints

$$a \leq u_t \leq b,$$

where $0 \leq a < b$.

The HJB equation for the optimal value function $V(t, n)$ is as follows.

$$\begin{cases} \dfrac{\partial V}{\partial t}(t, n) + \sup_{a \leq u \leq b} \{V(t, n+1) - V(t, n)\} u & = & 0, \\ V(T, n) & = & I\{n = K\}. \end{cases} \tag{11.25}$$

This is a recursive system of ODEs, for $n = 0, 1, 2, \ldots$, and we see immediately that the optimal control has the form

$$\hat{u}_t(n) = \begin{cases} a & \text{if} \quad V(t, n+1) < V(t, n) \\ b & \text{if} \quad V(t, n+1) > V(t, n) \end{cases}.$$

with a tie if $V(t, n) = V(t, n+1)$. The optimal control will thus switch between the extreme values a and b; in the control literature this is known as a **bang–bang** solution.

It is clear that for $n \geq K + 1$ we have $V(t, n) = 0$, so for $n = K$ we have the HJB equation

$$\begin{cases} \dfrac{\partial V}{\partial t}(t, K) + \sup_{a \leq u \leq b} \{-V(t, K)\} u & = & 0, \\ V(T, K) & = & 1. \end{cases}$$

Since $V \geq 0$, the supremum is realized by

$$\hat{u}_t(K) = a, \quad 0 \leq t \leq T,$$

so we obtain the ODE

$$\begin{cases} \dfrac{\partial V}{\partial t}(t,K) - aV(t,K)u &= 0, \\ V(T,K) &= 1, \end{cases}$$

with solution

$$V(t,K) = e^{-a(T-t)}.$$

Intuitively, the result $\hat{u}_t(K) = a$ and $V(t,K) = e^{-a(T-t)}$ is more or less obvious. If you have reached the target level, so that $N_t = K$, then you will of course set the intensity to a minimum, i.e. $\hat{u}_t(K) = a$, and then you do nothing but hope that the N process does not jump before the time horizon T. This implies that, given $N_t = K$, the N-process will be Poisson with constant intensity a on the interval $[t,T]$ and the expression $e^{-a(T-t)}$ is the probability of no jumps on that time interval.

Having thus computed $V(t,K)$, the system (11.25) can be solved recursively in n to obtain the functions $V(t,K-1), V(t,K-2), \ldots, V(t,1), V(t,0)$, as well as the corresponding optimal controls. This recursion is, however, rather messy.

11.10   Exercise

Exercise 11.1   Solve the scalar Wiener- and MPP-driven linear quadratic regulator problem where we want to minimize

$$E\left[\int_0^T \{QX_t^2 + Ru_t^2\}\, dt + HX_T^2\right],$$

given the dynamics

$$dX_t = \{AX_t + Bu_t\}\, dt + C\, dW_t + \int_R z\Psi(dt, dz).$$

Here X and u are scalars, Ψ is a marked point process with mark space R and a deterministic intensity measure $\lambda(dz)$ which is constant in time. Use the notation

$$\mu = \int_R z\lambda(dz), \qquad \sigma^2 = \int_R z^2\lambda(dz),$$

where we assume that both μ and σ^2 exist.

11.11   Notes

Optimal control for Wiener-driven SDEs are treated in great depth in the almost ency-clopedic monograph Pham (2010). For a deep study of optimal control and stopping for jump diffusions, see Øksendal & Sulem (2007).

Part IV

Non-Linear Filtering Theory

12 Non-Linear Filtering with Wiener Noise

In this chapter we will present some of the basic ideas and results of non-linear filtering. In order to prepare the ground for point-process observations, we start with filtering in the presence of Wiener noise.

12.1 The Filtering Model

We consider a filtered probability space $(\Omega, \mathcal{F}, P, \mathbf{F})$ where as usual the filtration $\mathbf{F} = \{\mathcal{F}_t; t \geq 0\}$ formalizes the idea of an increasing flow of information. The basic model of non-linear filtering consists of a pair of processes (X, Z) with dynamics as follows.

$$dX_t = a_t dt + dM_t, \tag{12.1}$$
$$dZ_t = b_t dt + dW_t. \tag{12.2}$$

In this model the processes a and b are allowed to be arbitrary $\mathbf{F}$-adapted processes, M is an $\mathbf{F}$-martingale and W is an $\mathbf{F}$-Wiener process. At the moment we also assume that M and W are independent. We will below consider models where X and Z are multidimensional and where there is correlation between M, and W; but for the moment we focus on this simple model.

Remark Note that M is allowed to be an arbitrary martingale, so it does not have to be a Wiener process. The assumption that W is Wiener is, however, very important.

The interpretation of this model is that we are interested in the *state process* X, but that we cannot observe X directly. What we can observe is instead the *observation process* Z, so our main problem is to draw conclusions about X, given the observations of Z. We would for example like to compute the condition expectation

$$\hat{X}_t = E\left[X_t | \mathcal{F}_t^Z\right],$$

where $\mathcal{F}_t^Z = \sigma\{Z_s, s \leq t\}$ is the information generated by Z on the time interval $[0, t]$ or, more ambitiously, we would like to compute $\mathcal{L}(X_t | \mathcal{F}_t^Z)$, i.e. the entire conditional distribution of X_t given observations of Z on $[0, t]$. We will sometimes use the alternative notation

$$\pi_t[X] = E\left[X_t | \mathcal{F}_t^Z\right],$$

and, for a function $f : R \to R$, we will use the extended notation

$$\pi_t[f] = E\left[f(X_t)| \mathcal{F}_t^Z \right].$$

A very common concrete example of the model above is given by a model of the form

$$dX_t = \mu(t, X_t)dt + \sigma(t, X_t)dV_t,$$
$$dZ_t = b(X_t)dt + dW_t$$

where V and W are independent Wiener processes. In this case we can thus observe X indirectly through the term $b(X_t)$, but the observations are corrupted by the noise generated by W.

12.2   The Innovation Process

Consider again the Z-dynamics

$$dZ_t = b_t dt + dW_t.$$

Our best guess of b_t given $\mathcal{F}_t^Z$ is obviously given by $\hat{b}_t = E\left[b_t | \mathcal{F}_t^Z \right]$, and W is a process with zero mean so, at least intuitively, we expect that we would have

$$E\left[dZ_t | \mathcal{F}_t^Z \right] = \hat{b}_t dt.$$

This would imply that the "detrended" process v, defined by $dv_t = dZ_t - \hat{b}_t dt$ should be an $\mathbf{F}^Z$-martingale. As we will see below, this conjecture is correct and we can even improve on it, but first the formal definition.

Definition 12.1   The **innovation process** v is defined by

$$dv_t = dZ_t - \hat{b}_t dt.$$

We now have the following central result.

Proposition 12.2   *The innovation process v is an $\mathbf{F}^Z$-Wiener process.*

Proof   We give a sketch of the proof. According to the Lévy theorem, it is enough to prove the following

(i) The process v is an $\mathbf{F}^Z$-martingale.
(ii) The process $v_t^2 - t$ is an $\mathbf{F}^Z$-martingale.

To prove (i) we use the definition of v to obtain

$$E_s^Z\left[v_t - v_s \right] = E_s^Z\left[Z_t - Z_s \right] - E_s^Z\left[\int_s^t \hat{b}_u du \right],$$

where we have used the shorthand notation

$$E_s^Z\left[\cdot \right] = E\left[\cdot | \mathcal{F}_s^Z \right].$$

From the Z-dynamics we have

$$Z_t - Z_s = \int_s^t b_u \, du + W_t - W_s,$$

so we can write

$$E_s^Z [v_t - v_s] = E_s^Z \left[\int_s^t \{b_u - \hat{b}_u\} \, du \right] + E_s^Z [W_t - W_s].$$

Using iterated expectations and the F-Wiener property of W we have

$$E_s^Z [W_t - W_s] = E_s^Z [E [W_t - W_s | \mathcal{F}_s]] = 0.$$

We also have

$$E_s^Z \left[\int_s^t \{b_u - \hat{b}_u\} \, du \right] = \int_s^t E_s^Z [b_u - \hat{b}_u] \, du = \int_s^t E_s^Z [E_u^Z [b_u - \hat{b}_u]] \, du$$

$$= \int_s^t E_s^Z [\hat{b}_u - \hat{b}_u] \, du = 0,$$

which proves (i).

To prove (ii) we use the Itô formula to obtain

$$dv_t^2 = 2v_t dv_t + (dv_t)^2.$$

Since $dv_t = (b_t - \hat{b}_t) + dW_t$ we see that $(dv_t)^2 = dt$, so we have

$$dv_t^2 - dt = 2v_t dv_t.$$

From (i) we know that v is a martingale so the term $2v_t dv_t$ should be a martingale increment, which proves (ii).   □

Remark   Note that we have, in a sense, cheated a little bit, since the proof of (ii) actually requires a stochastic calculus theory which covers stochastic integrals with respect to general martingales and not only Wiener processes. This is the case not only when we use the Itô formula on v^2 without knowing *a priori* that v is an Itô process, but also when we conclude that $v_t dv_t$ is a martingale increment without having an *a priori* guarantee that dv_t is a stochastic differential with respect to a Wiener process. Given this general stochastic calculus (see Protter, 2004), the proof is completely correct.

The innovation process v will play a very important role in the theory. To highlight this role we now reformulate the result above in a slightly different way.

Proposition 12.3   *The Z-dynamics can be written as*

$$dZ_t = \hat{b}_t dt + dv_t, \qquad (12.3)$$

where v is an $\mathbf{F}^Z$-Wiener process.

Remark   Note that we now have two expressions for the Z-dynamics. We have the original dynamics

$$dZ_t = b_t dt + dW_t,$$

and we also have

$$dZ_t = \hat{b}_t dt + dv_t.$$

It is important to realize that the Z process in the left-hand side of these equations is, trajectory-by-trajectory, *exactly the same process*. The difference is that the first equation gives us the Z-dynamics relative to the filtration $\mathbf{F}$, whereas the second equation gives us the Z-dynamics with respect to the $\mathbf{F}^Z$-filtration.

12.3   Filter Dynamics and the FKK Equations

We now go on to derive an equation for the dynamics of the filter estimate $\hat{X}$. From the X-dynamics, $dX_t = a_t dt + dM_t$, and from the previous argument concerning Z, the obvious guess is that the term $d\hat{X}_t - \hat{a}_t dt$ should be a martingale; and this is indeed the case.

Lemma 12.4   *The process m, defined by*

$$dm_t = d\hat{X}_t - \hat{a}_t dt,$$

is an $\mathbf{F}^Z$-martingale.

Proof   We have

$$E_s^Z [m_t - m_s] = E_s^Z [X_t - X_s] - E_s^Z \left[\int_s^t \hat{a}_u du \right]$$

$$= E_s^Z \left[\int_s^t \{a_u - \hat{a}_u\} du \right] + E_s^Z [M_t - M_s].$$

We also have

$$E_s^Z \left[\int_s^t \{a_u - \hat{a}_u\} du \right] = E_s^Z \left[\int_s^t E_u^Z [a_u - \hat{a}_u] du \right] = E_s^Z \left[\int_s^t \{\hat{a}_u - \hat{a}_u\} du \right] = 0,$$

and furthermore

$$E_s^Z [M_t - M_s] = E_s^Z [E[M_t - M_s | \mathcal{F}_s]] = 0. \qquad \square$$

We thus have the filter dynamics

$$d\hat{X}_t = \hat{a}_t dt + dm_t,$$

where m is an $\mathbf{F}^Z$-martingale, and it remains to see if we can say something more specific about m. From the definition of the innovation process v it seems reasonable to hope that we have the equality

$$\mathcal{F}_t^Z = \mathcal{F}_t^v, \tag{12.4}$$

and, if this conjecture is true, the martingale representation theorem for Wiener processes would guarantee the existence of an adapted process h such that

$$dm_t = h_t dv_t.$$

The conjecture (12.4) is known as the "innovations hypothesis" and in its time it was a minor industry. In discrete time the corresponding innovations hypothesis is more or less trivially true, but in continuous time the situation is much more complicated. As a matter of fact, the (continuous-time) innovations hypothesis is *not* generally true, but the good news is that Fujisaki, Kallianpur and Kunita proved the following result, which we quote from Fujisaki, Kallinapur & Kunita (1972).

Proposition 12.5 *There exists an adapted process h such that*

$$dm_t = h_t dv_t. \tag{12.5}$$

We thus have the filter dynamics

$$d\widehat{X}_t = \hat{a}_t dt + h_t dv_t, \tag{12.6}$$

and it remains to determine the precise structure of the *gain* process h. We have the following result.

Proposition 12.6 *The gain process h is given by*

$$h_t = \widehat{X_t b_t} - \widehat{X}_t \hat{b}_t. \tag{12.7}$$

Proof We give a slightly heuristic proof. The full formal proof (see Fujisaki et al., 1972) uses the same idea as below, but it is more technical. We start by noticing that, for $s < t$ we have (from iterated expectations)

$$E\left[X_t Z_t - \widehat{X}_t Z_t \middle| \mathcal{F}_s^Z\right] = 0.$$

This leads to the (informal) identity

$$E\left[d(XZ)_t - d(\widehat{X}Z)_t \middle| \mathcal{F}_s^Z\right] = 0. \tag{12.8}$$

From the Itô formula we have, using the independence between W and M.

$$d(XZ)_t = X_t b_t dt + X_t dW_t + Z_t a_t dt + Z_t dM_t.$$

The Itô formula applied to (12.6) and (12.3) gives us

$$d(\widehat{X}Z)_t = \widehat{X}_t \hat{b}_t dt + \widehat{X}_t dv_t + Z_t \hat{a}_t dt + Z_t h dv_t + h dt,$$

where the term $h dt$ comes from the equality $(dv_t)^2 = dt$, since v is a Wiener process. Plugging these expressions into the formula (12.8) and setting $s = t$ gives us the expression

$$\left(\widehat{X_t b_t} + Z_t \hat{a}_t - \widehat{X}_t \hat{b}_t - Z_t \hat{a}_t - h_t\right) dt = 0.$$

from which we conclude (12.7). □

If we collect our findings we have the main result of non-linear filtering, namely the Fujisaki–Kallianpur–Kunita filtering equations.

Theorem 12.7 (The FKK Filtering Equations)   *The filtering equations are*

$$d\widehat{X}_t = \hat{a}_t dt + \left\{ \widehat{X_t b_t} - \widehat{X}_t \hat{b}_t \right\} d\nu_t,$$ (12.9)

$$d\nu_t = dZ_t - \hat{b}_t dt.$$ (12.10)

A simple calculation shows that we can write the gain process h as

$$h_t = E\left[\left(X_t - \widehat{X}_t \right)\left(b_t - \hat{b}_t \right) \middle| \mathcal{F}_t^Z \right],$$ (12.11)

so we see that the innovations are amplified by the conditional error covariance between X and b.

12.4   The General FKK Equations

We now extend our filtering theory to a more general model.

Assumption   *We consider a filtered probability space $\{\Omega, \mathcal{F}, P, \mathbf{F}\}$, carrying a martingale M and a Wiener process W, where M and W are not assumed to be independent. On this space we have the model*

$$dX_t = a_t dt + dM_t,$$ (12.12)

$$dZ_t = b_t dt + \sigma_t dW_t,$$ (12.13)

where a and b are $\mathbf{F}$-adapted scalar processes. The process σ is assumed to be strictly positive and $\mathbf{F}^Z$-adapted.

We note that the assumption about σ being $\mathbf{F}^Z$-adapted is not so much an assumption as a result, since the quadratic variation property of W implies that we can in fact estimate σ_t^2 without error on a arbitrary short interval. The moral of this is that although the drift b will typically depend in some way on the state process X, we can *not* let σ be of the form $\sigma_t = \sigma(X_t)$, since then the filter would trivialize. We also note that we cannot allow σ to be zero at any point, since then the filter will degenerate.

In a setting like this it is more or less obvious that the natural definition of the innovation process ν is

$$d\nu_t = \frac{1}{\sigma_t}\{dZ_t - \pi_t[b]\,dt\},$$

and it is not hard to prove that ν is a Wiener process. We can now more or less copy the arguments in Section 12.3. After some work we end up with the following general FKK equations. See Fujisaki et al. (1972) for details.

Theorem 12.8 (Fujisaki–Kallianpur–Kunita)   *With assumptions as above we have the following filter equations.*

$$d\widehat{X}_t = \hat{a}_t dt + \left[\widehat{D}_t + \frac{1}{\sigma_t}\left\{ \widehat{X_t b_t} - \widehat{X}_t \hat{b}_t \right\} \right] d\nu_t,$$

$$d\nu_t = \frac{1}{\sigma_t}\left\{ dZ_t - \hat{b}_t dt \right\},$$

where

$$D_t = \frac{d\langle M, W \rangle_t}{dt}.$$

Furthermore, the innovation process v is an $\mathbf{F}^Z$-Wiener process.

A proper definition of the process D above requires a more general theory for semi-martingales, but for most applications the following results are sufficient.

- If M has continuous trajectories, then

$$dD_t = dM \cdot dW_t,$$

   with the usual Itô multiplication rules.
- If M is a pure jump process without a Wiener component, then

$$dD_t = 0.$$

12.5   Filtering a Markov Process

A natural class of filtering problems to study is obtained if we consider a time-homogeneous Markov process X, living on some state space $\mathcal{M}$, with generator $\mathcal{G}$, and we are interested in estimating $f(X_t)$ for some real-valued function $f : \mathcal{M} \to R$.

12.5.1   The Markov Filter Equations

If f is in the domain of $\mathcal{G}$ we can then apply the Dynkin theorem and obtain the dynamics

$$df(X_t) = (\mathcal{G}f)(X_t)dt + dM_t, \tag{12.14}$$

where M is a a martingale.

Remark   For the reader who is unfamiliar with Dynkin and general Markov processes we note that a typical example would be that X is governed by an SDE of the form

$$dX_t = \mu(X_t)dt + \sigma(X_t)dW_t^0,$$

where W^0 is a a Wiener process. In this case the Itô formula will give us

$$df(X_t) = \left\{ \mu(X_t)f'(X_t) + \frac{1}{2}\sigma^2(X_t)f''(X_t) \right\} dt + \sigma(X_t)f'(X_t)dW_t^0,$$

so in this case the generator $\mathcal{G}$ is given by

$$\mathcal{G} = \mu(x)\frac{\partial}{\partial x} + \frac{1}{2}\sigma^2(x)\frac{\partial^2}{\partial x^2}.$$

and

$$dM_t = \sigma(X_t)f'(X_t)dW_t^0.$$

Let us now assume that the observations are of the form

$$dZ_t = b(X_t)dt + dW_t,$$

where b is a function $b : M \to R$ and W is a Wiener process which, for simplicity, is independent of X. The filtration $\mathbf{F}$ is defined as

$$\mathcal{F}_t = \mathcal{F}_t^X \vee \mathcal{F}_t^W.$$

We thus have the filtering model

$$df(X_t) = (\mathcal{G}f)(X_t)dt + dM_t,$$
$$dZ_t = b(X_t)dt + dW_t,$$

and if we introduce the notation

$$\pi_t[g] = E\left[g(X_t)|\mathcal{F}_t^Z\right], \tag{12.15}$$

for any real-valued function $g : M \to R$, we can apply the FKK equations to obtain

$$d\pi_t[f] = \pi_t[\mathcal{G}f]\,dt + \{\pi_t[fb] - \pi_t[f]\cdot\pi_t[b]\}\,dv_t, \tag{12.16}$$
$$dv_t = dZ_t - \pi_t[b]\,dt. \tag{12.17}$$

12.5.2   On the Filter Dimension

We would now like to consider the equation (12.16) as an SDE driven by the innovation process v, but the problem is that the equation is not closed, since in the SDE for $\pi_t[f]$ we have the expressions $\pi_t[\mathcal{G}f]$, $\pi_t[fb]$ and $\pi_t[b]$ which all have to be determined in some way.

The obvious way to handle, for example, the term $\pi_t[b]$, is of course the following

- Use Dynkin on b to obtain

$$db(X_t) = (\mathcal{G}b)(X_t)dt + dM_t^b,$$

  where M^b is a martingale.
- Apply the FKK equations to the system

$$db(X_t) = (\mathcal{G}b)(X_t)dt + dM_t^b,$$
$$dZ_t = b(X_t)dt + dW_t,$$

  to obtain the filter equation

$$d\pi_t[b] = \pi_t[\mathcal{G}b]\,dt + \{\pi_t[b^2] - (\pi_t[b])^2\}\,dv_t.$$

We now have an equation for $d\pi_t[b]$, but this equation contains (among other things) the term $\pi_t[b^2]$, so we now need an equation for this. In order to derive that equation, we can of course apply Dynkin to the process $b^2(X_t)$ and again use the FKK equations, but this leads to the equation

$$d\pi_t[b^2] = \pi_t[\mathcal{G}b^2]\,dt + \{\pi_t[b^3] - \pi_t[b^2]\cdot\pi_t[b]\}\,dv_t,$$

and we now have to deal with the term $\pi_t[b^3]$ etc.

As the reader realizes, this procedure will in fact lead to an infinite number of filtering equations. This is in fact the generic situation for a filtering problem and the argument is roughly as follows.

- In general there will not exist a finite-dimensional sufficient statistic for the X process.
- In particular, an old estimate $\pi_t[X]$, plus the new information dv_t is not sufficient to allow us to determine an updated estimate $\pi_{t+dt}[X]$.
- In order to be able to update, even such a simple object as the conditional expectation $\pi_t[X]$, we will, in the generic case, need the *entire conditional distribution* $\mathcal{L}(X_t|\mathcal{F}_t^Z)$.
- The conditional distribution is typically an infinite-dimensional object.
- In the generic case we can therefore expect to have an *infinite-dimensional filter*.

An alternative way of viewing (12.16) is now to view it, not as a scalar equation for a fixed choice of f, but rather as an infinite number of equations, with one equation for each (say, bounded continuous) f. Viewed in this way, the filtering equation (12.16) represents an infinite-dimensional system for the determination of the entire condition distribution $\mathcal{L}(X_t|\mathcal{F}_t^Z)$. We will capitalize on this idea later when we derive the dynamics for the conditional density.

12.5.3   Finite-Dimensional Filters

From the discussion above it is now clear that the generic filter is infinite-dimensional, the reason being that we need the entire conditional distribution of X in order to update our filter estimates, and this distribution is typically an infinite-dimensional object. This is, in some sense, bad news, but there is really nothing we can do about the situation – it is simply a fact of life.

We can, however, also draw some more positive conclusions from the dimension argument: after a moment's reflection we have the following important idea.

Idea   If we know on *a priori* grounds that, for all t, the conditional distribution $\mathcal{L}(X_t|\mathcal{F}_t^Z)$ belongs to a class of probability distributions which is parameterized by a *finite number of parameters*, then we can expect the have a finite-dimensional filter. The filter equations should then provide us with the dynamics of the parameters for the conditional distribution.

There are in fact two well-known models when we have *a priori* information of the type above. They are known as the **Kalman** model and the **Wonham** model respectively. We will discuss them in detail later, but we introduce them already at this point.

The Kalman model:

The simplest case of a Kalman model is given by the linear system

$$dX_t = aX_t dt + cdW_t^0,$$
$$dZ_t = bX_t dt + dW_t,$$

where W^0 and W are, possibly correlated, Wiener processes. For this model it is easy to see that the pair (X, Z) will be jointly Gaussian. We then recall the standard fact that if (ξ, η) is a pair of Gaussian vectors, then the conditional distribution $\mathcal{L}\{\xi \mid \eta\}$ will also be Gaussian. This property can be shown to extend also to the process case, so we conclude that the conditional distribution $\mathcal{L}(X_t \mid \mathcal{F}_t^Z)$ is Gaussian. The Gaussian distribution is, however, determined by only two parameters - the mean and the variance, so for this model we expect to have a two-dimensional filter with one equation for the conditional mean and another for the conditional variance.

The Wonham model:

In the Wonham model, the process X is a continuous-time Markov chain which takes values in the finite state-space $\{1, 2, \ldots, n\}$, and where the observation process is of the form

$$dZ_t = b(X_t)dt + dW_t.$$

For this model it is immediately obvious that the conditional distribution $\mathcal{L}(X_t \mid \mathcal{F}_t^Z)$ is determined by a finite number of parameters, since it is in fact determined by the conditional probabilities $p_t^1, \ldots, p_t^n$ where $p_t^i = P\left(X_t = i \mid \mathcal{F}_t^Z\right)$. We thus expect to have an n-dimensional filter.

12.6   The Kalman Filter

In this section we will discuss the Kalman filter in some detail.

12.6.1   The Kalman Model

The basic Kalman model is as follows.

$$dX_t = AX_t dt + C dW_t^0,$$
$$dZ_t = BX_t dt + D dW_t,$$

All processes are allowed to be vector-valued with $X_t \in R^k$ $Z_t \in R^k$, $W_t \in R^n$ and $W^0 \in R^d$. The matrices A, C, B and D have the obvious dimensions and we need two basic assumptions.

Assumption   *We assume the following.*

1. *The $n \times n$ matrix D is invertible.*
2. *The Wiener processes W and W^0 are independent.*
3. *The distribution of X_0 is Gaussian with mean vector y_0 and covariance matrix R_0.*

The independence assumption, and also the Gaussian assumption concerning X_0, can be relaxed, but the invertibility of D is important and cannot be omitted.

This model can of course be treated using the FKK theory, but since the vector-valued case is a bit messy we only carry out the derivation in detail for the simpler scalar case in Section 12.6.2. In Section 12.6.3 we then state the general result without proof.

12.6.2   Deriving the Filter Equations in the Scalar Case

We will examine the special case of a scalar model of the form

$$dX_t = aX_t dt + cdW_t^0,$$
$$dZ_t = X_t dt + dW_t,$$

where all processes and constants are scalar. From the discussion in Section 12.5.3 we know that the conditional distribution $\mathcal{L}(X_t | \mathcal{F}_t^Z)$ is Gaussian, so it should be enough to derive filter equations for the conditional man and variance.

The FKK equation for the conditional mean is given by

$$d\pi_t[X] = a\pi_t[X] dt + \{\pi_t[X^2] - (\pi_t[X])^2\} dv_t, \qquad (12.18)$$
$$dv_t = dZ_t - b\pi_t[X] dt. \qquad (12.19)$$

This would be a closed system if we did not have the term $\pi_t[X^2]$. Along the lines of the discussion in Section 12.5.3 we therefore use Itô on the process X^2 to obtain

$$dX_t^2 = \{2aX_t^2 + c^2\} dt + 2cX_t dW_t^0.$$

This will give us the filter equation

$$d\pi_t[X^2] = \{2a\pi_t[X^2] + c^2\} dt + \{\pi_t[X^3] - \pi_t[X^2] \cdot \pi_t[X]\} dv_t. \qquad (12.20)$$

We now have the term $\pi_t[X^3]$ to deal with. A naive continuation of the procedure above will produce an infinite number of filtering equations for all conditional moments $\pi_t[X^k]$, where $k = 1, 2, \ldots$.

In this case, however, because of the particular dynamical structure of the model, we know that the conditional distribution $\mathcal{L}(X_t | \mathcal{F}_t^Z)$ is Gaussian. We therefore define the conditional variance process H by

$$H_t = E\left[\left(X_t - \hat{X}_t\right)^2 \middle| \mathcal{F}_t^Z\right] = \pi_t[X^2] - (\pi_t[X])^2.$$

In order to obtain a dynamical equation for H we apply Itô to (12.18) to give us

$$d(\pi_t[X])^2 = \{2a(\pi_t[X])^2 + H_t^2\} dt + 2\pi_t[X] H_t dv_t.$$

Using this and equation (12.20) we obtain the H-dynamics as

$$dH_t = \{2aH_t + c^2 - H_t^2\} dt + \{\pi_t[X^3] - 3\pi_t[X^2]\pi_t[X] + 2(\pi_t[X])^3\} dv_t.$$

We can now use the Gaussian structure of the problem. Recall that for any Gaussian variable ξ we have

$$E[\xi^3] = 3E[\xi^2]E[\xi] - 2(E[\xi])^3.$$

Since $\mathcal{L}(X_t | \mathcal{F}_t^Z)$ is Gaussian we thus conclude that

$$\pi_t[X^3] - 3\pi_t[X^2]\pi_t[X] + 2(\pi_t[X])^3 = 0,$$

so, as expected, H is in fact deterministic and we have the Kalman filter

$$d\widehat{X}_t = a\widehat{X}_t dt + H_t dv_t,$$
$$\dot{H}_t = 2aH_t + c^2 - H_t^2,$$
$$dv_t = dZ_t - \widehat{X}_t dt.$$

The first equation gives us the evolution of the conditional mean. The second equation, which is a so-called Riccati equation, gives us the evolution of the conditional variance and we note that this is deterministic. We also note that since the conditional distribution is Gaussian, the Kalman filter does in fact provide us with the entire conditional distribution, not just the conditional mean and variance.

12.6.3   The Full Kalman Model

We now return to the vector model:

$$dX_t = AX_t dt + C dW_t^0,$$
$$dZ_t = BX_t dt + D dW_t.$$

This can be treated very much along the lines of the scalar case in the previous section, but the calculations are a bit more complicated. We thus confine ourselves to stating the final result.

Proposition 12.9 (The Kalman Filter)   *With notation and assumptions as above we have the following filter equations, where $\star$ denotes transpose:*

$$d\widehat{X}_t = a\widehat{X}_t dt + R_t B^\star (DD^\star)^{-1} dv_t, \tag{12.21}$$
$$\dot{R}_t = AR_t + R_t A^\star - R_t B^\star (DD^\star)^{-1} BR_t + CC^\star, \tag{12.22}$$
$$dv_t = dZ_t - B\widehat{X}_t dt. \tag{12.23}$$

Furthermore, the conditional error covariance matrix is given by R above.

12.7   The Wonham Filter

We consider again the Wonham model. In this model, the X process is a time-homogeneous Markov chain on a finite state-space D and, without loss of generality, we may assume that $D = \{1, 2 \ldots, n\}$. We denote the intensity matrix of X by H, and the probabilistic interpretation is that

$$P(X_{t+h} = j \,|\, X_t = i) = H_{ij} h + o(h), \quad i \neq j,$$
$$H_{ii} = -\sum_{j \neq i} H_{ij}.$$

We cannot observe X directly, but instead we can observe the process Z, defined by

$$dZ_t = b(X_t) + dW_t, \tag{12.24}$$

where W is a Wiener process. In this model it is obvious that the conditional distribution of X will be determined by the conditional probabilities, so we define the indicator processes $\delta_t^1, \ldots, \delta_t^n$ by

$$\delta_i(t) = I\{X_t = i\}, \quad i = 1, \ldots, n,$$

where $I\{A\}$ denotes the indicator for an event A, so $I\{A\} = 1$ if A occurs and is zero otherwise. The Dynkin theorem, together with a simple calculation, gives us

$$d\delta_i(t) = \sum_{j=1}^{n} H_{ji}\delta_j(t)dt + dM_t^i, \quad i = 1, \ldots, n, \tag{12.25}$$

where $M^1, \ldots, M^n$ are martingales. In vector form we can thus write

$$d\delta(t) = H^\star \delta(t)dt + dM_t.$$

Applying the FKK theorem to the dynamics above gives us the filter equations

$$d\pi_t[\delta_i] = \sum_{j=1}^{n} H_{ji}\pi_t[\delta_j]\, dt + \{\pi_t[\delta_i b] - \pi_t[\delta_i]\,\pi_t[b]\}\, dv_t, \quad i = 1, \ldots, n,$$

$$dv_t = dZ_t - \pi_t[b]\, dt.$$

We now observe that, using the notation $b_i = b(i)$, we have the obvious relations

$$b(X_t) = \sum_{j=1}^{n} b_j\delta_j(t), \quad \delta_i(t)b(X_t) = \delta_i(t)b_i,$$

which gives us

$$\pi_t[b] = \sum_{j=1}^{n} b_j\widehat{\delta}_j(t), \quad \pi_t[\delta^i b] = \widehat{\delta}_i(t)b_i.$$

Plugging this into the FKK equations gives us the Wonham filter.

Proposition 12.10 (The Wonham Filter) *With assumptions as above, the Wonham filter is given by*

$$d\widehat{\delta}_i(t) = \sum_{j=1}^{n} H_{ji}\widehat{\delta}_j(t)dt + \left\{ b_i\widehat{\delta}_i(t) - \widehat{\delta}_i(t) \cdot \sum_{j=1}^{n} b_j\widehat{\delta}_j(t) \right\} dv_t, \tag{12.26}$$

$$dv_t = dZ_t - \sum_{j=1}^{n} b_j\widehat{\delta}_j(t)dt, \tag{12.27}$$

where

$$\widehat{\delta}_i(t) = P\left(X_t = i \,\big|\, \mathcal{F}_t^Z\right), \quad i = 1, \ldots, n.$$

12.8   Exercises

Exercise 12.1   Consider the filtering model

$$dX_t = a_t dt + dV_t$$
$$dZ_t = b_t dt + \sigma_t dW_t,$$

where

- the process σ is $\mathcal{F}_t^Z$ adapted and positive;
- W and V are, possibly correlated, Wiener processes.

Prove, along the lines above, that the filtering equations are given by

$$d\widehat{X}_t = \widehat{a}_t dt + \left[\widehat{D}_t + \frac{1}{\sigma_t} \left\{ \widehat{X_t b_t} - \widehat{X}_t \widehat{b}_t \right\} \right] dv_t,$$
$$dv_t = \frac{1}{\sigma_t} \left\{ dZ_t - \widehat{b}_t dt \right\},$$
$$D_t = \frac{d\langle V, W \rangle_t}{dt}.$$

Exercise 12.2   Consider the filtering model

$$dZ_t = X dt + dW_t,$$

where X is a random variable with distribution function F and W is a Wiener process which is independent of X. As usual we observe Z. Write down the infinite system of filtering equations for the determination of $\Pi_t[X] = E\left[X | \mathcal{F}_t^Z \right]$.

12.9   Notes

The original paper, Fujisaki et al. (1972), provides a very readable account of the FKK theory. The two-volume set Liptser & Shiryayev (2004) is a standard reference on filtering. It includes, as well the Wiener-driven FKK framework, also a deep theory of point processes and the related filtering theory. It is, however, not an easy read. A far-reaching account of Wiener-driven filtering theory is given in Bain & Crisan (2009).

13 The Conditional Density*

In this chapter we will present some further results from filtering theory, such as the stochastic PDE for the conditional density and the Zakai equation for the unnormalized density. These results are important but also more technical than the previous ones, so this chapter can be skipped in a first reading.

13.1 The Evolution of the Conditional Density

The object of this section is to derive a stochastic PDE describing the evolution of the conditional density of the state process X. This theory is, at some points, quite technical, so we only give the basic arguments. For details, see Bain & Crisan (2009). We specialize to the Markovian setting

$$dX_t = a(X_t)dt + c(X_t)dW_t^0, \tag{13.1}$$
$$dZ_t = b(X_t)dt + dW_t^1, \tag{13.2}$$

where W^0 and W^1 are independent Wiener processes. We recall from (12.16)–(12.17) that for any real-valued function $f : R \to R$, the filter equations for $\pi_t[f] = E\left[f(X_t)|\mathcal{F}_t^Z\right]$ are then given by

$$d\pi_t[f] = \pi_t[\mathcal{A}f]\,dt + \{\pi_t[fb] - \pi_t[f]\cdot\pi_t[b]\}\,dv_t, \tag{13.3}$$
$$dv_t = dZ_t - \pi_t[b]\,dt, \tag{13.4}$$

where the infinitesimal generator $\mathcal{A}$ is given by

$$\mathcal{A} = a(x)\frac{\partial}{\partial x} + \frac{1}{2}c^2(x)\frac{\partial^2}{\partial x^2}.$$

We now view f as a "test function" varying within a large class C of test functions. The filter equation (13.3) will then, as f varies over C, determine the entire conditional distribution $\mathcal{L}\left(X_t|\mathcal{F}_t^Z\right)$. We now make an important assumption.

Assumption *We assume that X has a conditional density process $p_t(y)$, with respect to Lebesgue measure, so that*

$$\pi_t[f] = E\left[f(X_t)|\mathcal{F}_t^Z\right] = \int_R f(x)p_t(x)dx.$$

In order to have a more suggestive notation we introduce a natural pairing (the inner product in L^2) denoted by $\langle , \rangle$ for any smooth real-valued functions g and f (where f has compact support). This is defined by

$$\langle g, f \rangle = \int_R f(x)g(x)dx.$$

We can thus write

$$\pi_t[f] = \langle p_t, f \rangle,$$

and with this notation, the filter equation takes the form

$$d\langle p_t, f \rangle = \langle p_t, \mathcal{A}f \rangle dt + \{\langle p_t, fb \rangle - \langle p_t, f \rangle \langle p_t, b \rangle\} dv_t.$$

We can now dualize this (see the exercises) to obtain

$$d\langle p_t, f \rangle = \langle \mathcal{A}^* p_t, f \rangle dt + \{\langle bp_t, f \rangle - \langle p_t, f \rangle \langle p_t, b \rangle\} dv_t,$$

where $\mathcal{A}^*$ is the adjoint operator:

$$\mathcal{A}^* p(x) = -\frac{\partial}{\partial y}[a(x)p(x)] + \frac{1}{2}\frac{\partial^2}{\partial x^2}\left[c^2(x)p(x)\right].$$

If this holds for all test functions f we are led to the following result. See Bain & Crisan (2009) for the full story.

Theorem 13.1 (Kushner–Stratonovich)   *Assume that X has a conditional density process $p_t(y)$ with respect to Lebesgue measure. Under suitable technical conditions, the density will satisfy the following stochastic partial differential equation (SPDE)*

$$dp_t(x) = \mathcal{A}^* p_t(x)dt + p_t(x)\left\{b(x) - \int_R b(x)p_t(x)dx\right\}dv(t), \qquad (13.5)$$

$$dv_t = dZ_t - \left(\int_R b(x)p_t(x)dx\right)dt. \qquad (13.6)$$

As we noted above, this is an SPDE for the conditional density. In order to connect to more familiar topics, we note that if the observation dynamics have the form

$$dZ_t = b(X_t)dt + \sigma dW_t,$$

then the SPDE would become

$$dp_t(y) = \mathcal{A}^* p_t(x)dt + \frac{p_t(x)}{\sigma}\left\{b(x) - \int_R b(x)p_t(x)dx\right\}dv(t),$$

$$dv_t = \frac{1}{\sigma}\left\{dZ_t - \left(\int_R b(x)p_t(x)dx\right)dt\right\}.$$

If we now let $\sigma \to +\infty$, which intuitively means that in the limit we only observe noise, then the filter equation degenerates to

$$\frac{\partial}{\partial t}p_t(x) = \mathcal{A}^* p_t(x),$$

which is the Fokker–Planck equation for the unconditional density.

13.2   Separation of Filtering and Detection

The main result in this section is Proposition 13.4 and Corollary 13.9, which will be important later on when we derive the Zakai equation. The reader who wishes to proceed directly to the Zakai equation can skip the present section and simply accept Proposition 13.4 and the corollary dogmatically. The theorem is, however, of considerable importance also in statistics so, for readers with more general interests, we provide the full story.

Consider a measurable space $\{\Omega, \mathcal{F}\}$ and two probability measures, P_0 and P_1 on this space. Let us now consider a case of hypothesis testing in this framework. We have two hypotheses, H_0 and H_1, with the interpretation

$$H_0 : P_0 \text{ holds,}$$
$$H_1 : P_1 \text{ holds.}$$

If $P_1 \ll P_0$, then we know from Neyman–Pearson that the appropriate test variable is given by the likelihood

$$L = \frac{dP_1}{dP_0}, \quad \text{on } \mathcal{F}$$

and that the optimal test is of the form

$$\text{Reject } H_0 \text{ if } L \geq R,$$

where R is a suitably chosen constant.

The arguments above hold as long as we really have access to all the information contained in $\mathcal{F}$. If, instead of $\mathcal{F}$, we only have access to the information in a smaller sigma-algebra $\mathcal{G} \subset \mathcal{F}$, then we cannot perform the test above, since L is not determined by the information in $\mathcal{G}$ (i.e. L is not $\mathcal{G}$-measurable).

From an abstract point of view, this problem is easily solved by simply applying Neyman–Pearson to the space $\{\Omega, \mathcal{G}\}$ instead of $\{\Omega, \mathcal{F}\}$. It is then obvious that the optimal test variable is given by

$$\widehat{L} = \frac{dP_1}{dP_0}, \quad \text{on } \mathcal{G},$$

and the question is now how $\widehat{L}$ is related to L.

This question is answered by the following standard result.

Proposition 13.2   *With notation as above, we have*

$$\widehat{L} = E^0[L|\mathcal{G}], \tag{13.7}$$

where E^0 denotes expectation under P_0.

We will now study the structure of L when L is generated by a Girsanov transformation and we also have a filtered space, so information increases over time. To this end we consider a filtered probability space $\{\Omega, \mathcal{F}, P_0, \mathbf{F}\}$ and two $\mathbf{F}$ adapted processes Z and h. We assume the following:

- Z is an **F**-Wiener process under P_0;

- the process h is **F**-adapted and P_0-independent of Z.

We now introduce some useful notation connected to Girsanov transformations.

Definition 13.3 For any **F**-adapted process g we define the process $L(g)$ by

$$L_t(g) = e^{\int_0^t g_s \, dZ_s - \frac{1}{2} \int_0^t g_s^2 \, ds}. \tag{13.8}$$

We see that $L(g)$ is simply the likelihood process for a Girsanov transformation if we use g as the the Girsanov kernel.

Assuming that $L(h)$ is a true martingale we may now define the measure P_1 by setting

$$L_t(h) = \frac{dP_1}{dP_0}, \quad \text{on } \mathcal{F}_t, \quad 0 \le t \le T.$$

where T is some fixed time horizon.

From the Girsanov theorem we know that we can write

$$dZ_t = h_t \, dt + dW_t^1,$$

where W^1 is an P_1-Wiener process. We thus see that we have the following situation.

- Under P_0, the process Z is a Wiener process without drift.
- Under P_1, the process Z has the drift h.

If we now interpret the process h as a signal, we have the following two hypotheses:

H_0 : We have no signal. We only observe Wiener noise.

H_1 : We observe the signal h, disturbed by the Wiener noise.

The task at hand is to test H_0 against H_1 sequentially over time, assuming that we can only observe the process Z. From Proposition 13.2 we know that, at time t, the optimal test statistics is given by $\widehat{L}_t(h)$, defined by

$$\widehat{L}_t(h) = E^0 \left[L_t(h) | \mathcal{F}_t^Z \right],$$

and the question is how we compute this entity. The answer is given by the following beautiful result.

Proposition 13.4 *With notation as above we have*

$$\widehat{L}_t(h) = L_t(\widehat{h}),$$

where

$$\widehat{h}_t = E^1 \left[h_t | \mathcal{F}_t^Z \right].$$

Note that $\widehat{L}_t(h)$ is a P_0-expectation, whereas $\widehat{h}_t$ is a P_1-expectation. Before going on to the proof, we remark that this is a separation result. It says that the filtered "detector" $\widehat{L}$ can be separated into two parts: The (unfiltered) detector L, and the P_1-filter estimate $\widehat{h}$.

Proof   From the definition of $L_t(h)$ we have

$$dL_t(h) = L_t(h)h_t \, dZ_t,$$

and we observe the P_0-Wiener process Z. From Theorem 12.8 we then have

$$d\widehat{L}_t(h) = \pi_t^0[L(h)h] \, dZ_t,$$

where superscript 0 denotes expectation under P_0: thus E^0 is expectation under P_0 and π^0 is conditional expectation under P_0. Similarly for E^1 and π^1. We can write this equation as

$$d\widehat{L}_t(h) = \widehat{L}_t(h)\eta_t \, dZ_t,$$

where

$$\eta_t = \frac{\pi_t^0[L(h)h]}{\pi_t^0[L(h)]},$$

so we have in fact

$$\widehat{L}_t(h) = L_t(\eta).$$

On the other hand we know

$$\eta_t = \frac{E^0\left[L_t(h)h_t \mid \mathcal{F}_t^Z\right]}{E^0\left[L_t(h) \mid \mathcal{F}_t^Z\right]},$$

so from the abstract Bayes formula we obtain

$$\eta_t = E^1\left[h_t \mid \mathcal{F}_t^Z\right] = \pi_t^1[h]. \qquad \square$$

For future use we note the following result which in fact is a a part of the proof above.

Corollary 13.5   *We have*

$$d\widehat{L}_t(h) = \pi_t^1[h]\,\widehat{L}_t(h)dZ_t. \tag{13.9}$$

13.3   The Unnormalized Filter Equation

We consider again a filtered probability space $\{\Omega, \mathcal{F}, P, \mathbf{F}\}$ and a Markovian filter model of the form

$$dX_t = a(X_t)dt + c(X_t)dW_t^0, \tag{13.10}$$

$$dZ_t = b(X_t)dt + dW_t, \tag{13.11}$$

where W^0 and W are independent. In Section 13.1 we studied the same model and we derived the Kushner–Stratonovich equation (13.5) for the density $p_t(y)$. We now present an alternative approach to the filtering problem along the following lines.

- Perform a Girsanov transformation from P to a new measure P_0 such that X and Z are independent under P_0.
- Compute filter estimates under P_0. This should be easy, due to the independence.

- Transform the results back to P using the abstract Bayes formula.
- This will lead to a study of the so-called **unnormalized** estimate $\sigma_t[f]$, and it turns out that the equation for $\sigma_t[f]$ is much simpler than the equation for $\pi_t[f]$.
- We will derive a result, known as the Zakai equation, for the unnormalized estimate $\sigma_t(f)$. This equation will lead us to an SPDE for the unnormalized density $q_t(y)$. The Zakai equation is, in many ways, much nicer than the Kushner–Stratonovich equation for the density $p_t(x)$.

13.3.1   The Basic Construction

Consider a probability space $\{\Omega, \mathcal{F}, P_0\}$ as well as two independent P_0-Wiener processes Z and W^0. We define filtration **F** by

$$\mathcal{F}_t = \mathcal{F}_t^Z \vee \mathcal{F}_\infty^{W^0},$$

and the process X by

$$dX_t = a(X_t)dt + c(X_t)dW_t^0.$$

We now define the likelihood process L by

$$dL_t = L_t b(X_t)dZ_t,$$
$$L_0 = 1,$$

and where P is given by

$$L_t = \frac{dP}{dP_0}, \quad \text{on } \mathcal{F}_t, \ 0 \le t \le T$$

for some fixed horizon T. From the Girsanov theorem it now follows that the we can write

$$dZ_t = b(X_t)dt + dW_t,$$

where W is a $(P, \mathbf{F})$-Wiener process. In particular, the process W is independent of $\mathcal{F}_0 = \mathcal{F}_\infty^{W^0}$, so W and W^0 are indenpendent under P. Since $L_0 = 1$ we have $P = P_0$ on $\mathcal{F}_0$, and since $\mathcal{F}_0 = \mathcal{F}_\infty^{W^0}$, we see that (W^0, X) has the same distribution under P as under P_0.

The end result of all this is that under P we have our standard model

$$dX_t = a(X_t)dt + c(X_t)dW_t^0,$$
$$dZ_t = b(X_t)dt + dW_t.$$

13.3.2   The Zakai Equation

We define $\pi_t[f]$ as usual by

$$\pi_t[f] = E^P\left[f(X_t)| \mathcal{F}_t^Z \right],$$

and from the abstract Bayes formula we then have

$$\pi_t[f] = \frac{E^0\left[L_t f(X_t)| \mathcal{F}_t^Z\right]}{E^0\left[L_t| \mathcal{F}_t^Z\right]}.$$

Definition 13.6   The **unnormalized estimate** $\sigma_t[f]$ is defined by

$$\sigma_t[f] = E^0\left[L_t f(X_t)| \mathcal{F}_t^Z\right].$$

We have thus derived the Kallianpur–Striebel formula.

Proposition 13.7 (Kallianpur–Striebel)   *The standard filter estimate $\pi_t[f]$ and the unnormalized estimate $\sigma_t[f]$ are related by the formula*

$$\pi_t[f] = \frac{\sigma_t[f]}{\sigma_t[1]}. \tag{13.12}$$

It turns out that $\sigma_t[f]$ is easier to study than $\pi_t[f]$, so we now go on to derive the dynamics of $\sigma_t[f]$.

This is in fact quite easy. From the Kallianpur–Striebel formula we have

$$\sigma_t[f] = \pi_t[f] \cdot \sigma_t[1].$$

From the FKK theorem we have

$$d\pi_t[f] = \pi_t[\mathcal{A}f]\, dt + \{\pi_t[fb] - \pi_t[f] \cdot \pi_t[b]\}\, dv_t,$$
$$dv_t = dZ_t - \pi_t[b]\, dt,$$

and from Corollary 13.5 have

$$d\sigma_t[1] = \sigma_t[1]\,\pi_t[b]\, dZ_t.$$

We can now apply the Itô formula to the product $\pi_t[f] \cdot \sigma_t[1]$. This leads to calculations which, at first sight, look rather forbidding. It turns out, however, that there are a surprisingly large number of cancellations in these calculations and in the end we obtain the following result.

Theorem 13.8 (The Zakai Equation)   *The unnormalized estimate $\sigma_t[f]$ satisfies the Zakai equation*

$$d\sigma_t[f] = \sigma_t[\mathcal{A}f]\, dt + \sigma_t[b]\, dZ_t. \tag{13.13}$$

We note that the Zakai equation has a much simpler structure than the corresponding FKK equation. First, it is driven directly by the observation process Z rather than by the innovation process v. Second, the non-linear (product) term $\pi_t[f] \cdot \pi_t[b]\, dv_t$ is replaced by the term $\sigma_t[b]\, dZ_t$ which does not involve f.

13.3.3   The SPDE for the Unnormalized Density

Let us now assume that there exists an unnormalized density process $q_t(y)$, with interpretation

$$\sigma_t[f] = \int_R f(x) q_t(x)\, dx.$$

We can then dualize the Zakai equation, exactly as we did in the derivation of the Kushner–Stratonovich equation, to obtain the following result.

Proposition 13.9   *Assuming that there exists an unnormalized density $q_t(y)$, we have the following SPDE.*

$$dq_t(y) = \mathcal{A}^\star q_t(x)dt + b(x)q_t(x)dZ_t. \tag{13.14}$$

Just as for the Zakai equation, we note that the SPDE above for the unnormalized density is much simpler than the Kushner–Stratonovich equation for the normalized density $p_t(x)$.

13.4   Exercise

Exercise 13.1   Consider a probability space $(\Omega, \mathcal{F}, P)$ and the filtering model

$$dZ_t = Xdt + dW_t,$$

where X is a random variable with distribution function F, and W is a Wiener process that is independent of X. As usual we observe Z. In Exercise 12.2 we saw that a naive application of the FKK equation produced an infinite-dimensional filter. The object of this exercise is to show that we can do better.

Define, therefore, the functions $f : R \times R_+ \to R$ and $g : R \times R_+ \to R$ by

$$f(t,z) = \int_R x e^{-\frac{t}{2}\left(x - \frac{z}{t}\right)^2} dF(x),$$

$$g(t,z) = \int_R e^{-\frac{t}{2}\left(x - \frac{z}{t}\right)^2} dF(x),$$

and show that

$$E^P \left[X | \mathcal{F}_t^Z \right] = \frac{f(t, Z_t)}{g(t, Z_t)}$$

Hint: Perform a Girsanov transformation from P to a new measure Q so that Z and X are Q-independent. You may then use (a generalization of) the fact that if ξ and η are independent random variables where ξ has distribution function F then, for any function $H : R^2 \to R$, we have

$$E\left[H(\xi, \eta) | \sigma\{\eta\}\right] = h(\eta),$$

where h is defined by

$$h(y) = \int_R H(x, y)dF(x).$$

13.5   Notes

A very complete account of Wiener-driven FKK theory as well as unnormalized filtering is given in Bain & Crisan (2009).

14 Non-Linear Filtering with Counting-Process Observations

In this chapter we will present the basic ideas and results of non-linear filtering when the state process modulates the intensity of a point process. The ideas are very similar to the Wiener case but, because of the presence of jumps, the situation is a bit more complicated. We start with the simplest filtering model where we only observe a single counting process, but later we extend the theory to the k-variate situation. In Section 14.5.1 we will treat the case when the observation process is driven by a Wiener process in addition to a marked point process.

14.1 A Model with Counting-Process Observations

We consider a filtered probability space $(\Omega, \mathcal{F}, P, \mathbf{F})$ where as usual the filtration $\mathbf{F} = \{\mathcal{F}_t; \ t \geq 0\}$ formalizes the idea of an increasing flow of information. Our basic model consists of a pair of processes (X, Z) with dynamics as follows:

$$dX_t = a_t dt + dM_t, \tag{14.1}$$

$$dZ_t = dN_t. \tag{14.2}$$

The process a is allowed to be an arbitrary integrable $\mathbf{F}$-adapted processes, M is an $\mathbf{F}$-martingale and N is a scalar $\mathbf{F}$-counting process with $\mathbf{F}$-intensity process λ.

Remark Note that M is allowed to be an arbitrary martingale, so it does not have to be a Wiener process.

We need an essential assumption.

Assumption 14.1 *We assume that there exists an $\mathbf{F}$-optional process D such that*

$$E\left[\Delta M_t \Delta N_t \mid \mathcal{F}_{t-}\right] = D_t dt. \tag{14.3}$$

In most concrete cases this assumption is satisfied. The two typical situations are the following.

- Assume that M and N have no common jumps. Then we obviously have

$$D_t = 0.$$

- Assume that

$$dM_t = \sigma_t dW_t + \beta_t \{dN_t - \lambda dt\},$$

where σ is F-optional and β is F-predictable. In this case an easy calculation gives us

$$E\left[\Delta M_t \Delta N_t | \mathcal{F}_{t-}\right] = E\left[\beta_t \Delta N_t \Delta N_t | \mathcal{F}_{t-}\right] = \beta_t E\left[\Delta N_t | \mathcal{F}_{t-}\right] = \beta_t \lambda_t dt,$$

so we have

$$D_t = \beta_t \lambda_t.$$

The interpretation of the model above is that we are interested in the *state process* X, but that we cannot observe X directly. What we can observe is instead the *observation process* Z, so our main problem is to draw conclusions about X, given the observations of Z. For example, we might like to compute the conditional expectation

$$\pi_t[X] = E\left[X_t | \mathcal{F}_t^Z\right],$$

where $\mathcal{F}_t^Z$ is the information generated by Z on the time interval $[0,t]$ or, more ambitiously, we would like to compute $\mathcal{L}(X_t \mid \mathcal{F}_t^Z)$, i.e. the entire conditional distribution of X_t given observations of Z on $[0,t]$.

A more concrete example of a model of this form could be as follows.

- The state process X solves a jump-diffusion SDE of the form

$$dX_t = \mu(X_t)dt + \sigma(X_t)dW_t + \beta(X_{t-})dN_t^0,$$

  with driving Wiener process W and, say, a Poisson process N^0 with constant intensity γ.
- The state process X modulates the intensity λ of the observations point process N, through the formula

$$\lambda_t = \lambda(t, X_t),$$

  where λ in the right-hand side denotes a deterministic function.

14.2   Optional and Predictable Projections

In Chapter 12 on filtering for Wiener-driven models, we used the notations $\pi_t[X]$ and $\widehat{X}_t$ to denote the same object, namely

$$E\left[X_t | \mathcal{F}_t^Z\right].$$

In our present setting where we have jumps present, the situation is more delicate so we have to be more careful. We therefore make the following definitions.

Definition 14.2   For any F-optional process X, we define the **optional projection** $\pi_t[X]$ and the **predictable projection**, $\widehat{X}$, of X onto F^Z by

$$\pi_t[X] = E\left[X_t | \mathcal{F}_t^Z\right], \tag{14.4}$$

$$\widehat{X}_t = E\left[X_t | \mathcal{F}_{t-}^Z\right]. \tag{14.5}$$

Strictly speaking, the definition above is not a proper definition at all. If, for example, we look at $\pi_t[X] = E\left[X_t | \mathcal{F}_t^Z\right]$ for a fixed t, then the conditional expectation is really an equivalence class of random variables. In order to obtain a *bona fide* process we have to select one member of the equivalence class for each t and then glue these random variables together in order to obtain an optional process. This is by no means trivial and we have the same problem with the predictable projection.

This problem, and many others, are solved by using the far-reaching (but far from simple) "general theory of processes" which can be found in Dellacherie & Meyer (1972) or any other book on semimartingale theory. In the general theory, the projections above are given in a more abstract form but for most practical purposes the definitions above are good enough. The reader of the present text can take comfort in the following facts

- The projections above do indeed exist.
- The optional projection $\pi_t[X]$ is $\mathbf{F}^Z$-optional.
- The predictable projection $\widehat{X}_t$ is $\mathbf{F}^Z$-predictable.

14.3 Filter Equations for Counting-Process Observations

We recall that the state dynamics are given by

$$dX_t = a_t \, dt + dM_t,$$

where a is $\mathbf{F}$-optional and M is an $\mathbf{F}$-martingale. We now recall from Lemma 12.4 the following result on the filter dynamics.

Lemma 14.3 *The process $\pi_t[X]$ admits the representation*

$$d\pi_t[X] = \pi_t[a] \, dt + dM_t^{\star}, \tag{14.6}$$

where $M^{\star}$ is an $\mathbf{F}^Z$-martingale.

Since $M^{\star}$ is a martingale under the internal filtration generated by N, we have access to the martingale representation theorem, but first we need to find the correct intensity process for N. We recall that λ is the $\mathbf{F}$-intensity of N; from Proposition 4.4 we deduce that the $\mathbf{F}^Z$-predictable intensity of N is given by $\widehat{\lambda}_t = E\left[\lambda_t | \mathcal{F}_{t-}^Z\right]$. We can now, very much like in the Wiener case, define the innovations process.

Definition 14.4 The innovations process ν is defined by

$$d\nu_t = dN_t - \widehat{\lambda}_t dt.$$

Since $M^{\star}$ is an $\mathbf{F}^Z$-martingale, the martingale representation theorem 5.1 gives us an $\mathbf{F}^Z$-predictable process h such that

$$dM^{\star} = h_t \, d\nu_t.$$

We refer to h as the **gain** process, and we thus have the filter dynamics

$$d\pi_t[X] = \pi_t[a] \, dt + h_t \, d\nu_t. \tag{14.7}$$

This leaves us with the problem of determining h, for which we have the following result.

Proposition 14.5   *The gain process h is given by*

$$h_t = \frac{1}{\widehat{\lambda}_t} \left\{ \widehat{(X\lambda)}_t - \widehat{X}_t \widehat{\lambda}_t + \widehat{D}_t \right\},$$

(14.8)

where D is defined in (14.3).

Proof   We give a slightly heuristic proof. The full proof is based on the same ideas, but is a bit more technical.

We start by introducing the shorthand notation $E_t^Z [\cdot] = E [\cdot | \mathcal{F}_t^Z]$. By iterated expectations we then have

$$E_s^Z [Z_t X_t - Z_s X_s] = E_s^Z [Z_t \pi_t[X] - Z_s \pi_s[X]].$$

On the infinitesimal scale, this should (intuitively) imply that

$$E \left[d(Z_t X_t) - d(Z_t \pi_t[X]) | \mathcal{F}_{t-}^Z \right] = 0,$$

(14.9)

and we now proceed to compute the differentials. Since Z is of bounded variation we can use the product differentiation rule, and the X-dynamics, to obtain

$$d(Z_t X_t) = Z_{t-} dX_t + X_{t-} dZ_t + \Delta Z_t \Delta X_t$$
$$= Z_{t-} a_t dt + Z_{t-} dM_t + X_{t-} dN_t + \Delta M_t \Delta N_t$$
$$= Z_{t-} a_t dt + Z_{t-} dM_t + X_{t-} [dN_t - \lambda_t dt] + X_{t-} \lambda_t dt + \Delta M_t \Delta N_t.$$

Using iterated expectations, and the fact that dM_t and $[dN_t - \lambda_t dt]$ are **F**-martingale increments, we obtain

$$E \left[d(Z_t X_t) | \mathcal{F}_{t-}^Z \right] = \left\{ Z_{t-} \widehat{a}_t + \widehat{(X\lambda)}_t + \widehat{D}_t \right\} dt,$$

(14.10)

where D is defined in (14.3). From the filter dynamics, (14.7), we obtain

$$d(Z_t \pi_t[X]) = Z_{t-} d\pi_t[X] + \pi_{t-}[X] dZ_t + \Delta Z_t \Delta \pi_t[X]$$
$$= Z_{t-} \pi_t[a] dt + Z_{t-} dM_t^\star + \pi_{t-}[X] dN_t + h_t \Delta v_t \Delta N_t.$$

Furthermore we have $\Delta v_t = \Delta N_t$ and $\Delta N_t \Delta N_t = dN_t$, so we obtain

$$d(Z_t \pi_t[X]) = Z_{t-} \pi_t[a] dt + Z_{t-} dM_t^\star + \pi_{t-}[X] dN_t + h_t \Delta N_t$$
$$= Z_{t-} \pi_t[a] dt + Z_{t-} dM_t^\star + \pi_{t-}[X] \left[dN_t - \widehat{\lambda}_t dt \right]$$
$$+ \pi_t[X] \widehat{\lambda}_t dt + h_t \left[dN_t - \widehat{\lambda}_t dt \right] + h_t \widehat{\lambda}_t dt,$$

where we have used the fact that $\pi_{t-}[X] \widehat{\lambda}_t dt = \pi_t[X] \widehat{\lambda}_t dt$. From iterated expectations we have

$$E \left[\pi_t[a] | \mathcal{F}_{t-}^Z \right] = E \left[E \left[a_t | \mathcal{F}_t^Z \right] | \mathcal{F}_{t-}^Z \right] = E \left[a_t | \mathcal{F}_{t-}^Z \right] = \widehat{a}_t,$$

as well as

$$E \left[\pi_t[X] | \mathcal{F}_{t-}^Z \right] = \widehat{X}_t.$$

Since $dM^\star$ as well as $\left[dN_t - \widehat{\lambda}_t dt\right]$ are $\mathbf{F}^Z$-martingale increments, we obtain

$$E\left[d(Z_t \pi_t[X]) \mid \mathcal{F}^Z_{t-}\right] = \left\{Z_{t-}\widehat{a}_t + \widehat{X_t \lambda}_t + h_t \widehat{\lambda}_t\right\} dt. \tag{14.11}$$

Plugging (14.10) and (14.11) into (14.9) gives us (14.8). □

We can finally state the main result.

Theorem 14.6 (Filtering Equations)   *Under Assumption 14.1, the filter dynamics for the model (14.1)–(14.2) are given by*

$$d\pi_t[X] = \pi_t[a]\, dt + \frac{1}{\widehat{\lambda}_t}\left\{(\widehat{X\lambda})_t - \widehat{X}_t\widehat{\lambda}_t + \widehat{D}_t\right\}\left[dN_t - \widehat{\lambda}_t dt\right], \tag{14.12}$$

where D is determined by

$$E\left[\Delta M_t \Delta N_t \mid \mathcal{F}_{t-}\right] = D_t dt. \tag{14.13}$$

We now have some comments on this result.

Concerning the gain process, a simple calculation shows that

$$(\widehat{X\lambda})_t - \widehat{X}_t\widehat{\lambda}_t = \left[\widehat{(X_t - \widehat{X}_t)(\lambda_t - \widehat{\lambda}_t)}\right] = E\left[(X_t - \widehat{X}_t)(\lambda_t - \widehat{\lambda}_t)\mid \mathcal{F}^Z_{t-}\right].$$

so this part of the gain process has the interpretation of being a conditional error covariance.

It is perhaps a bit disappointing that the right-hand side of the filter equations contains predictable projections, such as $\widehat{\lambda}_t$, instead of optional projections, such as $\pi_t[\lambda]$ (which would give us a more closed system of SDEs) We do, however, have the following simple but useful result.

Proposition 14.7   *Suppose that, within the filtering framework above, X is a process which can be written*

$$X_t = A^0_t + M^0_t, \tag{14.14}$$

where A^0 is $\mathbf{F}$-optional with continuous trajectories and M^0 is a (cadlag) $\mathbf{F}$-martingale. Defining $\widehat{X}_t$, as before, by $\widehat{X}_t = E\left[X_t \mid \mathcal{F}^N_{t-}\right]$, the following hold:

$$\widehat{X}_t = E\left[X_{t-}\mid \mathcal{F}^N_{t-}\right], \tag{14.15}$$

$$\widehat{X}_t = \pi_{t-}[X]. \tag{14.16}$$

Proof   Concerning (14.15), we see that, given the assumptions above, we have

$$X_{t-} = A^0_{t-} + M^0_{t-} = A^0_t + M^0_{t-}$$

and thus

$$E\left[X_{t-}\mid \mathcal{F}^N_{t-}\right] = E\left[A^0_t \mid \mathcal{F}^N_{t-}\right] + E\left[M^0_{t-}\mid \mathcal{F}^N_{t-}\right] = \widehat{A^0}_t + E\left[M^0_{t-}\mid \mathcal{F}^N_{t-}\right].$$

By iterated expectations we have

$$E\left[M^0_{t-}\mid \mathcal{F}^N_{t-}\right] = E\left[E\left[M^0_t \mid \mathcal{F}_{t-}\right]\mid \mathcal{F}^N_{t-}\right] = E\left[M^0_t \mid \mathcal{F}^N_{t-}\right] = \widehat{M^0}_t$$

which yields

$$\widehat{X}_t = E\left[X_{t-}|\mathcal{F}_{t-}^N\right].$$

Moving on to (14.16), it follows from (14.14) that we can apply the filtering formulas, so $\pi_t[X]$ has the structure

$$\pi_t[X] = A_t + m_t, \tag{14.17}$$

where A is $\mathbf{F}^Z$-optional with continuous trajectories and m is an $\mathbf{F}^Z$-martingale. We now want to compute $\widehat{X}_t$ and by iterated expectations, and the dynamics above, we have

$$\widehat{X}_t = \widehat{\pi_t[X]} = \widehat{A}_t + \widehat{m}_t.$$

Since A is $\mathbf{F}^Z$-optional with continuous trajectories it is in fact also $\mathbf{F}^Z$-predictable, so we have $\widehat{A}_t = A_t$ and, because of continuity, we also have $A_t = A_{t-}$. Furthermore, since m is an $\mathbf{F}^Z$-martingale, we obtain

$$\widehat{m}_t = E\left[m_t|\mathcal{F}_{t-}^Z\right] = m_{t-},$$

where we have used the fact that m is cadlag. As a result we see that, given the assumption above, we have

$$\widehat{X}_t = \pi_{t-}[X]. \tag*{□}$$

Proposition 14.8 (Filtering Equations) *Assume that, as well as the process X, the processes λ and D also have dynamics of the structural form (14.14). Then the filter equations in Theorem 14.6 can be written as*

$$d\pi_t[X] = \pi_t[a]\, dt + h_t\left[dN_t - \pi_{t-}[\lambda]\, dt\right], \tag{14.18}$$

where

$$h_t = \frac{1}{\pi_{t-}[\lambda]}\left\{\pi_{t-}[X\lambda] - \pi_{t-}[X]\,\pi_{t-}[\lambda] + \pi_{t-}[D]\right\}.$$

Proof Apply (14.16) to X, λ and D, and plug the results into the filtering equations. Since X and λ has the form (14.1), the Itô formula implies that this also $X\lambda$ has the structure (14.1) so we can apply (14.16) also to $X\lambda$. □

14.4   Filtering a Finite-State Markov Chain

As we observed already in Section 12.5.2, the filtering equation (14.12) is not closed, in the sense that the objects in the right-hand side are not determined within the equation, and we recap our idea from Section 12.5.2:

Idea If we know on *a priori* grounds that, for all t, the conditional distribution $\mathcal{L}(X_t \mid \mathcal{F}_t^Z)$ belongs to a class of probability distributions which is parameterized by a *finite number of parameters*, then we can expect the have a finite-dimensional filter. The filter equations should then provide us with the dynamics of the parameters for the conditional distribution.

An obvious situation where the conditional distribution $\mathcal{L}(X_t|\mathcal{F}_t^Z)$ is determined by a finite number of parameters, is of course when X is a finite-state Markov chain in continuous time. In Section 12.7 we studied such a model with observations under Wiener noise and derived the Wonham filter. We will now study the same model but with point-process observations.

14.4.1   The Model

In our model, living on a filtered space $(\Omega, \mathcal{F}, P, \mathbf{F})$, we consider a process X, which is a time-homogeneous Markov chain on a finite state-space D. Without loss of generality we may assume that $D = \{1, 2, \ldots, n\}$. We denote the intensity matrix of X by H. Then the probabilistic interpretation is that

$$P(X_{t+h} = j \,|\, X_t = i) = H_{ij}h + o(h), \quad i \neq j,$$

$$H_{ii} = -\sum_{j \neq i} H_{ij}.$$

We cannot observe X directly, but we can instead observe the process Z, which is defined as follows.

Definition 14.9   The observations process Z is a Cox counting process N with $\mathbf{F}$-intensity given by

$$\lambda_t = c(X_t). \tag{14.19}$$

where c is a non-negative function.

We note that the existence of such a model is covered by the existence result for Cox processes in Section 5.5.

14.4.2   The Filter

In this model it is obvious that the conditional distribution of X will be determined by the conditional probabilities, so we define the indicator processes $\delta_t^1, \ldots, \delta_t^n$ by

$$\delta_t^i = I\{X_t = i\}, \quad i = 1, \ldots, n,$$

where $I\{A\}$ denotes the indicator for an event A, so $I\{A\} = 1$ if A occurs, and zero otherwise. The Dynkin theorem, together with a simple calculation, gives us the equation

$$d\delta_t^i = \sum_{j=1}^n H_{ji}\delta_t^j\, dt + dM_t^i, \quad i = 1, \ldots, n, \tag{14.20}$$

where $M^1, \ldots, M^n$ are martingales. In vector form this we can thus write

$$d\delta_t = H^\star \delta_t\, dt + dM_t, \tag{14.21}$$

where $\star$ denotes transpose. From the model construction and (14.20) we now note that:

- the model satisfies the conditions of Proposition 14.8;

- for every i, the processes δ^i and Z have no common jumps.

We thus obtain the filter equations

$$d\pi_t[\delta^i] = \sum_{j=1}^n H_{ji}\pi_t[\delta^j]\, dt + \{\pi_{t-}[\delta^i \lambda] - \pi_{t-}[\delta^i]\, \pi_{t-}[\lambda]\}\, dv_t, \quad i = 1,\ldots,n,$$

$$dv_t = dN_t - \pi_{t-}[\lambda]\, dt.$$

We now observe that, using the notation $c_i = c(i)$, we have the obvious relations

$$\lambda_t = c(X_t) = \sum_{j=1}^n c_j \delta_t^j, \quad \delta_t^i c(X_t) = \delta_t^i c_i,$$

which gives us

$$\pi_t[\lambda] = \sum_{j=1}^n c_j \pi_t[\delta^j], \quad \pi_t[\delta^i c] = \pi_t[\delta_i]\, c_i.$$

Plugging this into the filtering equations gives us the filter.

Proposition 14.10 (The finite-state Markov chain filter)   *With assumptions as above, the filter is given by*

$$d\pi_t[\delta^i] = \sum_{j=1}^n H_{ji}\pi_t[\delta^j]\, dt + \left\{c_i\pi_{t-}[\delta^i] - \pi_{t-}[\delta^i]\cdot \sum_{j=1}^n c_j\pi_{t-}[\delta^j]\right\} dv_t,$$

$$dv_t = dN_t - \sum_{j=1}^n c_j\pi_{t-}[\delta^j]\, dt,$$

where

$$\pi_t[\delta^i] = P\left(X_t = i \,\big|\, \mathcal{F}_t^Z\right), \quad i = 1,\ldots,n,$$

and

$$\pi_{t-}[\delta^i] = P\left(X_t = i \,\big|\, \mathcal{F}_{t-}^Z\right), \quad i = 1,\ldots,n.$$

14.5   Unnormalized Filter Estimates

In Chapter 13 we developed a theory for the unnormalized filter equation (the Zakai equation) in the case of Wiener-driven observations. We will now develop the corresponding theory for the case of point-process observations and, as the reader will observe, the the theories are parallel. Several arguments below are thus more or less copied from Chapter 13, but they are included in order to keep the present chapter reasonably self contained. We start by briefly discussing the filter problem for a Markovian state process.

14.5.1   Filtering a Markov Process

In this section we consider a filtered space $(\Omega, \mathcal{F}, P, \mathbf{F})$, carrying a time-homogeneous cadlag Markov process X, living on some state space $\mathcal{M}$, with generator G, and we are interested in estimating $f(X_t)$ for some real-valued function $f : \mathcal{M} \to R$. If f is in the domain of G we can then apply the Dynkin theorem and obtain the dynamics

$$df(X_t) = (Gf)(X_t)dt + dM_t, \tag{14.22}$$

where M is an **F**-martingale. The observations process Z is assumed to be a counting process N with an **F**-intensity of the form

$$\lambda_t = \lambda(X_{t-}), \tag{14.23}$$

where λ in the right-hand side denotes a deterministic mapping $\lambda : \mathcal{M} \to R_+$.

We can now apply Theorem 14.6 to obtain the filter equation

$$d\pi_t[f] = \pi_t[Gf]\,dt + \frac{1}{\widehat{\lambda}_t} \left\{ (\widehat{f\lambda})_t - \widehat{f}_t\widehat{\lambda}_t + \widehat{D}_t \right\} \left[dN_t - \widehat{\lambda}_t dt \right], \tag{14.24}$$

where D as usual is defined by

$$E\left[\Delta M_t \Delta N_t \,|\, \mathcal{F}_{t-}\right] = D_t dt.$$

The filter equation (14.24) is a complex highly non-linear equation. In order to obtain something simpler, the idea is now, exactly as in Section 13.3, to proceed as follows.

- Perform a Girsanov transformation from P to a new measure P_0 such that X and Z are independent under P_0.
- Compute filter estimates under P_0. This should be easy, thanks to the independence.
- Transform the results back to P using the abstract Bayes formula.
- This will lead to a study of the so-called *unnormalized* estimate $\sigma_t[f]$.
- We will derive an equation known as the Zakai equation, for the unnormalized estimate $\sigma_t[f]$: the equation for $\sigma_t[f]$ turns out to be much simpler than that for $\pi_t[f]$.

First, however, we need a result concerning separation of filtering and detection.

14.5.2   Separation of Filtering and Detection

The setup below is parallel to that of Section 13.2, so the reader is referred to that section for the necessary statistical background.

We consider, on the time interval $[0, T]$, a filtered space $(\Omega, \mathcal{F}, P_0, \mathbf{F})$ carrying the following exogenously given objects:

- an $(P_0, \mathbf{F})$-Poisson process N with unit intensity;
- a non-negative **F**-predictable process λ.

We will perform some Girsanov transformations on the space above, and to this end we need the following definition.

Definition 14.11   For any non-negative **F**-predictable process h, we define the process $L(h)$ by

$$
\begin{cases}
dL_t(h) &= L_{t-}(h)\,(h_t - 1)\,\{dN_t - dt\} \\
L_0(h) &= 1.
\end{cases}
\tag{14.25}
$$

We now perform a change of measure from P_0 to P using $L(\lambda)$ as the likelihood process, so that

$$
L_t(\lambda) = \frac{dP}{dP_0}, \quad \text{on } \mathcal{F}_t,
$$

and we assume that $E^P\left[L_T(\lambda)\right] = 1$. The Girsanov theorem now implies the following:

- under P_0, the counting process N has the **F**-intensity 1;
- under P, the counting process N has the **F**-intensity λ.

Within a statistical framework, we can thus form two hypotheses:

- H_0: N is governed by P_0.
- H_1: N is governed by P.

The natural test statistic, based on the information flow **F**, is (by Neyman–Pearson) given by the likelihood process $L(\lambda)$. See Section 13.2 for a more detailed discussion.

Suppose now that, instead of the information flow **F**, we can only observe the process N. We would then need the likelihood process L^N, defined by

$$
L_t^N = \frac{dP}{dP_0}, \quad \text{on } \mathcal{F}_t^N,
$$

and we have the following standard result.

Lemma 14.12   *With notation as above, we have*

$$
L_t^N = E^{P_0}\left[L_t(\lambda)\mid \mathcal{F}_t^N\right].
$$

This implies that L_t^N is the filter estimate of L_t under P. We now need some notation.

Definition 14.13   For any sufficiently integrable **F**-optional process X we write

$$
\pi_t[X] = E^{P_0}\left[X_t\mid \mathcal{F}_t^N\right],
$$
$$
\widehat{X}_t = E^{P_0}\left[X_t\mid \mathcal{F}_{t-}^N\right],
$$
$$
\widehat{X}_t^P = E^P\left[X_t\mid \mathcal{F}_{t-}^N\right].
$$

Note the different measures used above. Written in this notation, we see that our test statistic L_t^N can be written as

$$
L_t^N = \pi_t[L(\lambda)],
$$

and we now have the following nice result.

Proposition 14.14 *With notation as above we have*

$$\pi_t[L(\lambda)] = L_t(\widehat{\lambda}^P),$$ (14.26)

or alternatively

$$L_t^N = L_t(\widehat{\lambda}^P).$$ (14.27)

In particular we have the dynamics

$$d\pi_t[L(\lambda)] = \pi_{t-}[L(\lambda)] \left[\widehat{\lambda}_t^P - 1\right] \{dN_t - dt\}.$$ (14.28)

Proof Since the filter estimate $\pi_t[L(\lambda)]$ is under P_0, we start by recalling the $L(\lambda)$-dynamics under P_0 as

$$dL_t(\lambda) = 0 \cdot dt + L_{t-}(\lambda)(\lambda_t - 1)\{dN_t - dt\}.$$

We also recall that N has intensity $\lambda^{P_0} = 1$ under P_0. Theorem 14.6 then gives us the dynamics for $L_t^N = \pi_t^0[L(\lambda)]$ as

$$dL_t^N = 0 \cdot dt + \frac{1}{1}\left\{\widehat{D}_t\right\}\left[dN_t - \widehat{1}dt\right]$$

or

$$dL_t^N = \widehat{D}_t \left[dN_t - dt\right],$$

where D is defined by

$$E^{P_0}\left[\Delta M_t \Delta N_t | \mathcal{F}_{t-}\right] = D_t dt.$$

The martingale M in our case is given by

$$dM_t = L_{t-}(\lambda)(\lambda_t - 1)\{dN_t - dt\}.$$

We thus see that

$$D_t = L_{t-}(\lambda)(\lambda_t - 1),$$

so that

$$\widehat{D}_t = \widehat{[L_{t-}(\lambda)(\lambda_t - 1)]} = E^P\left[L_{t-}(\lambda)(\lambda_t - 1)|\mathcal{F}_{t-}^N\right].$$

We thus obtain

$$dL_t^N = \widehat{[L_{t-}(\lambda)(\lambda_t - 1)]} \cdot [dN_t - dt]$$

which we can write as

$$dL_t^N = L_{t-}^N g_t \left[dN_t - dt\right].$$

with

$$g_t = \frac{\widehat{[L_{t-}(\lambda)(\lambda_t - 1)]}}{L_{t-}^N} = \frac{\widehat{[L_{t-}(\lambda)\lambda_t]}}{L_{t-}^N} - 1.$$

We can thus write

$$dL_t^N = L_{t-}^N(h_t - 1)[dN_t - dt],$$ (14.29)

where

$$h_t = \frac{[L_{t-}(\lambda)\lambda_t]\widehat{}}{L_{t-}^N} = \frac{E^P\left[L_{t-}(\lambda)\lambda_t \mid \mathcal{F}_{t-}^N\right]}{E^P\left[L_{t-}(\lambda)\mid \mathcal{F}_{t-}^N\right]}.$$

It now follows from Bayes' formula that

$$h_t = E^P\left[\lambda_t \mid \mathcal{F}_{t-}^N\right] = \widehat{\lambda}_t^P.$$

Comparing (14.29) with (14.25) we see that $L_t^N = L_t(\widehat{\lambda}^P)$ which proves our proposition.

□

14.5.3   A Model with a Driving MPP

In this section we will present a way of constructing a model (X, Z) like the one in Section 14.5.1, by using Girsanov theory. To that end we consider a probability space $(\Omega, \mathcal{F}, P_0,)$ carrying the following *independent* objects.

- A scalar process X satisfying the SDE

$$dX_t = \mu(X_t)dt + \sigma(X_t)dW_t + \int_E \beta(X_{t-}, z)\Psi(dt, dz), .$$

  where $\mu(x)$, $\sigma(x)$ and $\beta(x, z)$ are given deterministic functions, and where Ψ has a predictable intensity λ^Ψ of the functional form

$$\lambda_t^\Psi = \lambda^\Psi(X_{t-}, dz).$$

- A Poisson process N with unit intensity.
- A deterministic mapping $\lambda : M \to R_+$.

Let us now define a filtration **F** by

$$\mathcal{F}_t = \mathcal{F}_t^N \vee \mathcal{F}_\infty^X,$$

and change measure from P_0 to a new measure P by a Girsanov transformation where the likelihood process L is defined by

$$\begin{cases} dL_t &= L_{t-}[\lambda(X_{t-}) - 1]\{dN_t - dt\} \\ L_0 &= 1 \end{cases}$$

and we assume that $E^{P_0}[L_T] = 1$, for all $T \geq 0$. It now follows from the Girsanov theorem that the process N will have a P-intensity λ_t given by

$$\lambda_t = \lambda(X_{t-}).$$

Furthermore, since $\mathcal{F}_\infty^X \subseteq \mathcal{F}_0$ and $L_0 = 1$, we see that the distribution of X is the same under P as under P_0. In this way we have thus constructed a model of a form that we considered in Section 14.5.1. We now move on to filtering, but first we have an important observations.

Note 14.15   Since Ψ and N are independent under P_0, they have no common jumps under P_0 and since $P \ll P_0$, we see that this also holds under P.

14.5.4   The Zakai Equation

We define $\pi_t[f]$ as usual by

$$\pi_t[f] = E^P\left[f(X_t)|\mathcal{F}_t^Z\right],$$

and from the abstract Bayes formula we then have

$$\pi_t[f] = \frac{E^0\left[L_t f(X_t)|\mathcal{F}_t^Z\right]}{E^0\left[L_t|\mathcal{F}_t^Z\right]},$$

where E^0 is shorthand for E^{P_0}, and this leads us to the following definition.

Definition 14.16   The **unnormalized estimate** $\sigma_t[f]$ is defined by

$$\sigma_t[f] = E^0\left[L_t f(X_t)|\mathcal{F}_t^Z\right].$$

We have thus derived the Kallianpur–Striebel formula.

Proposition 14.17 (Kallianpur–Striebel)   *The standard filter estimate $\pi_t[f]$ and the unnormalized estimate $\sigma_t[f]$ are related by the formula*

$$\pi_t[f] = \frac{\sigma_t[f]}{\sigma_t[1]}. \tag{14.30}$$

It turns out that $\sigma_t[f]$ is easier to study than $\pi_t[f]$, so we now go on to derive the dynamics of $\sigma_t[f]$. This is in fact quite easy. From the Kallianpur–Striebel formula we have

$$\sigma_t[f] = \pi_t[f] \cdot \sigma_t[1].$$

From the Itô formula we obtain

$$df(X_t) = (\mathcal{G}f)(X_t)dt + dM_t,$$

where M is a martingale, and we recall from Proposition 9.5 that the infinitesimal generator of the process X above is given (under P_0 as well as under P) by

$$\mathcal{G}f(x) = \mu_t f_x(x) + \frac{1}{2}\sigma_t^2 f_{xx}(x) + \int_E f_\beta(x,z)\lambda_t^\Psi(x,dz), \tag{14.31}$$

where

$$f_\beta(x,z) = f(x + \beta(x,z)) - f(x). \tag{14.32}$$

From (14.24) we have

$$d\pi_t[f] = \pi_t[\mathcal{G}f]\,dt + \frac{1}{\widehat{\lambda}_t}\left\{(\widehat{f\lambda})_t - \widehat{f}_t\widehat{\lambda}_t\right\}\left[dN_t - \widehat{\lambda}_t dt\right], \tag{14.33}$$

where, because of no common jumps (see Note 14.15), the term $\widehat{D}$ is not present, and from (14.28) we have

$$d\sigma_t[1] = \sigma_{t-}[1]\,(\pi_{t-}[\lambda] - 1)\,\{dN_t - dt\}.$$

We can now apply the Itô formula to the product $\pi_t[f] \cdot \sigma_t[1]$. As in the pure Wiener case, there are a surprisingly large number of cancellations in these calculations and in the end we obtain the following result.

Theorem 14.18 (The Zakai Equation)   *The unnormalized estimate $\sigma_t[f]$ satisfies the Zakai equation*

$$d\sigma_t[f] = \sigma_t[\mathcal{G}f]\,dt + \sigma_{t-}[f(\lambda - 1)]\,\{dN_t - dt\}. \tag{14.34}$$

We note that, as in the Wiener case, the Zakai equation for $\sigma_t[f]$ has a much simpler structure than the corresponding equation for $\pi_t[f]$. First, the Zakai equation is driven by the compensated Poisson process $dN_t - dt$, rather than by the innovation process $dN_t - \widehat{\lambda}_t dt$. Second, the gain term, $\sigma_t[f(\lambda - 1)]$, in the Zakai equation has a much simpler structure than the corresponding term in the equation for $\pi_t[f]$.

Let us now assume that there exists an unnormalized density process $q_t(x)$, for X with respect to some dominating measure m (where m often is Lebesgue measure or a counting measure) so that we have the interpretation

$$\sigma_t[f] = \int_R f(x)q_t(x)dm(x).$$

We can then dualize the Zakai equation, exactly like we did in the derivation of the Kushner–Stratonovich equation, to obtain the following result, where $\mathcal{G}^\star$ denotes the adjoint operator.

Proposition 14.19   *The unnormalized density $q_t(x)$ satisfies the following SPIDE*

$$dq_t(x) = \mathcal{G}^\star q_t(x)dt + q_{t-}(x)\,[\lambda(x) - 1]\,\{dN_t - dt\}. \tag{14.35}$$

15 Filtering with k-Variate Counting-Process Observations

The results of the previous chapters easily extend to the case of a k-variate observations process $N = (N^1, \ldots, N^k)$.

15.1 Filtering

We consider again a filtered probability space $(\Omega, \mathcal{F}, P, \mathbf{F})$. Our basic model consists of a pair of processes (X, Z) with dynamics as follows.

$$dX_t = a_t dt + dM_t, \tag{15.1}$$

$$dZ_t = dN_t. \tag{15.2}$$

The process a is allowed to be an arbitrary integrable $\mathbf{F}$-adapted process, and M is assumed to be an $\mathbf{F}$-martingale. The novelty is that N is now a k-variate $\mathbf{F}$-counting process $N = (N^1, \ldots, N^k)$ with $\mathbf{F}$-intensity processes $(\lambda^1, \ldots, \lambda^k)$, so $Z = (Z^1, \ldots, Z^k)$ is a k-dimensional process. We recall that, for any process X, we use the notation

$$\pi_t[X] = E\left[X_t | \mathcal{F}_t^Z\right],$$
$$\widehat{X}_t = E\left[X_t | \mathcal{F}_{t-}^Z\right].$$

Exactly like in the scalar case we can now cite Lemma 14.3 to deduce that the process $\pi_t[X]$ admits the representation

$$d\pi_t[X] = \pi_t[a] \, dt + dM_t^\star,$$

where $M^\star$ is an $\mathbf{F}^Z$-martingale. It now follows from the martingale representation theorem that we can write

$$dM_t^\star = \sum_{i=1}^k h_t^i \left[dN_t^i - \widehat{\lambda}_t^i dt\right],$$

where the gain process $h = (h^1, \ldots, h^k)$ is $\mathbf{F}^N$-predictable. The remaining problem is thus to identify the gain process h and to this end note that

$$E\left[X_t Z_t^i - X_s Z_s^i \big| \mathcal{F}_s^Z\right] = E\left[X_t Z_t^i - X_s Z_s^i \big| \mathcal{F}_s^Z\right],$$

which after a small calculation leads us to the (informal) relation

$$E\left[d(Z_t^i X_t) - d(Z_t^i \pi_t[X]) \big| \mathcal{F}_{t-}^Z\right] = 0.$$

This leaves us with the problem of determining the gain process h, and for that we have the following result.

Proposition 15.1 *The gain process h is given by*

$$h_t^i = \frac{1}{\widehat{\lambda}_t^i} \left\{ (\widehat{X\lambda})_t^i - \widehat{X}_t \widehat{\lambda}_t^i + \widehat{D}_t^i \right\},$$ (15.3)

where D is defined by

$$E\left[\Delta M_t \Delta N_t^i \big| \mathcal{F}_{t-}\right] = D_t^i dt.$$ (15.4)

Proof   The proof follows exactly the proof of Proposition 14.5.   □

We thus have the following main result.

Theorem 15.2 (Filtering Equations)   *With assumptions as above, the filter equations are*

$$d\pi_t[X] = \pi_t[a]\,dt + \sum_{i=1}^{k} \frac{1}{\widehat{\lambda}_t^i} \left\{ (\widehat{X\lambda^i})_t - \widehat{X}_t \widehat{\lambda}_t^i + \widehat{D}_t^i \right\} \left[dN_t^i - \widehat{\lambda}_t^i dt \right],$$ (15.5)

where D^i is determined by

$$E\left[\Delta M_t \Delta N_t^i \big| \mathcal{F}_{t-}\right] = D_t^i dt.$$ (15.6)

As for the univariate case we can simplify these equations as follows.

Proposition 15.3 (Filtering Equations)   *Assume that, apart from the process X, also the processes λ and D have dynamics of the structural form (14.14). Then the filter equations in Theorem 15.2 can be written as*

$$d\pi_t[X] = \pi_t[a]\,dt + h_t^i \sum_{i=1}^{k} \left[dN_t^i - \pi_{t-}[\lambda^i]\,dt \right],$$ (15.7)

where

$$h_t^i = \frac{1}{\pi_{t-}[\lambda^i]} \left\{ (\pi_{t-}[X\lambda^i] - \pi_{t-}[X]\pi_{y-}[\lambda^i] + \pi_{t-}[D^i] \right\}.$$ (15.8)

15.2   Separation of Filtering and Detection

We now move on to extend the separation result from Section 14.5.2, so we consider a filtered space $(\Omega, \mathcal{F}, P_0, \mathbf{F})$ carrying the following exogenously given objects:

- a k-variate counting process $N = (N^1, \ldots, N^k)$ with all components being $(P_0, \mathbf{F})$-Poisson processes with unit intensity;
- a non-negative $\mathbf{F}$-predictable process $\lambda = (\lambda^1, \ldots, \lambda^k)$.

We then extend the definition of the likelihood process L in the obvious way.

Definition 15.4 For any non-negative **F**-predictable process $h = (h^1, \ldots, h^k)$, we define the process $L(h)$ by

$$\begin{cases} dL_t(h) = L_{t-}(h) \sum_{i=1}^k (h_t^i - 1) \{dN_t^i - dt\} \\ L_0(h) = 1. \end{cases} \qquad (15.9)$$

We now perform a change of measure from P_0 to P by using $L(\lambda)$ as the likelihood process, so

$$L_t(\lambda) = \frac{dP}{dP_0}, \quad \text{on } \mathcal{F}_t,$$

and we assume that $E^P[L_T(\lambda)] = 1$. The Girsanov theorem now implies the following:

- under P_0, the counting process N has the **F**-intensity 1;
- under P, the counting process N has the **F**-intensity λ.

We recall our definition from the univariate case. Note the different measures.

Definition 15.5 For any sufficiently integrable **F**-optional process X we use the following notation

$$\pi_t[X] = E^{P_0}\left[X_t | \mathcal{F}_t^N\right],$$
$$\widehat{X}_t = E^{P_0}\left[X_t | \mathcal{F}_{t-}^N\right],$$
$$\widehat{X}_t^P = E^P\left[X_t | \mathcal{F}_{t-}^N\right].$$

The following result then holds.

Proposition 15.6 *With notation as in Definition 14.13 above we have*

$$\pi_t[L(\lambda)] = L_t(\widehat{\lambda}^P), \qquad (15.10)$$

or alternatively

$$L_t^N = L_t(\widehat{\lambda}^P). \qquad (15.11)$$

In particular we have the dynamics

$$d\pi_t[L(\lambda)] = \pi_{t-}[L(\lambda)] \sum_{i=1}^k (\widehat{\lambda}_t^{iP} - 1) \{dN_t^i - dt\}. \qquad (15.12)$$

15.2.1   The Zakai Equation

The obvious extension of the model in Section 14.5.3 is that we consider a probability space $(\Omega, \mathcal{F}, P_0,)$ carrying the following *independent* objects.

- A scalar process X satisfying the SDE

$$dX_t = \mu(X_t)dt + \sigma(X_t)dW_t + \int_E \beta(X_{t-}, z)\Psi(dt, dz),$$

where $\mu(x)$, $\sigma(x)$ and $\beta(x, z)$ are given deterministic functions, and where Ψ has a predictable intensity λ^Ψ of the form

$$\lambda_t^\Psi = \lambda^\Psi(X_{t-}).$$

- A k-variate Poisson process N with unit intensity.
- A deterministic mapping $\lambda : M \to R_+^k$.

We then define a filtration $\mathbf{F}$ by

$$\mathcal{F}_t = \mathcal{F}_t^N \vee \mathcal{F}_\infty^X,$$

and change measure from P_0 to a new measure P by a Girsanov transformation where the likelihood process L is defined by

$$\begin{cases} dL_t &= L_{t-} \sum_{i=1}^k \left[\lambda^i(X_{t-}) - 1 \right] \left\{ dN_t^i - dt \right\} \\ L_0 &= 1 \end{cases}$$

and where we assume that $E^{P_0}[L_T] = 1$, for all $T \geq 0$. It now follows from the Girsanov theorem that the process N will have an $\mathbf{F}$-predictable P-intensity process $\lambda = (\lambda^1, \ldots, \lambda^k)$ given by

$$\lambda_t = \lambda(X_{t-}).$$

We again define the unnormalized estimate $\sigma_t[f]$ by

$$\sigma_t[f] = E^0 \left[L_t f(X_t) | \mathcal{F}_t^Z \right].$$

and the Kallianpur–Striebel formula (14.30) still holds. The conclusion is expected and reads as follows.

Theorem 15.7 (The Zakai Equation)   *The unnormalized estimate $\sigma_t[f]$ satisfies the Zakai equation*

$$d\sigma_t[f] = \sigma_t[\mathcal{G}f] \, dt + \sum_{i=10}^k \sigma_{t-}\left[f(\lambda^i - 1) \right] \left\{ dN_t^i - dt \right\}. \tag{15.13}$$

The unnormalized density $q_t(x)$ satisfies the SPIDE

$$dq_t(x) = \mathcal{G}^\star q_t(x) dt + q_{t-}(x) \sum_{i=1}^k \left[\lambda^i(x) - 1 \right] \left\{ dN_t^i - dt \right\}. \tag{15.14}$$

15.3   Filtering with Marked Point-Process Observations

We now move on to study the filtering problem when we can observe a marked point process. The general ideas are exactly the same as earlier in this chapters, but there is an added problem which makes it necessary to study this case in some detail. We consider again a filtered probability space $(\Omega, \mathcal{F}, P, \mathbf{F})$. The model is defined as follows.

Assumption 15.8   *We assume the following.*

- *The state process X has the usual dynamics*

$$dX_t = a_t dt + dM_t, \tag{15.15}$$

where a is $\mathbf{F}$-optional and M is an $\mathbf{F}$-martingale.

- We can observe the filtration $\mathbf{F}^{\Psi}$, where Ψ is a marked point process on the mark space $(E, \mathcal{E})$.
- The point process Ψ admits a predictable $\mathbf{F}$-intensity measure process $\lambda_t(dz)$.

For the convenience of the reader we recall the definition of $\mathbf{F}^{\Psi}$ as

$$\mathcal{F}_t^{\Psi} = \sigma\{\Psi([0,s] \times A) : A \in \mathcal{E}, \ s \le t\}.$$

For any process X we now define the filter estimate $\pi_t[X]$ by

$$\pi_t[X] = E\left[X_t | \mathcal{F}_t^{\Psi}\right]$$

and we go on to derive the dynamics of $\pi_t[X]$. The general strategy is exactly like the case of observing a single counting process, and we can cite Lemma 14.3 to deduce that the process $\pi_t[X]$ admits the representation

$$d\pi_t[X] = \pi_t[a]\,dt + dM_t^{\star}, \tag{15.16}$$

where $M^{\star}$ is an $\mathbf{F}^{\Psi}$-martingale.

This is where the story starts to get a bit technical. In the case of observing a single counting process N with $\mathbf{F}$-intensity λ, we could at this point rely on the martingale representation theorem which allowed us to write

$$dM_t^{\star} = h_t\left[dN_t - \widehat{\lambda}_t\,dt\right],$$

where $\widehat{\lambda}$ was the $\mathbf{F}^N$-predictable projection of λ, given by

$$\widehat{\lambda}_t = E\left[\lambda_t | \mathcal{F}_{t-}^N\right].$$

This could easily be generalized to the case of a k-variate observations process $N = (N^1, \ldots, N^k)$, by simply projecting each intensity process λ^i separately. The problem we now are facing is that we no longer have a finite number of individual intensity processes, but instead an intensity measure-valued process $\lambda_t(dz)$. We would thus like to project (in some sense) the measure-valued process $\lambda_t(dz)$ onto the predictable σ-algebra generated by $\mathbf{F}^{\Psi}$. This can in fact be done, and we cite the following result from Last & Brandt (1995).

Proposition 15.9 *There exists a measure-valued process $\widehat{\lambda}_t(dz)$, referred to as the* **predictable projection of** $\lambda_t(dz)$, *such that the following hold.*

- *For every $t \ge 0$ and every $A \in \mathcal{E}$ we have*

$$\widehat{\lambda}_t(A) = E\left[\lambda_t(A) | \mathcal{F}_{t-}^{\Psi}\right].$$

- *For every fixed $A \in \mathcal{E}$ the process $\widehat{\lambda}_t(A)$ is $\mathbf{F}^{\Psi}$-predictable.*
- *The process $\widehat{\lambda}_t(dz)$ is the $\mathbf{F}^{\Psi}$-predictable intensity process of Ψ.*

Given this result, it now follows from the martingale representation theorem 9.1 that there exists a $\mathcal{P}^{\Psi}$-predictable process $H_t(z)$ such that

$$dM_t^{\star} = \int_E H_t(z)\left\{\Psi(dt, dz) - \widehat{\lambda}_t(dz)dt\right\}. \tag{15.17}$$

From (15.16) we thus have the filter dynamics

$$d\pi_t[X] = \pi_t[a]\,dt + \int_E H_t(z)\left\{\Psi(dt,dz) - \hat{\lambda}_t(dz)dt\right\},\qquad(15.18)$$

and it remains to determine the gain process H.

We now face another problem. In the case of counting-process observations we noted that we had the relation

$$E\left[N_t^i X_t - N_s^i X_s\,|\,\mathcal{F}_s^N\right] = E\left[N_t^i\pi_t[X] - N_s^i\pi_s[X]\,|\,\mathcal{F}_s^N\right]$$

for each $i = 1,\dots,k$, and this allowed us to determine the gain process. In the present setting there is nothing like the ith component of Ψ, so instead we proceed as follows.

Definition 15.10   For an arbitrary $A \in \mathcal{E}$, we define the process N^A by

$$N_t^A = \Psi([0,t],A).$$

It is then clear that N^A is a counting process with **F**-intensity $\lambda_t(A)$, and $\mathbf{F}^\Psi$-predictable intensity $\hat{\lambda}_t(A)$.

We also need an assumption about the common jumps of N^A and M.

Assumption 15.11   *For each $A \in \mathcal{E}$ there exists an **F**-optional process $D(A)$ such that*

$$E\left[\Delta M_t \Delta N_t^A\,|\,\mathcal{F}_{t-}\right] = D_t(A)dt.\qquad(15.19)$$

It is easy to see that, for fixed t, $D_t(\cdot)$ is a random measure on $(E,\mathcal{E})$, so, as in Proposition 15.9, we can project this to obtain a measure-valued process $\hat{D}$ such that

$$\hat{D}_t(A) = E\left[D_t(A)\,|\,\mathcal{F}_{t-}^\Psi\right].\qquad(15.20)$$

We can now more or less copy the arguments from the single counting-process case. We note that, from iterated expectations, we have

$$E\left[N_t^A X_t - N_s^A X_s\,|\,\mathcal{F}_s^\Psi\right] = E\left[N_t^A\pi_t[X] - N_s^A\pi_s[X]\,|\,\mathcal{F}_s^\Psi\right].$$

On the infinitesimal scale, this should (intuitively) imply that

$$E\left[d(N_t^A X_t) - d(N_t^A\pi_t[X])\,|\,\mathcal{F}_{t-}^\Psi\right] = 0,\qquad(15.21)$$

and we now go on to compute the differentials. Since N^A is of bounded variation we can use the product differentiation rule, and the X-dynamics, to obtain

$$
\begin{aligned}
d(N_t^A X_t) &= N_{t-}^A dX_t + X_{t-}dN_t^A + \Delta N_t^A \Delta X_t\\
&= N_{t-}^A a_t\,dt + N_{t-}^A dM_t + X_{t-}dN_t^A + \Delta M_t \Delta N_t^A\\
&= N_{t-}^A a_t\,dt + N_{t-}^A dM_t + X_{t-}\left[dN_t^A - \lambda_t(A)dt\right] + X_{t-}\lambda_t(A)dt\\
&\quad + \Delta M_t \Delta N_t^A.
\end{aligned}
$$

Using iterated expectations, and the fact that dM_t and $\left[dN_t^A - \lambda_t(A)dt\right]$ are **F**-martingale increments, we obtain

$$E\left[d(N_t^A X_t)\,|\,\mathcal{F}_{t-}^\Psi\right] = \left\{N_{t-}^A\hat{a}_t + \widehat{X_t\lambda_t}(A) + \hat{D}_t(A)\right\}dt.\qquad(15.22)$$

From the filter dynamics, (15.16), we obtain

$$d(N_t^A \pi_t[X]) = N_{t-}^A d\pi_t[X] + \pi_{t-}[X] dN_t^A + \Delta N_t^A \Delta \pi_t[X]$$
$$= N_{t-}^A \pi_t[a] dt + N_{t-}^A dM_t^\star + \pi_{t-}[X] dN_t^A + \Delta \pi_t[X] \Delta N_t^A.$$

Furthermore, from (15.18) we deduce

$$\Delta \pi_t[X] \Delta N_t^A = \int_A H_t(z) \Psi(dt, dz).$$

We thus obtain

$$d(N_t^A \pi_t[X]) = N_{t-}^A \pi_t[a] dt + N_{t-}^A dM_t^\star + \pi_{t-}[X] dN_t^A + \int_A H_t(z) \Psi(dt, dz)$$
$$= N_{t-}^A \pi_t[a] dt + N_{t-}^A dM_t^\star + \pi_{t-}[X] \left[dN_t^A - \widehat{\lambda}_t(A) dt \right]$$
$$+ \pi_t[X] \widehat{\lambda}_t(A) dt + \int_A H_t(z) \left[\Psi(dt, dz) - \widehat{\lambda}_t(dz) dt \right]$$
$$+ \int_A H_t(z) \widehat{\lambda}_t(dz) dt.$$

From iterated expectations we have

$$E\left[\pi_t[a] \mid \mathcal{F}_{t-}^\Psi \right] = E\left[E\left[a_t \mid \mathcal{F}_t^\Psi \right] \mid \mathcal{F}_{t-}^\Psi \right] = E\left[a_t \mid \mathcal{F}_{t-}^\Psi \right] = \widehat{a}_t,$$

as well as

$$E\left[\pi_t[X] \mid \mathcal{F}_{t-}^\Psi \right] = \widehat{X}_t.$$

Taking expectations we obtain

$$E\left[d(N_t^A \pi_t[X]) \mid \mathcal{F}_{t-}^\Psi \right] = \left\{ N_{t-}^A \widehat{a}_t + \widehat{X}_t \widehat{\lambda}_t(A) + \int_A H_t(z) \widehat{\lambda}_t(dz) \right\} dt. \qquad (15.23)$$

Equations (15.21), (15.22) and (15.23) gives us the relation

$$\widehat{X_t \lambda_t}(A) + \widehat{D_t}(A) = \widehat{X}_t \widehat{\lambda}_t(A) + \int_A H_t(z) \widehat{\lambda}_t(dz)$$

and, since this holds for every $A \in \mathcal{E}$, we should now be able to identify H. We then note that $\widehat{X_t \lambda_t}(\cdot)$, $\widehat{D_t}(\cdot)$ and $\widehat{X}_t \widehat{\lambda}_t(\cdot)$ are random measures, and that all of them are absolutely continuous with respect to $\widehat{\lambda}_t(dz)$. We therefore make the following definition.

Definition 15.12 The processes $R_t^{X\lambda}(z)$ and $R_t^D(z)$ are defined as the Radon–Nikodym derivatives

$$R_t^{X\lambda}(z) = \frac{d(\widehat{X\lambda})_t}{d\widehat{\lambda}_t}(z), \qquad (15.24)$$

$$R_t^D(z) = \frac{d(\widehat{D})_t}{d\widehat{\lambda}_t}(z). \qquad (15.25)$$

With this notation the relation above becomes

$$\int_A H_t(z) \widehat{\lambda}_t(dz) = \int_A \left[R_t^{X\lambda}(z) - \widehat{X}_t + R_t^D(z) \right] \widehat{\lambda}_t(dz),$$

and, since this holds for all $A \in \mathcal{E}$, we have

$$H_t(z) = R_t^{X\lambda}(z) - \widehat{X}_t + R_t^D(z), \quad \widehat{\lambda}_t(dz)\text{-a.e.}$$

Modulo a number of technical conditions we have thus proved the following result.

Theorem 15.13   *Under Assumptions 15.8 and 15.11, the filter dynamics are given by*

$$d\pi_t[X] = \pi_t[a] \, dt + \int_E H_t(z) \left\{ \Psi(dt, dz) - \widehat{\lambda}_t(dz)dt \right\}, \tag{15.26}$$

where

$$H_t(z) = R_t^{X\lambda}(z) - \widehat{X}_t + R_t^D(z), \quad \widehat{\lambda}_t(dz)\text{-a.e.} \tag{15.27}$$

and where $R_t^{X\lambda}$ and R_t^D are given in Definition 15.12.

Part V

Applications in Financial Economics

16 Basic Arbitrage Theory

16.1 Portfolios

In this chapter we recall some central concepts and results from general arbitrage theory. For details the reader is referred to Björk (2020) or any other standard textbook on the subject.

We consider a filtered probability space $(\Omega, \mathcal{F}, P, \mathbf{F})$ and a model of a financial market consisting of $N + 1$ financial assets (without dividends). We assume that the market is perfectly liquid, that there is no credit risk, no bid–ask spread and that prices are not affected by our portfolios.

We denote by S_t^i the price at time t of one unit of asset number i, for $i = 0, \ldots, N$. We let S denote the corresponding N-dimensional column-vector process, and assume that all asset price processes are adapted. The asset S^0 will play a special role below as the **numeraire asset** and we assume that S^0 is strictly positive with probability one. In general the price processes are allowed to be semimartingales but in our applications the prices will be restricted to being jump diffusions.

We now go on to define the concept of a portfolio strategy, in particular, the concept of a "self-financing portfolio". Intuitively this is a portfolio strategy where there is no external withdrawal from, or infusion of money into, the portfolio. It is far from trivial how this should be formalized in continuous time, but a careful discretization argument leads to the following formal definition, where we let h_t^i denote the number of units of asset number i that are held in the portfolio.

Definition 16.1 We introduce the following concepts.

- A **portfolio strategy** is an $(N + 1)$-dimensional predictable (row-vector) process $h = (h^1, \ldots, h^N)$. For a given strategy h, the corresponding **value process** V^h is defined by

$$V_t^h = \sum_{i=0}^{N} h_t^i S_t^i, \tag{16.1}$$

or equivalently

$$V_t^h = h_t S_t. \tag{16.2}$$

- The strategy is said to be **self-financing** if

$$dV_t^h = \sum_{i=0}^{N} h_t^i \, dS_t^i, \tag{16.3}$$

or equivalently

$$dV_t^h = h_t \, dS_t. \tag{16.4}$$

- For a given strategy h, the corresponding **relative portfolio** $u = (u^1, \ldots, u^N)$ is defined by

$$u_t^i = \frac{h_t^i S_{t-}^i}{V_{t-}^h}, \quad i = 0, \ldots, N, \tag{16.5}$$

and we will have

$$\sum_{i=0}^{N} u_t^i = 1.$$

Note that, by the definition above, the relative portfolio will be predictable. We should, in all honesty, also require some minimal integrability properties for our admissible portfolios, but we will suppress these and some other technical conditions. The reader is referred to the specialist literature for details.

It is often easier to work with the relative portfolio u than with the portfolio h, and we immediately have the following obvious result.

Proposition 16.2 *If u is the relative portfolio corresponding to a self-financing portfolio h, then we have*

$$dV_t^h = V_{t-}^h \sum_{i=0}^{N} u_t^i \frac{dS_t^i}{S_{t-}^i}. \tag{16.6}$$

In most market models we have a (locally) risk-free asset: the formal definition is as follows.

Definition 16.3 Suppose that one of the asset price processes, henceforth denoted by B, has dynamics of the form

$$\begin{cases} dB_t &= r_{t-}B_{t-}dt, \\ B_0 &= 1, \end{cases} \tag{16.7}$$

where r is some adapted random process, and the initial condition $B_0 = 1$ is a normalizing factor of no importance. In such a case we say that the asset B is (locally) **risk-free** and we refer to B as the **bank account**. The process r is referred to as the corresponding **short rate**.

The term "locally risk-free" is more or less obvious. We first note that since B has continuous trajectories, it will be predictable. If we (loosely speaking) are standing at time $t-$ then, since r is adapted, we know the value of r_{t-}. Since b is predictable we also know B_{t-}, which implies that already at time $t-$ we know the value $B_t = B_t$ of B at time t. The asset B is thus risk free on the local (infinitesimal) time scale, even if the short rate r is random. The interpretation is the usual, i.e. we can think of B as the value of a bank account where we have the short rate r. Typically we will choose B as the asset S^0. Since we integrate with respect to Lebesgue measure we can (and will) write the B-dynamics as

$$dB_t = r_t B_t dt, \tag{16.8}$$

and we have

$$S_t = e^{\int_0^t r_s \, ds}. \tag{16.9}$$

16.2   Arbitrage

The definition of arbitrage is standard.

Definition 16.4   A portfolio strategy h is an **arbitrage** strategy on the time interval $[0, T]$ if the following conditions are satisfied.

1. The strategy h is self financing
2. The initial cost of h is zero, i.e.

$$V_0^h = 0.$$

3. At time T we have

$$P\left(V_T^h \geq 0\right) = 1,$$
$$P\left(V_T^h > 0\right) > 0.$$

An arbitrage strategy is thus a money-making machine which, with positive probability, produces a positive amount of money out of nothing. The economic interpretation is that the existence of an arbitrage opportunity indicates a serious case of mispricing in the market, and a minimal requirement of market efficiency is that there are no arbitrage opportunities. The single most important result in mathematical finance is the "First Fundamental Theorem" below which connects absence of arbitrage to the existence of a martingale measure.

Definition 16.5   Consider a market model consisting of $N + 1$ assets $S^0, \ldots, S^N$, and assume that the *numeraire asset* S^0 has the property that $S_t^0 > 0$ with probability one for all t. An equivalent *martingale measure*, often referred to as an EMM, is a probability measure Q with the properties that

1. Q is equivalent to P, i.e. $Q \sim P$.
2. The normalized price processes $Z_t^0, \ldots, Z_t^N$, defined by

$$Z_t^i = \frac{S_t^i}{S_t^0}, \quad i = 0, \ldots, N,$$

are (local) martingales under Q.

We can now state the main abstract result of arbitrage theory. The proof is beyond the scope of this book but can be found in Delbaen & Schachermayer (1994).

Theorem 16.6 (First Fundamental Theorem)   *The market model is free of arbitrage possibilities if and only if there exists a martingale measure Q.*

Remark   The First Fundamental Theorem as stated above is a "folk theorem" in the sense that it is not stated with all necessary technical conditions. The statement above will however do nicely for our purposes. For a precise statement and an outline of the full proof, see Björk (2020). For the full (extremely difficult) proof see Delbaen & Schachermayer (1994).

We note that if there exists a martingale measure Q, then it will depend upon the choice of the numeraire asset S^0, so we should really index Q as Q^0. In most cases the numeraire asset will be the bank account B and in this case the measure Q, which more precisely should be denoted by Q^B, is known as the **risk-neutral martingale measure**.

We end this section by citing the following useful result. We will later prove it within a jump-diffusion setting, but we state it in full generality.

Proposition 16.7   *Assume absence of arbitrage and let Q denote a (not necessarily unique) risk-neutral martingale measure with B as numeraire. Let S denote the arbitrage-free price process of an arbitrary asset, underlying or derivative. The following statements are then equivalent.*

- *The normalized process*

$$Z_t = \frac{S_t}{B_t}$$

  is a Q-martingale.
- *The Q-dynamics of S are of the form*

$$dS_t = r_t S_t dt + S_t dM_t^Q,$$

  where M^Q is a Q-martingale.

What this result says is that, under Q, the local mean rate of return of any traded asset (without dividends) must equal the short rate. In particular we have the following easy but important result.

Proposition 16.8   *Assume that V is the value of a self-financing portfolio or any other asset in an arbitrage-free market, and suppose that V has dynamics*

$$dV_t = k_t V_t dr, \tag{16.10}$$

where K is some adapted process. Then we have

$$k_t = r_t, \tag{16.11}$$

for all t.

Proof Define Z by $Z_t = V_t/B_t$. This gives us the P-dynamics

$$dZ_t = Z_t(k_t - r_t)dt,$$

and this process is not a martingale under any $Q \sim P$ unless $k - t = r_t$.   □

Intuitively this is also obvious: the process V above is a "synthetic bank" with short rate k. If $k > r$ we borrow in the normal bank at market short rate r, invest in V and make a risk-free profit. If $k < r$ we shortsell V, invest in the normal bank at market short rate r and make a risk-free profit.

16.3   Martingale Pricing

We now study the possibility of pricing contingent claims. The formal definition of a claim is as follows.

Definition 16.9   Given a a stochastic basis $(\Omega, \mathcal{F}, P, \mathbf{F})$ and a specified point in time T, often referred to as "the exercise date") a **contingent T-claim** is a random variable $\mathcal{Z} \in \mathcal{F}_T$.

The interpretation is that the holder of the claim will obtain the random amount $\mathcal{Z}$ of money at time T. We now consider the "primary" or "underlying" market $S^0, S^1, \ldots, S^N$ as given *a priori*, and we fix a T-claim $\mathcal{Z}$. Our task is that of determining a "reasonable" price-process $\Pi_t[\mathcal{Z}]$ for $\mathcal{Z}$ and to that end we assume that the primary market is arbitrage-free. A main idea is the following.

The derivative should be priced in a way that is *consistent* with the prices of the underlying assets. More precisely we should demand that the extended market $\Pi[\mathcal{Z}], S^0, S^1, \ldots, S^N$ is free of arbitrage possibilities.

We thus demand that there should exist a martingale measure Q^0 for the extended market $\Pi[\mathcal{Z}], S^0, S^1, \ldots, S^N$. This implies the following facts.

- The martingale measure Q^0 is an EMM also for the underlying market $S^0, \ldots, S^N$.
- The normalized price process $\Pi_t[\mathcal{Z}]/S_t^0$ is a Q^0-martingale.

Assuming enough integrability, and using the martingale property of the normalized derivatives price, we obtain

$$\frac{\Pi_t[\mathcal{Z}]}{S_t^0} = E^{Q^0}\left[\left.\frac{\Pi_T[\mathcal{Z}]}{S_T^0}\right|\mathcal{F}_t\right] = E^{Q^0}\left[\left.\frac{\mathcal{Z}}{S_T^0}\right|\mathcal{F}_t\right], \qquad (16.12)$$

where we have used the fact that, in order to avoid arbitrage at time T, we must have $\Pi_T[X] = X$. We thus have the following result.

Theorem 16.10 (General Pricing Formula)   *The arbitrage-free price process for the T-claim $\mathcal{Z}$ is given by*

$$\Pi_t[\mathcal{Z}] = S_t^0 E^{Q^0}\left[\left.\frac{\mathcal{Z}}{S_T^0}\right|\mathcal{F}_t\right], \qquad (16.13)$$

where Q^0 is the (not necessarily unique) martingale measure for the a priori given market $S^0, S^1, \ldots, S^N$, with S^0 as the numeraire.

Note that different choices of Q will generically give rise to different price processes. In particular we note that if we assume that if S^0 is the money account B, so

$$S_t^0 = S_0^0 \cdot e^{\int_0^t r(s)ds},$$

where r is the short rate, then (16.13) reduced to the familiar "risk-neutral valuation formula".

Theorem 16.11 (Risk-Neutral Valuation Formula) *Assuming the existence of a short rate, the pricing formula takes the form*

$$\Pi_t[\mathcal{Z}] = E^Q \left[e^{-\int_t^T r(s)ds} \mathcal{Z} \middle| \mathcal{F}_t \right]. \tag{16.14}$$

where Q is a (not necessarily unique) martingale measure with the money account as the numeraire.

The pricing formulas (16.13) and (16.14) are very nice, but it is clear that if there exists more than one martingale measure (for the chosen numeraire), then the formulas do not provide a unique arbitrage-free price for a given claim $\mathcal{Z}$. It is thus natural to ask under what conditions the martingale measure is unique; turns out to be closely linked to the possibility of hedging contingent claims.

16.4   Hedging

Consider a market model $S^0, \ldots, S^N$ and a contingent T-claim $\mathcal{Z}$.

Definition 16.12   If there exists a self-financing portfolio h such that the corresponding value process V^h satisfies the condition

$$V_T^h = \mathcal{Z}, \quad P\text{-a.s.}, \tag{16.15}$$

then we say that h **replicates** $\mathcal{Z}$, that h is a **hedge** against $\mathcal{Z}$, or that $\mathcal{Z}$ is **attained** by h. If, for every T, all T-claims can be replicated, then we say that the market is **complete**.

Given the hedging concept, we now have a second approach to pricing. Let let us assume that $\mathcal{Z}$ can be replicated by h. Since the holding of the derivative contract and the holding of the replicating portfolio are equivalent from a financial point of view, we see that price of the derivative must be given by the formula

$$\Pi_t[\mathcal{Z}] = V_t^h, \tag{16.16}$$

since otherwise there would be an arbitrage possibility (why?).

We now have two obvious problems.

- What will happen in a case when $\mathcal{Z}$ can be replicated by two different portfolios g and h?

• How is the formula (16.16) connected to the previous pricing formula (16.13)?

To answer these question, let us assume that the market is free of arbitrage and let us also suppose that, at the T-claim, X is replicated by the portfolios g and h. We choose the bank account B as the numeraire and consider a fixed martingale measure Q. Since Q is a martingale measure for the underlying market $S^0, \ldots, S^N$, it is easy to see that this implies that Q is also a martingale measure for V^g and V^h in the sense that V^h/B and V^g/B are Q-martingales. Using this we obtain

$$\frac{V_t^h}{B_t} = E^Q \left[\frac{V_T^h}{B_T} \bigg| \mathcal{F}_T \right]$$

and similarly for V^g. Since, by assumption, we have $V_T^h = Z$ we thus have

$$V_t^h = E^Q \left[Z \frac{B_t}{B_T} \bigg| \mathcal{F}_T \right],$$

which will hold for any replicating portfolio and for any martingale measure Q. Assuming absence of arbitrage we have thus proved the following.

• If Z is replicated by g and h, then

$$V_t^h = V_t^g, \quad t \geq 0.$$

• For an attainable claim, the value of the replicating portfolio coincides with the risk-neutral valuation formula, i.e.

$$V_t^h = E^Q \left[e^{-\int_t^T r_s \, ds} Z \bigg| \mathcal{F}_T \right].$$

From (16.14) it is obvious that every claim Z will have a unique price if and only if the martingale measure Q is unique. On the other hand, it follows from the alternative pricing formula (16.16) that there will exist a unique price for every claim if every claim can be replicated. The following result is therefore not surprising.

Theorem 16.13 (Second Fundamental Theorem) *Given a fixed numeraire S^0, the corresponding martingale measure Q^0 is unique if and only if the market is complete.*

Proof We have already seen above that if the market is complete, then the martingale measure is unique. The other implication is a very deep result: the reader is referred to Delbaen & Schachermayer (1994) for the proof. □

16.5 Heuristic Results

In this section we will provide a very useful and general rule-of-thumb which can be used to determine whether a certain model is complete and/or free of arbitrage. The arguments will be purely heuristic.

Let us consider a model with N traded underlying assets *plus* the risk-free asset (i.e. $N + 1$ assets in total). We assume that the price processes of the underlying assets are driven by R "random sources". We cannot give a precise definition of what constitutes a "random source" here, but the following informal rules will be enough for our purposes.

- Every independent Wiener process counts as one source of randomness. Thus, if we have five independent Wiener processes, then $R = 5$.
- Every independent Poisson process counts as one source of randomness. Thus, if we have five independent Wiener processes and three independent Poisson processes, then $R = 5 + 3 = 8$.
- If we have a driving point process with random jump size, then *every possible jump size* counts as one source of randomness. This implies in particular that if we have a driving marked point process with the real line as the mark space, then then $R = \infty$.

When discussing completeness and absence of arbitrage it is important to realize that these concepts work in opposite directions. Let the number of random sources R be fixed. Then every new underlying asset added to the model (without increasing R) will of course give us a potential opportunity of creating an arbitrage portfolio, so in order to have an arbitrage-free market the number N of underlying assets must be small in comparison to the number of random sources R.

On the other hand we see that every new underlying asset added to the model gives us new possibilities of replicating a given contingent claim, so completeness requires N to be large in comparison to R.

We cannot formulate and prove a precise result here, but the following rule is nevertheless extremely useful. In concrete cases it can in fact be given a precise formulation and a precise proof.

Rule-of-Thumb 16.14   *Let N denote the number of underlying* traded *assets in the model,* excluding *the risk-free asset, and let R denote the number of random sources. Generically we then have the following relations.*

1. *The model is arbitrage-free if and only if $N \le R$.*
2. *The model is complete if and only if $N \ge R$.*
3. *The model is complete and arbitrage-free if and only if $N = R$.*

As an example we take the Black–Scholes model, where we have one underlying asset S plus the risk-free asset so $M = 1$. We have one driving Wiener process, giving us $R = 1$, so in fact $M = R$. Using the Rule-of-Thumb above we thus expect the Black–Scholes model to be arbitrage-free as well as complete and this is indeed the case.

16.6   Change of Numeraire

In many applications it is very convenient to change from one numeraire asset price process, say S^0, to another, say S^1, while staying within the same pricing system. To study this let us consider an arbitrary T-claim $\mathcal{Z}$ and price it using S^0 and S^1 as numeraires.

Using (16.13) we obtain obtain

$$\Pi_0[Z] = S_0^0 E^{Q^0} \left[\frac{Z}{S_T^0} \right]$$

and

$$\Pi_0[Z] = S_0^1 E^{Q^1} \left[\frac{Z}{S_T^1} \right].$$

In a complete market the measures Q^0 and Q^1 are unique, but if the market is incomplete there are many martingale measures. We therefore assume (very reasonably) that the market uses a unique price system Π for all claims. If this is the case then we must have

$$S_0^1 E^{Q^1} \left[\frac{Z}{S_T^1} \right] = S_0^0 E^{Q^0} \left[\frac{Z}{S_T^0} \right]$$

and, introducing the likelihood process $L^{0,1}$ by

$$L_t^{0,1} = \frac{dQ^1}{dQ^0}, \quad \text{on } \mathcal{F}_t,$$

we obtain

$$E^{Q^0} \left[L_T^{0,1} S_0^1 \frac{Z}{S_T^1} \right] = E^{Q^0} \left[S_0^0 \frac{Z}{S_T^0} \right].$$

Since this holds for all $Z \in \mathcal{F}_T$ we must have

$$L_T^{0,1} = \frac{S_T^1}{S_T^0} \cdot \frac{S_0^0}{S_0^1},$$

and we have proved the following result.

Proposition 16.15   *The likelihood process $L^{0,1} = \frac{dQ^1}{dQ^0}$ is given by*

$$L_t^{0,1} = \frac{S_t^1}{S_t^0} \cdot \frac{S_0^0}{S_0^1}. \tag{16.17}$$

In particular, if $S^1 = S$ and $S^0 = B$ we use the notation $L_t^S = L_t^{0,1}$; we then have

$$L_t^S = \frac{S_t}{B_t} \cdot \frac{1}{S_0}. \tag{16.18}$$

16.7   Stochastic Discount Factors

In previous sections we have seen that we can price a contingent T-claim Z by using the formula

$$\Pi_0[Z] = E^Q \left[e^{-\int_t^T r_s \, ds} Z \right], \tag{16.19}$$

where Q is a martingale measure with the bank account as a numeraire. In many applications of the theory, in particular in equilibrium theory, it is very useful to write this expected value directly under the objective probability measure P instead of under Q. This can easily be obtained by using the likelihood process L, where a usual L is defined on the interval $[0, T]$ through

$$L_t = \frac{dQ}{dP}, \quad \text{on } \mathcal{F}_t. \tag{16.20}$$

We can then write (16.19) as

$$\Pi_0[\mathcal{Z}] = E^P\left[B_T^{-1} L_T \mathcal{Z}\right],$$

which naturally leads us to the following definition.

Definition 16.16   Assume the existence of a short rate process r. For any fixed martingale measure Q, let the likelihood process L be defined by (16.20). The **stochastic discount factor** (SDF) process **M**, corresponding to Q, is defined as

$$\mathbf{M}_t = B_t^{-1} L_t. \tag{16.21}$$

We thus see that there is a one-to-one correspondence between martingale measures and stochastic discount factors so we can write the pricing formula as

$$\Pi_0[\mathcal{Z}] = E^P[\mathbf{M}_T \mathcal{Z}].$$

This gives us the arbitrage-free price at $t = 0$ as a P-expectation, but the result can easily be generalized to arbitrary times t.

Proposition 16.17   *Assume absence of arbitrage. With notation as above, the following hold.*

- *For any sufficiently integrable T-claim $\mathcal{Z}$, the arbitrage-free price is given by*

$$\Pi_t[\mathcal{Z}] = \frac{1}{\mathbf{M}_t} E^P[\mathbf{M}_T \mathcal{Z} | \mathcal{F}_t]. \tag{16.22}$$

- *For any arbitrage-free asset price process S (derivative or underlying) the process*

$$\mathbf{M}_t S_t \tag{16.23}$$

  is a (local) P-martingale.

Proof   Use the abstract Bayes' formula.   □

Note   We observe that, as far as pricing is concerned we only need to know **M** up to a multiplicative constant. In finance and economics the expression

$$\frac{\mathbf{M}_t}{\mathbf{M}_s}$$

is often referred to as "the stochastic discount factor for the interval $[s, t]$". In discrete time it is natural to study the one-step SDF $\mathbf{m}_n$ defined as

$$\mathbf{m}_n = \frac{\mathbf{m}_n}{\mathbf{m}_{n-1}}.$$

In many applications, and in particular in equilibrium theory, it turns out that the dynamics of M are very important, so we will now derive these. Assume therefore that we have a filtered space $(\Omega, \mathcal{F}, P, \mathbf{F})$ carrying a d-dimensional Wiener process W and a marked point process $\Psi(dt, dz)$ with intensity measure $\lambda_t(dz)$. Assume furthermore that the filtration is the internal one, so $\mathbf{F} = \mathbf{F}^{W, \Psi}$. We then know from the martingale representation theorem that the likelihood P-dynamics have the form

$$dL_t = L_t \gamma_t^* dW_t + \int_E \varphi_t(z) \{\Psi(dt, dz) - \lambda_t(dz)dt\}, \qquad (16.24)$$

where γ is a column-vector process, $*$ denotes transpose and φ is $\mathcal{P}$-predictable. Applying the Itô formula to (16.21) give us the following easy but important result.

Proposition 16.18 *With a Girsanov transformation of the form (16.24), the P-dynamics of the stochastic discount factor M are given by*

$$dL_t = -r_t L_t dt + L_t \gamma_t^* dW_t + L_{t-} \int_E \varphi_t(z) \{\Psi(dt, dz) - \lambda_t(dz)dt\}. \qquad (16.25)$$

The point of this result is that it allows us to identify the Girsanov kernels γ and φ, as well as the short rate r from the M-dynamics. This technique will be used when we study dynamic market equilibrium models in Chapter 26.

Note Although SDFs and martingale measures are logically equivalent, it is often convenient to be able to switch from one to the other. In asset pricing and equilibrium theory it is often natural to use the SDF formalism, whereas it seems more convenient to use the language of martingale measures in connection with arbitrage theory and pricing of contingent claims.

16.8 Dividends

So far we have assumed that all assets are non-dividend paying. We now extend the theory and to that end we consider a market consisting of the usual bank account B as well as N risky assets. The novelty is that all assets are allowed to pay dividends so we need to formalize this idea. As usual we denote by S_t^i the price at time t of asset number i, and we denote by D^i the **cumulative dividend process** of asset number i. The interpretation is that D_t is the total amount of dividends that have been paid by holding one unit of the asset over the time period $[0, t]$. Intuitively this means that over an infinitesimal interval $[t, t + dt]$ the holder of the asset will obtain the amount $dD_t = D_{t+dt} - D_t$. For simplicity we also assume that the trajectory of D^i is continuous.

A market of this kind thus consist of the bank account $S^0 = B$ and a collection of **price dividend pairs** $(S^1, D^1), \ldots, (S^N, D^N)$. We now have the following extension of the previous theory. For details, see Björk (2020).

Definition 16.19 A **portfolio strategy** is an $(N + 1)$-dimensional predictable (row vector) process $h = (h^0, h^1, \ldots, h^N)$. For a given strategy h, the corresponding **value**

process V^h is defined by

$$V_t^h = \sum_{i=0}^{N} h_t^i S_t^i, \qquad (16.26)$$

The strategy is said to be **self-financing** if

$$dV_t^h = \sum_{i=0}^{N} h_t^i dG_t^i, \qquad (16.27)$$

where the **gain process** G is defined by

$$dG_t^i = dS_t^i + dD_t^i.$$

For a given strategy h, the corresponding **relative portfolio** $u = (u^0, \ldots, u^N)$ is defined by

$$u_t^i = \frac{h_t^i S_{t-}^i}{V_{t-}^h}, \qquad i = 0, \ldots, N. \qquad (16.28)$$

The interpretation of the self-financing condition (16.27) is clear. If you hold h units of an asset over the interval $[t - dt, t]$ then you gain (or lose) $h_t dS_t$, because of the price change, and you get $h_t dD_t$ in dividends.

Proposition 16.20 *If u is the relative portfolio corresponding to a self-financing portfolio h, then we have*

$$dV_t^h = V_{t-}^h \sum_{i=0}^{N} u_t^i \frac{dS_t^i + dD_t^i}{S_{t-}^i}. \qquad (16.29)$$

The definition of arbitrage is exactly as in the non-dividend case, but for the concept of a martingale measure we need a small variation.

Definition 16.21 An equivalent **martingale measure** is a probability measure Q with the properties that

1. Q is equivalent to P, i.e. $Q \sim P$.
2. The **normalized gain processes** $G_t^{Z0}, \ldots, G_t^{ZN}$, defined by

$$G_t^{Zi} = \frac{S_t^i}{B_t} + \int_0^t \frac{1}{B_s} dD_s^i, \qquad i = 0, \ldots, N,$$

are (local) martingales under Q.

The martingale property of G^Z has a very natural interpretation. For a price–dividend pair (S, D) we obtain, after some reshuffling of terms.

$$S_t = E^Q \left[\int_t^T e^{-\int_t^s r_u du} dD_s + e^{-\int_t^T r_s ds} S_T \,\middle|\, \mathcal{F}_t \right].$$

This is a risk-neutral valuation formula, which says that the stock price at time t equals the arbitrage-free value at t of the stock price S_T plus the sum (or rather integral) of the arbitrage-free value at t of all dividends over the interval $[t, T]$.

In the present setting the first fundamental theorem takes the following form.

Theorem 16.22 (First Fundamental Theorem)   *The market model is free of arbitrage possibilities if and only if there exists a martingale measure Q.*

The pricing formulas are as expected.

Proposition 16.23   *The arbitrage-free price of a T-claim $\mathcal{Z}$ is given by*

$$\Pi_t[\mathcal{Z}] = E^Q \left[e^{-\int_t^T r_s \, ds} \mathcal{Z} \middle| \mathcal{F}_t \right].$$

16.9   Consumption

We now introduce consumption into the theory. For simplicity we will only consider consumption programs which can be represented in terms of a **consumption rate process** c. The interpretation is that over an infinitesimal interval $[t, t + dt]$ the economic agent consumes the amount $c_t dt$, so the dimension of c_t is "consumption per unit time". We define portfolios $h^0, \ldots, h^N$, portfolio value V and martingale measure, as in the previous section.

Given a market of the form $B, (S^1, D^1), \ldots, (S^N, D^N)$ we say that a portfolio h is self-financing for the consumption process c if

$$dV_t^h = \sum_{i=0}^N h_t^i \left\{ dS_t^i + dD_t^i \right\} - c_t dt,$$

or, in terms of relative weights

$$dV_t^h = V_{t-}^h \sum_{i=0}^N u_t^i \frac{dS_t^i + dD_t^i}{S_{t-}^i} - c_t dt.$$

The pricing theory for contingent claims can easily be extended to include consumption.

Proposition 16.24   *The arbitrage-free price at time t for the consumption process c restricted to the interval $[t, T]$ is given by*

$$\Pi_t[c] = E^Q \left[\int_t^T e^{-\int_t^T r_s \, ds} c_s \, ds \middle| \mathcal{F}_t \right].$$

Using the stochastic discount factor $M = B^{-1} L$ we can write this as

$$\Pi_t[c] = \frac{1}{M_t} E^P \left[\int_t^T M_s c_s \, ds \middle| \mathcal{F}_t \right].$$

Furthermore, if the market is complete, then every consumption stream can be replicated by a self-financing portfolio.

16.10   Replicating a Consumption Process

Consider a market $B,(S^1,D^1),\ldots,(S^N,D^N)$ and a consumption process c. If the market is complete, then it follows from Proposition 16.24 that c can be replicated with a self-financing portfolio. This result, however, is an abstract existence result, but in a concrete case it will not tell us what the replicating portfolio looks like. We will now present a proposition which can be used to construct the replicating portfolio.

We start with a small lemma which show that, as could be expected, the property of being self-financing is invariant under normalization.

Lemma 16.25 (Invariance Lemma)   *Consider a market as above, a consumption process c and a portfolio process $h = (h^0,h^1,\ldots,h^N)$. Let V denote the corresponding value process. Then it holds that h is self-financing for c if and only if*

$$dV_t^Z = \sum_{i=1}^N h_t^i dG_t^{Zi} - c_t^z dt,$$

where the normalized value process V^Z and the normalized consumption process c^Z are defined by

$$V_t^Z = \frac{V_t}{B_t},$$

$$c_t^z = \frac{c_t}{B_t}.$$

Proof   The proof is left to the reader as an exercise.   □

Proposition 16.26   *Consider the market $B,(S^1,D^1),\ldots,(S^N,D^N)$ and a consumption process c. Define the process K by*

$$K_t = E^Q\left[\int_0^T c_s^z ds \,\middle|\, \mathcal{F}_t\right],$$

with c^z as in the lemma above. Assume that there exist predictable processes $h^1,\ldots,h^N$ such that

$$dK_t = \sum_{i=1}^N h_t^i dG_t^{zi}.$$

Then c can be replicated by the portfolio $h^0,h^1,\ldots,h^N$, with $h^1,\ldots,h^N$ as above and h^0 defined by

$$h_t^0 = K_t - \sum_{i=1}^N h_t^i Z_t^i - \int_0^t c_s^z ds.$$

Proof   The normalized portfolio value of $h^0,h^1,\ldots,h^N$ is, by definition, given by

$$V_t^Z = h_0 \cdot 1 + \sum_{i=}^N h_t^i Z_t^i$$

so, using the definition of h^0, we have

$$V_t^Z = K_t - \int_0^t c_s^z ds.$$

From the assumption $dK_t = \sum_i h_t^i dG_t^{zi}$ we thus obtain

$$dV_t^Z = \sum_i h_t^i dG_t^{zi} - c_t^z dt.$$

This shows that h is self-financing in the normalized economy and, by Lemma 16.25, it is also self-financing in nominal terms.   □

16.11   Exercise

Exercise 16.1   Prove Lemma 16.25.

16.12   Notes

The martingale approach to arbitrage pricing was developed in Harrison & Kreps (1979), Kreps (1981) and Harrison & Pliska (1981). It was then extended by Duffie & Huang (1986) and Delbaen & Schachermayer (1994), among others. Stochastic discount factors are treated in most textbooks on asset pricing, for example Cochrane (2001) and Duffie (2001).

17 Poisson-Driven Stock Prices

17.1 Introduction

In this chapter and the next we will study arbitrage pricing in two concrete and very simple models. We do this as a warm up exercise, so the reader who wants to study a more general model may progress directly to Chapter 19. The reason for studying these simple models is that, in these cases, we can connect the abstract martingale methods to the classical "delta-hedging approach" used by Black, Scholes and Merton, and others, during the 1970s. We mainly use these models as laboratory examples, but using the delta-hedging technique is in fact quite instructive (although quite messy).

The first case is very similar to the standard Black–Scholes model, the only difference being that while the stock price in the Black–Scholes model is driven by a Wiener process, the stock price in our model will be driven by a Poisson process. The Wiener-driven and the Poisson-driven models are structurally very close, so we hope the reader will recognize concepts and techniques from the Wiener case. Exactly as in that case, we have three different methods for pricing financial derivatives.

1. Construction of locally risk-free portfolios.
2. Construction of replicating portfolios.
3. Construction of equivalent martingale measures.

As one may expect, the martingale approach is the most general one, but it is still very instructive (and a good exercise) to see how far it is possible to go using the first two techniques above.

As usual we consider a filtered space $(\Omega, \mathcal{F}, P, \mathbf{F})$ and we assume that the space carries a Poisson process N with constant intensity λ. The filtration is the internal one generated by N. The market we will study is very simple. It consists of two assets, namely a risky asset with price process S and the usual bank account B. Given known constants α, β and r, the dynamics are as follows:

$$dS_t = \alpha S_{t-} dt + \beta S_{t-} dN_t, \tag{17.1}$$

$$dB_t = r B_t dt. \tag{17.2}$$

In order to have an economic interpretation of the stock price dynamics (17.1) we note that *between jumps* the price evolves according to the deterministic ODE

$$\frac{dS_t}{dt} = \alpha S_t,$$

so between jumps the stock price grows exponentially with the factor α. We thus see that the constant α is the local mean rate of return of the stock *between jumps*.

The question is now to get a grip on the overall mean rate of return and in order to do this we recall that if we define the process M by

$$M_t = N_t - \lambda t, \tag{17.3}$$

then M is a martingale and we can write (17.3) as

$$dN_t = \lambda dt + dM_t. \tag{17.4}$$

If we now plug (17.4) into (17.1) we obtain, after some reshuffling,

$$dS_t = S_{t-}(\alpha + \beta\lambda)\,dt + \beta S_{t-}dM_t, \tag{17.5}$$

and we see that the overall mean rate of return of the stock, *including jumps* is given by $\alpha + \beta\lambda$.

We see that we have two equivalent ways of viewing the stock price dynamics, since can write the dynamics either as (17.1) or as (17.5). From a probabilistic point of view, (17.5) is the most natural one, since it decomposes the dynamics into a predictable *drift* part (the dt term) and a *martingale* part (the dM term). This is known as "the semimartingale decomposition" of the S-dynamics. The representation (17.1), on the other hand, is often easier to use when we want to apply the Itô formula. As we will see below we will often switch between the two representations.

If we now move to the dN term in (17.1) we see that if N has a jump at time t, then the induced jump size of S is given by

$$\Delta S_t = \beta S_{t-},$$

so β is the **relative jump size** of the stock price; we will sometimes refer to β as the "jump volatility" of S. We also see that the sign of β determines the sign of the jump: assuming $S_t > 0$, if $\beta > 0$ then all jumps are upwards whereas if $\beta < 0$ all jump are downwards. In particular we see that if $\beta = -1$ then, if there is a jump at t, we obtain $\Delta S_t = (-1) \cdot S_{t-}$, i.e.

$$S_t = S_{t-} + \Delta S_t = S_{t-} - S_{t-} = 0.$$

In other words, if $\beta = -1$, then the stock price will jump to zero at the first jump of N (and the stock price will stay forever at the value zero). This indicates that we may use counting processes in order to model bankruptcy phenomena.

We can now collect our findings so far.

Proposition 17.1 *If S is given by (17.1) we have the following interpretation.*

- *The constant α is the local mean rate of return of the stock between jumps.*
- *Denoting the overall mean rate of return of the stock, including jumps under the measure P by μ^P, we have*

$$\mu^P = \alpha + \beta\lambda. \tag{17.6}$$

- *The relative jump size is given by β.*

Before we go on to pricing in this simple model, let us informally discuss conditions of no arbitrage.

Let us first assume that $\beta > 0$ and that $S_0 > 0$. Then all jumps are positive and it is clear that a necessary condition for no arbitrage is that $r > \alpha$, since otherwise the stock return would dominate the return of the bank between jumps, and dominate even more at a jump time (because of the positive jumps). In other words, if $r < \alpha$ then we would have an arbitrage by borrowing in the bank and investing in the stock.

If, on the other hand, $\beta < 0$, then all jumps are negative, and a necessary condition for no arbitrage is that $\alpha > r$, since otherwise the bank would dominate the stock between jumps and even more so at a jump time. We can summarize the findings as follows.

Proposition 17.2   *A necessary condition for absence of arbitrage is given by*

$$\frac{r - \alpha}{\beta} > 0. \tag{17.7}$$

17.2   The Classical Approach to Pricing

We now turn the the problem of pricing derivatives in the model above. In this section we will basically follow the "classical" Black–Scholes delta-hedging methodology and to this end we consider a contingent T-claim X of the form

$$X = \Phi(S_T),$$

where Φ is some given contract function. A typical example would be a European call option with exercise date T and strike price K, in which case Φ would have the form

$$\Phi(s) = \max\left[s - K, 0\right].$$

We now assume that the derivative is traded on a liquid market and that the price $\Pi_t[X]$ is of the form

$$\Pi_t[X] = F(t, S_t),$$

for some smooth function $F(t, s)$. Our job is to find out what the pricing function F must look like in order to avoid arbitrage on the extended market (S, B, F). To this end we carry out the following program.

1. Form a self-financing portfolio based on the stock S and the derivative F and denote the corresponding value process of the portfolio by V.
2. Choose the portfolio weights such that the dN terms in the V-dynamics cancel.
3. The V-dynamics will then be of the form

$$dV_t = V_t k_t dt,$$

for some random process k.
4. Thus V is a *risk-free* portfolio and in order to avoid arbitrage possibilities between V and the bank account we must, according to Proposition 16.8 have the equation

$$k_t = r, \quad t \geq 0.$$

5. This equation turns out to be a PIDE for determining the pricing function F.

We now go on to carry out this small program. Denoting the relative weights on the underlying stock and the derivative by u^S and u^F respectively, we have the portfolio dynamics

$$dV_t = V_{t-} \left\{ u_t^S \frac{dS_t}{S_{t-}} + u_t^F \frac{dF(t, S_t)}{F(t, S_{t-})} \right\},$$

which we write more compactly as

$$dV = V^- \left\{ u^S \frac{dS}{S^-} + u^F \frac{dF}{F^-} \right\},$$

where S^- is shorthand for S_{t-} etc. From the Itô formula we immediately have

$$dF(t, S_t) = \left\{ \frac{\partial F}{\partial t}(t, S_t) + \alpha S_t \frac{\partial F}{\partial s}(t, S_t) \right\} dt + F_\beta(t, S_{t-}) dN_t,$$

where

$$F_\beta(t, s) = F(t, s + \beta s) - F(t, s),$$

and we can rewrite this in shorthand as

$$dF = \alpha_F F dt + \beta_F^- F^- dN_t,.$$

where

$$\alpha_F(t, s) = \frac{\frac{\partial F}{\partial t}(t, s) + \alpha s \frac{\partial F}{\partial s}(t, s)}{F(t, s)} \qquad \beta_F(t, s) = \frac{F_\beta(t, s)}{F(t, s)}.$$

With this notation the portfolio dynamics takes the form

$$dV = V^- \left\{ u^S (\alpha dt + \beta dN) + u^F (\alpha_F dt + \beta_F^- dN) \right\},$$

or alternatively

$$dV = V^- \left\{ u^S \alpha + \alpha_F \right\} dt + V^- \left\{ u^S \beta + u^F \beta_F^- \right\} dN.$$

We thus see that if we choose the relative portfolio u such that

$$u^S \beta + u^F \beta_F^- = 0,$$

then the driving Poisson noise in the portfolio dynamics will vanish. Recalling the the portfolio weights must sum to unity, we thus define the relative portfolio by the system

$$u^S \beta + u^F \beta_F = 0$$
$$u^S + u^F = 1.$$

This is a simple 2×2 system of linear equations with the solution

$$u^S = \frac{\beta_F^-}{\beta_F^- - \beta},$$

$$u^F = -\frac{\beta}{\beta_F^- - \beta}.$$

Using this portfolio, the V-dynamics take the form

$$dV = V^- \left\{ \frac{\alpha \beta_F}{\beta_F - \beta} - \frac{\alpha_F \beta}{\beta_F - \beta} \right\} dt,$$

which represents the dynamics of a risk-free asset. From Proposition 16.8 we thus see that, in order to avoid arbitrage, the condition

$$\frac{\alpha \beta_F}{\beta_F - \beta} - \frac{\alpha_F \beta}{\beta_F - \beta} = r,$$

must be satisfied P-a.s. Substituting the definitions for α_F and β_F and reshuffling this equation will finally give us the equation

$$\frac{\partial F}{\partial t} + \alpha s \frac{\partial F}{\partial s} + \frac{r - \alpha}{\beta} F_\beta - rF = 0.$$

This is the required no-arbitrage condition for the pricing function F. Recalling that we have the obvious (why?) boundary condition $F(T, s) = \Phi(s)$, we have our first main pricing result.

Proposition 17.3 *Consider the pure Poisson model (17.1)–(17.2) and a T-claim X of the form $X = \Phi(S_T)$. Assume that the price process $\Pi_t[X]$ is of the form $\Pi_t[X] = F(t, S_t)$. Then, in order to avoid arbitrage, the pricing function F must satisfy the following PIDE on the time interval $[0, T]$:*

$$\begin{cases} \dfrac{\partial F}{\partial t}(t, s) + \alpha s \dfrac{\partial F}{\partial s}(t, s) + \dfrac{r - \alpha}{\beta} F_\beta(t, s) - rF(t, s) &= 0, \\[2mm] F(T, s) &= \Phi(s), \end{cases} \tag{17.8}$$

where F_β is defined by

$$F_\beta(t, s) = F(t, s + \beta s) - F(t, s). \tag{17.9}$$

Comparing the PIDE above with our results in Chapter 6 we see (with great satisfaction) that it is precisely of the form which allows for a Feynman–Kac representation. Using the Feynman–Kac theorem 6.8 we thus obtain

$$F(t, s) = e^{-r(T - t)} E^Q_{t,s} [\Phi(S_T)],$$

where S has dynamics

$$dS_t = \alpha S_t dt + \beta S_t dN_t, \tag{17.10}$$

and N is Poisson with intensity $(r - \alpha)/\beta$ under the measure Q. This looks quite nice, but at this point we have to be a bit careful since, in order to apply the relevant Feynman–Kac theorem, we need to assume that the condition

$$\frac{r - \alpha}{\beta} > 0 \tag{17.11}$$

is satisfied – for otherwise we are dealing with a Poisson process with negative intensity and such animals do not exist. This condition is, however, exactly the necessary condition for absence of arbitrage that we encountered in Proposition 17.2. We can summarize our finding as follows.

Proposition 17.4 *Consider the pure Poisson model (17.1)–(17.2), where N is Poisson with constant intensity λ under the objective measure P, and where we assume that the no-arbitrage condition (17.11) is satisfied. Consider a T claim X of the form $X = \Phi(S_T)$ and assume that the price process $\Pi_t[X]$ is of the form $\Pi_t[X] = F(t, S_t)$. Absence of arbitrage will then imply that F has the representation*

$$F(t, s) = e^{-r(T-t)} E_{t,s}^{Q} [\Phi(S_T)], \tag{17.12}$$

where the S-dynamics are given by (17.2), but where the process N under the measure Q is Poisson with intensity

$$\lambda^{Q} = \frac{r - \alpha}{\beta}.$$

An explicit formula for F is given by

$$F(t, s) = e^{-r(T-t)} \sum_{n=0}^{\infty} \Phi\left(s(1 + \beta)^n e^{\alpha(T-t)}\right) \frac{(r - \alpha)^n (T - t)^n}{\beta^n n!} e^{-\frac{r-\alpha}{\beta}(T-t)}. \tag{17.13}$$

We end this section by discussing how F depends on the various model parameters. The most striking fact, which we see from Propositions 17.3 and 17.4 is that while F depends on the parameters α, β and r, it does *not* depend on the parameter λ, which is the Poisson intensity λ under the objective measure P. This is the point-process version of the fact that in a Wiener-driven model, the pricing function does not depend on the local mean rate of return.

We also see that the dynamics of S is given by

$$dS_t = \alpha S_t dt + \beta S_{t-} dN_t,$$

under the objective measure P as well as under the martingale measure Q. The difference between P and Q is that while N is Poisson with intensity λ under P, it is Poisson with intensity $\lambda^{Q} = (r - \alpha)/\beta$ under Q.

We can also easily compute the local rate of return of S under Q. We can write

$$dN_t = \left(\frac{r - \alpha}{\beta}\right) dt + dM_t^{Q},$$

where M^{Q} is a Q-martingale. Inserting this into the S-dynamics above gives us

$$dS_t = r S_t dt + \beta S_{t-} dM_t^{Q},$$

which shows that the local mean rate of return under Q equals the short rate r.

This shows, as was expected, that Q is a risk-neutral martingale measure for S, i.e. that the process S/B is a Q-martingale. It is also easy to see that F/S is a Q-martingale.

17.3 The Martingale Approach to Pricing

In this section we will study in terms of martingales the simple Poisson market described above. We recall the P-dynamics of the stock price:

$$dS_t = \alpha S_t dt + \beta S_{t-} dN_t. \tag{17.14}$$

Our first task is to find the relevant no-arbitrage conditions. By elementary arguments we have already derived the *necessary* condition (17.11); we now want to find a *sufficient* condition too, and to this end we now determine the class of equivalent martingale measures (with the bank account as numeraire) for our market model on the compact interval $[0,T]$.

Since the filtration is the internal one, we know from the converse of the Girsanov theorem that every measure $Q \sim P$, regardless of whether Q is a martingale measure or not, is obtained by a Girsanov transformation of the form

$$L_t = \frac{dQ}{dP}, \quad \text{on } \mathcal{F}_t, \quad 0 \le t \le T,$$

where the likelihood process L has the dynamics

$$\begin{cases} dL_t &= L_{t-}\varphi_t \{dN_t - \lambda_t dt\}, \\ L_0 &= 1, \end{cases}$$

for some predictable process φ with $\varphi > -1$. From the Girsanov theorem we know that the Q-intensity of N is given by

$$\lambda_t^Q = (1 + \varphi_t)\lambda,$$

so we can write

$$dN_t = (1 + \varphi_t)\lambda dt + dM_t^Q,$$

where M^Q is a Q-martingale. Substituting this into the S-dynamics gives us the semi-martingale decomposition of S under Q as

$$dS_t = S_t \{\alpha + \beta(1 + \varphi_t)\lambda\} dt + \beta S_{t-} dM_t^Q,$$

so the local mean rate of return under Q is given by

$$\mu_t^Q = \alpha + \beta(1 + \varphi_t)\lambda.$$

A martingale measure with B as numeraire is characterized by the fact that $\mu_t^Q = r$, so Q is a risk-neutral martingale measure if and only if

$$\alpha + \beta(1 + \varphi_t)\lambda = r. \tag{17.15}$$

The first fundamental theorem now says that the market model is free of arbitrage if and only if there exists an equivalent martingale measure, so we see that we have absence of arbitrage if and only if equation (17.15) has a solution φ such that $\varphi_t > -1$. Since (17.15) has the simple solution

$$\varphi_t = \frac{r - \alpha}{\lambda \beta} - 1,$$

and $\lambda > 0$ we have more or less proved the following result.

Proposition 17.5 *The pure Poisson market model (17.1)–(17.2) is free of arbitrage if and only if the condition*

$$\frac{r - \alpha}{\beta} > 0, \tag{17.16}$$

is satisfied. If the condition is satisfied, then the market is also complete and the process N is Poisson under Q with intensity given by

$$\lambda_t^Q = \frac{r - \alpha}{\beta}. \tag{17.17}$$

Given this result, pricing of derivatives is very easy and we have the following.

Proposition 17.6 *Assume that the no-arbitrage condition* (17.16) *is satisfied and consider any contingent T-claim X. Then X will have a unique arbitrage-free price process $\Pi_t[X]$ given by*

$$\Pi_t[X] = e^{-r(T-t)} E^Q[X|\mathcal{F}_t]. \tag{17.18}$$

Furthermore, if X is of the form $X = \Phi(S_T)$ for some deterministic contract function Φ, then we have

$$\Pi_t[X] = F(t, S_t),$$

where the pricing function F satisfies the PIDE

$$\begin{cases} \dfrac{\partial F}{\partial t}(t,s) + \alpha s \dfrac{\partial F}{\partial s}(t,s) + \dfrac{r - \alpha}{\beta} F_\beta(t,s) - rF(t,s) &= 0, \\ F(T,s) &= \Phi(s), \end{cases} \tag{17.19}$$

Proof The risk-neutral valuation formula (17.18) is just a special case of Proposition 16.11. If X is of the form $X = \Phi(S_T)$ then, since S is Markov under Q, we can write

$$e^{-r(T-t)} E^Q[X|\mathcal{F}_t] = e^{-r(T-t)} E^Q[X|S_t] = F(t, S_t),$$

and the PIDE (17.19) is the Kolmogorov backward equation. □

17.4 The Martingale Approach to Hedging

As we saw in the previous section, the martingale measure is unique, so the second fundamental theorem guarantees that the Poisson model is complete. In this simple case we can in fact also provide a self-contained proof of market completeness. Let us thus consider a T-claim X.

Our formal job is to construct three processes, V, u^B and u^S such that the following hold.

- The (prospective) weights u^B and u^S are predictable and sum to unity, i.e.

$$u_t^B + u_t^S = 1.$$

- The V-dynamics have the form

$$dV_t = V_{t-} \left\{ u_t^S \frac{dS_t}{S_{t-}} + u_t^B \frac{dB_t}{B_{t-}} \right\}. \tag{17.20}$$

- V replicates X at maturity, i.e.

$$V_T = X.$$

In order to do this we consider the associated price process

$$\Pi_t[X] = e^{-r(T-t)} E^Q[X|\mathcal{F}_t].$$ (17.21)

Furthermore, we hope that for the replicating relative portfolio u we will have

$$V_t^u = \Pi_t[X].$$

The idea is now to use a martingale representation result to obtain the dynamics for $\Pi_t[X]$ from (17.21), compare these dynamics with (17.20) and thus identify the portfolio weights.

We start by noting that $\Pi_t[X]$, henceforth abbreviated as π_t, can be written as

$$\pi_t = e^{-r(T-t)} X_t,$$

where X is defined by

$$X_t = E^Q[X|\mathcal{F}_t].$$

It is clear that X is a Q-martingale, so by a slight variation of the martingale representation theorem 5.1 we deduce the existence of a predictable process g such that

$$dX_t = g_t X_{t-} \{dN_t - \lambda^Q dt\},$$ (17.22)

where

$$\lambda^Q = \frac{r-\alpha}{\beta}.$$ (17.23)

From the product formula we then obtain, after some reshuffling of terms,

$$d\pi_t = \pi_{t-} \{r - g_t \lambda^Q\} dt + g_t \pi_{t-} dN_t.$$ (17.24)

The V-dynamics above can be written in more detail as

$$dV_t = V_{t-} \{u_t^S \alpha + u_t^B r\} dt + V_{t-} u_t^S \beta dN_t,$$ (17.25)

and, comparing (17.24) with (17.25), we can now identify u^S from the dN term. More formally, let us define u^S by

$$u_t^S = \frac{g_t}{\beta}.$$

We can then write (17.24) as

$$d\pi_t = \pi_{t-} \left\{r - g_t \frac{r-\alpha}{\beta}\right\} dt + \pi_{t-} u_t^S \beta dN_t,$$

i.e.

$$d\pi_t = \pi_{t-} \{u_t^S \alpha + (1 - u_t^S)r\} dt + \pi_{t-} u_t^S \beta dN_t.$$

Then we see that if we define u^B by

$$u_t^B = 1 - u_t^S,$$

and define the process V by $V_t = \pi_t$ we obtain

$$dV_t = V_{t-} \{u_t^S \alpha + u_t^B r\} dt + V_{t-} u_t^S \beta dN_t,$$

which are the dynamics of a self-financing portfolio with weights u^B and u^S (summing to unity). We then obviously have $V_T = X$ and so have proved the following.

Proposition 17.7 *The model is complete and the replicating portfolio weights are given by*

$$u_t^S = \frac{g_t}{\beta},$$ (17.26)

$$u_t^B = 1 - u_t^S,$$ (17.27)

where g is given by the martingale representation theorem in (17.22).

18 The Simplest Jump-Diffusion Model

18.1 Introduction

In this chapter we will discuss the simplest possible jump-diffusion model for a financial market. We will use both the Black–Scholes classical hedging as well as the martingale argument. The calculations will sometimes be somewhat messy and from a logical development point of view this entire chapter can be skipped. In the next chapter we will consider a much more general model and derive much more general results with much less effort. The point of the present chapter is thus not one of logic, but one of enhancing economic intuition.

Formally we consider a stochastic basis $(\Omega, \mathcal{F}, P, \mathbf{F})$ carrying a standard Wiener process W as well as an independent Poisson process N with constant P-intensity λ. The market consists of a risky asset S and a bank account B with constant short rate. The asset dynamics are given by

$$dS_t = \alpha S_t dt + \sigma S_t dW_t + \beta S_{t-} dN_t, \tag{18.1}$$
$$dB_t = r B_t dt, \tag{18.2}$$

where α, σ, β and the short rate r are known constants.

We can interpret α as the local mean rate of return *between jumps*, and β as the relative jump size. In particular this implies that if S is the price of a common stock and thus non-negative, then we must have $\beta \geq -1$. To obtain the rate or return including jumps we write the S-dynamics in semimartingale form as

$$dS_t = (\alpha + \lambda\beta) S_t dt + \sigma S_t dW_t + \beta S_{t-} [dN_t - \lambda dt]. \tag{18.3}$$

Since the terms dW and $dN_t - \lambda dt$ are martingale increments, this implies that the local mean rate of return *including jumps* under P is given by

$$\mu^P = \alpha + \beta\lambda.$$

Before we go on to concrete calculations let us informally discuss what we may expect from the model above. Referring to the Rule-of-Thumb 16.14 we see that we have one risky asset S, so $N = 1$, and two sources of randomness, W and N, so $R = 2$. From the Rule-of-Thumb we thus conclude that we may expect the model to be arbitrage-free but *not complete*. In particular we should expect that the martingale measure is not unique, that there will not be unique arbitrage-free prices for financial derivatives, and that it will not be possible to form a risk-free portfolio based on a derivative and the underlying asset.

18.2 Classical Techniques

We now go on to study derivatives pricing using the classical methodology of risk-free portfolios developed by Black & Scholes (1973) and Merton (1973). To this end we consider a T-claim X of the form

$$X = \Phi(S_T),$$

and we assume that this derivative asset is traded on a liquid market with a price process of the form

$$\Pi_t[X] = F(t, S_t).$$

Our job is to see what we can say about the pricing function F, given the requirement of an arbitrage-free market, and the idea is, like in the previous chapter, roughly as follows.

- We form a portfolio based on S and F.
- We try to choose the portfolio weights such that the portfolio becomes risk-free, i.e. such that the corresponding value process V has dynamics of the form

$$dV_t = k_t V_t dt,$$

  with no driving noise terms.
- If this can be done, then absence of arbitrage implies that $k_t = r$ for all t, and this should give us some information about the pricing function F.

To carry out this program we need the price dynamics of the derivative. From the Itô formula we have

$$dF = \alpha_F F dt + \sigma_F F dW_t + \beta_F^- F^- dN_t, \tag{18.4}$$

where the uppercase index, as in F^-, denotes evaluation at $(t-, S_{t-})$. The coefficients are given by

$$\alpha_F(t, s) = \frac{F_t(t, s) + \alpha s F_s(t, s) + \frac{1}{2}\sigma^2 s^2 F_{ss}(t, s)}{F(t, s)}, \tag{18.5}$$

$$\sigma_F(t, s) = \frac{\sigma s F_s(t, s)}{F(t, s)}, \tag{18.6}$$

$$\beta_F(t, s) = \frac{F_\beta(t, s)}{F(t, s)}. \tag{18.7}$$

Here we have used the notation $F_t = \frac{\partial F}{\partial t}$ and similarly for F_s and F_{ss}. The function F_β is given by

$$F_\beta(t, s) = F(t, s + \beta s) - F(t, s). \tag{18.8}$$

If we now form a self-financing portfolio based on S and F, we obtain the following dynamics of the value process V.

$$dV = V^- \left\{ u^S \frac{dS}{S^-} + u^F \frac{dF}{F^-} \right\}.$$

Inserting the expression for dF above and collecting terms we obtain

$$dV = V^- \left\{ u^S \alpha + u^F \alpha_F^- \right\} dt + V^- \left\{ u^S \sigma + u^F \sigma_F^- \right\} dW$$
$$+ V^- \left\{ u^S \beta + u^F \beta_F^- \right\} dN.$$

We now want to balance the portfolio in such a way that it becomes locally risk-free, i.e. we want to choose the portfolio weights u^S and u^F such that the dW and the dN terms vanish. Recalling that the weights must sum to unity we then have the following system of equations

$$u^S \sigma + u^F \sigma_F^- = 0,$$
$$u^S \beta + u^F \beta_F^- = 0,$$
$$u^S + u^F = 1.$$

This system is, however, overdetermined since we have two unknowns and three equations, so in general it will not have a solution.

The economic reason for this is clear. If we want to hedge the derivative by using the underlying asset, then we have only one instrument (the stock) to hedge two sources of randomness (W and N).

We can summarize the situation as follows.

- The price of a particular derivative Φ will *not* be completely determined by the specification of the S-dynamics and the requirement that the market (B, S, F) is free of arbitrage.
- The reason for this fact is that arbitrage pricing is always a case of pricing a derivative *in terms of* the price of some underlying assets. In our market we do not have sufficiently many underlying assets.

Thus we will not obtain a unique price of a particular derivative. This fact does not mean, however, that prices of various derivatives can take any form whatsoever. From the discussion above we see that the reason for the incompleteness is that we do not have enough underlying assets, so if we adjoin one more asset to the market, without introducing any new Wiener or Poisson processes, then we expect the market to be complete. This idea can be expressed in the following ways.

- We cannot say very much about the price of any **particular** derivative.
- The requirement of an arbitrage-free derivative market implies that *prices of different derivatives* (i.e. claims with different contract functions or different times of expiration) will have to satisfy certain *internal consistency relations* in order to avoid arbitrage possibilities on the derivatives market.

We now go on to investigate these internal consistency relations so we assume that, apart from the claim $X = \Phi(S_T)$, there is also another T-claim, $X = \Gamma(S_T)$, traded on the market. We further suppose that the price of X is of the form

$$\Pi_t[X] = G(t, S_t),$$

for some pricing function G. We assume too that the market (B, S, F, G) is free of arbitrage so form a portfolio based on these assets, with predictable weights u^B, u^S, u^F and u^G. Since the weights must sum to unity we can write

$$u^B = 1 - u^S - u^F - u^G,$$

where u^S, u^F and u^G can be chosen without constraints. The corresponding value dynamics are then given by

$$dV = V^- \left\{ (1 - u^S - u^F - u^G)\frac{dB}{B} + u^S\frac{dS}{S^-} + u^F\frac{dF}{F^-} + u^G\frac{dG}{G^-} \right\}.$$

The differential dF is already given by (18.4)–(18.7), and we will of course have exactly the same structure for the differential dG. Collecting the various terms, we obtain

$$dV = V^- \left\{ r + (\alpha - r)u^S + (\alpha_F - r)u^F + (\alpha_G - r)u^G \right\} dt$$
$$+ V^- \left\{ \sigma u^S + \sigma_F u^F + \sigma_G u^G \right\} dW_t$$
$$+ V^- \left\{ \beta u^S + \beta_F u^F + \beta_G u^G \right\} dN_t.$$

If we now choose the weights such that the dW and dN terms vanish we obtain the system

$$\sigma u^S + \sigma_F u^F + \sigma_G u^G = 0,$$
$$\beta u^S + \beta_F u^F + \beta_G u^G = 0.$$

With such a a choice of weights, the portfolio becomes locally risk-free, and absence of arbitrage now implies that we must also have

$$r + (\alpha - r)u^S + (\alpha_F - r)u^F + (\alpha_G - r)u^G = r$$

or, equivalently,

$$(\alpha - r)u^S + (\alpha_F - r)u^F + (\alpha_G - r)u^G = 0.$$

The result of all this is that absence of arbitrage on the derivatives market implies that the system

$$(\alpha - r)u^S + (\alpha_F - r)u^F + (\alpha_G - r)u^G = 0,$$
$$\sigma u^S + \sigma_F u^F + \sigma_G u^G = 0,$$
$$\beta u^S + \beta_F u^F + \beta_G u^G = 0$$

admits a *non-trivial* solution. (The trivial solution $u^S = u^F = u^G = 0$ corresponds to putting all the money in the bank.) This is equivalent to saying that the coefficient matrix

$$\begin{bmatrix} \alpha - r & \alpha_F - r & \alpha_G - r \\ \sigma & \sigma_F & \sigma_G \\ \beta & \beta_F & \beta_G \end{bmatrix}$$

is singular. This, in turn, implies that the rows must be linearly dependent, so there must exist functions $\varphi_0(t, s)$ (the notation will become clear in the next section) and $\gamma_m(t, s)$ such that

$$\alpha - r = \gamma_m \sigma + \varphi_0 \beta, \tag{18.9}$$

$$\alpha_F - r = \gamma_m \sigma_F + \varphi_0 \beta_F, \tag{18.10}$$

$$\alpha_G - r = \gamma_m \sigma_G + \varphi_0 \beta_G. \tag{18.11}$$

This system allows a natural economic interpretation, and to see this we recall that α is the local mean rate of return for the stock *excluding jumps*. The local mean rate of return for the stock *including jumps* is given by

$$\mu^P = \alpha + \beta \lambda,$$

and in the same way the local mean rates of return for the F and G contracts are given by μ_F^P and μ_G^P, where

$$\mu_F^P = \alpha_F + \beta_F \lambda, \tag{18.12}$$

$$\mu_G^P = \alpha_G + \beta_G \lambda. \tag{18.13}$$

We can thus rewrite (18.9)–(18.11) as

$$\mu^P - r = \gamma_m \sigma + (\varphi_0 + \lambda)\beta, \tag{18.14}$$

$$\mu_F^P - r = \gamma_m \sigma_F + (\varphi_0 + \lambda)\beta_F, \tag{18.15}$$

$$\mu_G^P - r = \gamma_m \sigma_G + (\varphi_0 + \lambda)\beta_G. \tag{18.16}$$

Defining the function φ_m by

$$\varphi_0(t, s) + \lambda = \lambda \varphi_m, \tag{18.17}$$

we can write (18.9)–(18.11) as

$$\mu^P - r = \gamma_m \sigma + \varphi_m \lambda \beta, \tag{18.18}$$

$$\mu_F^P - r = \gamma_m \sigma_F + \varphi_m \lambda \beta_F, \tag{18.19}$$

$$\mu_G^P - r = \gamma_m \sigma_G + \varphi_m \lambda \beta_G. \tag{18.20}$$

These equations allow a very natural economic interpretation. On the left-hand side we have the **risk premium** for the assets S, F and G. On the right-hand side we have a sum of the diffusion and (mean) jump volatilities multiplied by the coefficients φ_m and γ_m respectively. The main point to observe is that whereas the risk premium, the diffusion volatility and the jump volatility vary from asset to asset, the coefficients φ_m and γ_m are *the same for all assets*. We can thus interpret γ_m as "the risk premium per unit of diffusion volatility", and φ_m as "the risk premium per unit of jump volatility". In the economics and finance literature they are referred to as

$$\gamma_m = \text{"the market price of diffusion risk"},$$

$$\varphi_m = \text{"the market price of jump risk"}.$$

We have now proved our first result.

Proposition 18.1 *Consider the model (18.1)–(18.2) and assume absence of arbitrage on the derivatives market. Then there will exist functions $\varphi_m(t,s)$ and $\gamma_m(t,s)$ such that, for any claim of the form $\Phi(S_T)$ with pricing function $F(t,s)$, the we have condition*

$$\mu_F^P - r = \gamma_m \sigma_F + \varphi_m \lambda \beta_F, \tag{18.21}$$

where the local mean rate of return, μ_F^P, the diffusion volatility, σ_F and the jump volatility, β_F, are defined by (18.5)–(18.7) and (18.21). The functions φ_m and γ_m are universal in the sense that they do not depend on the particular choice of the derivative. In particular we have

$$\alpha + \lambda\beta - r = \gamma_m \sigma + \varphi_m \lambda \beta, \tag{18.22}$$

so we have one degree of freedom for the choice of φ_m and γ_m.

We may also use the relations above to obtain an equation for the function F. Plugging the definitions of μ_F^P, α_F, σ_F and β_F into (18.12) gives us the following pricing result.

Proposition 18.2 *The pricing function F for the contract $\Phi(S_T)$ will satisfy the PIDE*

$$\begin{cases} F_t + (\alpha - \gamma_m\sigma)sF_s + \dfrac{1}{2}\sigma^2 s^2 F_{ss} + \lambda(1 - \varphi_m)F_\beta - rF &= 0, \\[2mm] \hspace{5cm} F(T,s) &= \Phi(s), \end{cases} \tag{18.23}$$

subject to the constraint 18.22.

This is the pricing PIDE for F, but in order to provide an explicit solution we need to know the market price of Wiener risk γ_m as well as the market price of jump risk φ_m. These objects are not given *a priori* nor are they determined within the model. The reason for this is of course that our model is incomplete, so there are infinitely many market prices of risk which are consistent with no-arbitrage. Thus there are also potentially infinitely many arbitrage-free price processes for the contract Φ. In a concrete market, exactly one of these price processes will be chosen by the market, and this process will be determined, not only on the requirement of absence of arbitrage, but also by the preferences towards risk on the market. These preferences are then codified in the the market choice of the market prices of the risks γ_m and φ_m. Note again that since γ_m and φ_m have to satisfy (18.22), we have one degree of freedom.

It is important to remember that pricing a derivative by no-arbitrage is always pricing a derivative *in terms of* already existing underlying assets. These underlying assets could be stocks or other derivatives. To illustrate this, suppose that, somehow, we are given the pricing function G exogenously. Then γ_m and φ_m are uniquely determined by (18.18) and (18.20), so we can solve the PIDE (18.23) for F. We thus obtain a unique pricing function F, but the point is that F is determined by our exogenous specification of G. We have thus determined F in terms of G.

We can of course also apply the Feynman–Kac representation formula to (18.23). In order to do that it is convenient to introduce the notation

$$\gamma = -\gamma_m, \tag{18.24}$$

$$\varphi = -\varphi_m. \tag{18.25}$$

We then have the PIDE

$$F_t + (\alpha + \gamma\sigma)sF_s + \frac{1}{2}\sigma^2 s^2 F_{ss} + \lambda(\varphi + 1)F_\beta - rF = 0, \qquad (18.26)$$

$$F(T, s) = \Phi(s), \qquad (18.27)$$

and Feynman–Kac gives us the following representation of F.

Proposition 18.3   *With notations as above, the pricing function F can be represented by the following risk-neutral valuation formula:*

$$F(t, s) = e^{-r(T-t)} E^Q_{t,s} [\Phi(S_T)] . \qquad (18.28)$$

The dynamics of S under Q are given by

$$dS_t = \{\alpha + \gamma\sigma\} S_t dt + S_t \sigma dW^Q_t + \beta S_{t-} dN_t, \qquad (18.29)$$

under the constraint

$$\mu^P - r = \gamma_m \sigma + \varphi_m \beta, \qquad (18.30)$$

where W^Q is Q-Wiener and N is Poisson under Q with intensity $\lambda(\varphi + 1)$.

We obviously expect the measure Q to be a martingale measure so, under Q, the mean local rate of return of S should be the short rate. To see this we compensate the dN term in the S-dynamics under Q to obtain the semimartingale dynamics

$$dS_t = \{\alpha + \gamma\sigma + \beta\lambda(1 + \varphi)\} S_t dt + S_t \sigma dW^Q_t + \beta S_{t-} \{dN_t - \lambda(1 + \varphi)dt\} . \quad (18.31)$$

It now follows from the (18.30) that in fact $\alpha + \gamma\sigma + \beta\lambda(1 + \varphi) = r$ so we have the semimartingale Q-dynamics

$$dS_t = rS_t dt + S_t \sigma dW^Q_t + \beta S_{t-} \{dN_t - \lambda(1 + \varphi)dt\},$$

which shows that Q, as expected, is a martingale measure. We also see from (18.31) that Q is obtained from P by a Girsanov transformation with kernels γ and φ, so the likelihood dynamics are

$$dL_t = L_t \gamma_t dW_t + L_{t-} \varphi \{dN_t - \lambda dt\} . \qquad (18.32)$$

18.3   Martingale Analysis

We now go on to study the jump-diffusion model

$$dS_t = \alpha S_t dt + \sigma S_t dW_t + \beta S_{t-} dN_t, \qquad (18.33)$$

$$dB_t = r B_t dt, \qquad (18.34)$$

from the point of view of martingale measures. This turns out to be very easy, and as usual we start looking for a potential martingale measure by applying the Girsanov theorem. We thus choose two predictable processes γ and φ with $\varphi > -1$, define the likelihood process L by

$$\begin{cases} dL_t &= L_t\gamma_t dW_t + L_{t-}\varphi_t \{dN_T - \lambda dt\}, \\ L_0 &= 1 \end{cases}$$ (18.35)

and define a new measure Q by

$$L_t = \frac{dQ}{dP}, \quad \text{on } \mathcal{F}_t, \ 0 \le t \le T.$$

From the Girsanov theorem we know that N has Q-intensity

$$\lambda_t^Q = (1 + \varphi_t)\lambda,$$ (18.36)

and that we can write

$$dW_t = \gamma_t dt + dW_t^Q,$$ (18.37)

where W^Q is Q-Wiener. Plugging (18.37) into the S-dynamics (18.33) and compensating N under Q gives us the Q-dynamics of S as

$$dS_t = \{\alpha + \gamma_t\sigma + (1 + \varphi_t)\beta\lambda\} S_t dt + \sigma S_t dW_t^Q$$ (18.38)
$$+\beta S_{t-} \{dN_t - (1 + \varphi_t)\lambda dt\}.$$ (18.39)

From this we see that Q is a martingale measure if and only if the relation

$$\alpha + \gamma_t\sigma + (1 + \varphi_t)\beta\lambda = r$$ (18.40)

is satisfied. We can write this as

$$\alpha + \beta\lambda - r = -\gamma_t\sigma - \varphi_t\lambda\beta,$$ (18.41)

which is of course the same as (18.13) and so we have the following result.

Proposition 18.4   *The measure Q above is a martingale measure if and only if the following conditions are satisfied:*

$$\varphi_t > -1,$$ (18.42)
$$\alpha + \beta\lambda - r = -\gamma_t\sigma - \varphi_t\lambda\beta,$$ (18.43)
$$E^P [L_T] = 1.$$ (18.44)

Furthermore, the Girsanov kernels γ and φ are related to the market price of the diffusion risk γ_m and the market price of jump risk φ_m by

$$\gamma_{mt} = -\gamma_t,$$ (18.45)
$$\varphi_{mt} = -\varphi_t.$$ (18.46)

19 A General Jump-Diffusion Model

We now move on to study a more interesting jump-diffusion model for the case of a non-dividend-paying aset. To that end we consider a filtered space $(\Omega, \mathcal{F}, P, \mathbf{F})$ carrying a d-dimensional Wiener process W, and a marked point process Ψ with mark space $(E, \mathcal{E})$ and intensity measure $\lambda_t(dz)$. The case of a dividend-paying asset is very similar, but a little bit messy so we treat the cases separately. See Section 19.4

19.1 Market Dynamics and Basic Concepts

In the first round we will consider a financial market consisting of (at least) two underlying assets: a risky non-dividend-paying asset with price process S and a locally risk-free asset (bank account) with price process B.

Assumption 19.1 *We assume that the P-dynamics of the price process are given by*

$$dS_t = \alpha_t S_t dt + S_t \sigma_t dW_t + S_{t-} \int_E \beta_t(z) \Psi(dt, dz), \tag{19.1}$$

$$dB_t = r_t B_t dt, \tag{19.2}$$

where α and r are scalar optional process, σ is an optional d-dimensional row-vector process and $\beta \geq -1$ is $\mathcal{P}$-predictable.

The reason why we must have $\beta \geq -1$ is that if we have a jump at t with mark z then

$$S_t = S_{t-} + \Delta S_t = S_{t-} + S_{t-}\beta_t(z) = S_{t-}[\beta(z) + 1] \tag{19.3}$$

so the condition $\beta \geq -1$ stops the price process from going negative.

There could also be other underlying assets in the market, but for the moment we concentrate on S and B. Our first task is to understand the economic significance of the stock price dynamics in terms of mean rate of return, volatility and risk premium. We then note that (19.1) is not in semimartingale form. In order to obtain the semimartingale decomposition we do what we always do, namely compensate the point process to obtain

$$dS_t = \left(\alpha_t + \int_E \beta_t(z) \lambda(dz) \right) S_t dt + S_t \sigma_t dW_t + S_{t-} \int_R \beta_t(z) \widetilde{\Psi}(dt, dz). \tag{19.4}$$

where

$$\widetilde{\Psi}(dt, dz) = \Psi(dt, dz) - \lambda_t(dz) dt. \tag{19.5}$$

We now go on to define return and volatility.

Definition 19.2   The **return process** R is defined by the relation

$$dR_t = \frac{dS_t}{S_{t-}}$$

so, with S defined by (19.4), we have

$$dR_t = \left(\alpha_t + \int_E \beta_t(z)\lambda(dz)\right) dt + \sigma_t dW_t + \int_R \beta_t(z)\widetilde{\Psi}(dt, dz). \qquad (19.6)$$

The total (squared) **volatility** v_t^2 is defined by the relation

$$v_t^2 = \frac{d\langle R, R\rangle_t}{dt}. \qquad (19.7)$$

where the angular bracket $\langle R, R\rangle_t$ was defined in Section 8.3.

Starting with the return, we have the following easy observation.

Lemma 19.3   *The following hold.*

- *The mean rate of return between jumps is given by α_t.*
- *The total mean rate of return μ_t (including jumps) is given by*

$$\mu_t = \alpha_t + \int_E \beta_t(z)\lambda(dz). \qquad (19.8)$$

We will often write the P-dynamics directly in semimartingale form as

$$dS_t = \mu_t S_t dt + S_t \sigma_t dW_t + S_{t-}\int_R \beta_t(z)\widetilde{\Psi}(dt, dz). \qquad (19.9)$$

Now, it may seem that the volatility definition above is a bit abstract, but from Intuition 8.8 we have the following interpretation.

Intuition 19.4   *The squared volatility admits the following informal interpretation:*

$$v_t^2 dt = \mathrm{Var}\left[dR_t \mid \mathcal{F}_{t-}\right],$$

so v_t^2 is the conditional variance of return per unit time.

We can finally use Proposition 8.9 to compute the volatility for the process S.

Proposition 19.5   *For the asset price process S with dynamics (19.1), the volatility is given by*

$$v_t^2 = \|\sigma_t\|^2 + \int_E \beta_t^2(z)\lambda_t(dz), \qquad (19.10)$$

where $\|\cdot\|$ denotes the Euclidian norm.

We see that the squared-volatility process can also be expressed as

$$v_t^2 = \|\sigma_t\|_{R^d}^2 + \|\beta t\|_{\lambda_t}^2, \qquad (19.11)$$

where $\|\cdot\|_{\lambda_t}$ denotes the norm in the Hilbert space $L^2[X, \lambda_t(dz)]$. We can also, for future use, express the volatility v as

$$v_t = \|(\sigma_t, \beta_t)\|_{\mathcal{H}}, \qquad (19.12)$$

where we view (σ_t, β_t) as a vector in the Hilbert space $\mathcal{H} = R^d \times L^2[E, \lambda_t(dz)]$. We finish this section by defining two important concepts of financial economics.

Definition 19.6 Denote, as above, the mean rate of return by μ, the risk-free short rate by r and the volatility by v.

- The **excess rate of return** or **risk premium** RP is defined by

$$RP_t = \mu_t - r_t. \tag{19.13}$$

- The **Sharpe ratio** SR is defined by

$$SR_t = \frac{RP_t}{v_t}; \tag{19.14}$$

so

$$SR_t = \frac{\mu_t - r_t}{v_t}. \tag{19.15}$$

The risk premium is a measure of how well an asset (or fund) is doing, in terms of the difference between expected rate of return and the risk-free rate of return. This, however, does not take into account how risky the asset is. The Sharpe ratio, on the other hand, normalizes the risk premium by dividing with the volatility of the asset (as a measure of riskiness). In words, the Sharpe ratio is thus the risk premium per unit of risk (in terms of volatility).

19.2   Equivalent Martingale Measures

Given the process S above we now search for an equivalent martingale measure Q, and for any Q equivalent to P (martingale measure or not) we define the the likelihood process L by

$$\frac{dQ}{dP} = L_t, \quad \text{on } \mathcal{F}_t; \quad 0 \leq t \leq T. \tag{19.16}$$

Since L is always a P-martingale and since, by Theorem 9.1, every martingale within the present framework admits a stochastic integral representation we know that L must have dynamics of the form

$$\begin{cases} dL_t = L_t \gamma_t^* dW_t + L_{t-} \int_E \varphi_t(z) \{\Psi(dt, dz) - \lambda_t(dz)dt\}, \\ L_0 = 1, \end{cases} \tag{19.17}$$

where the **Girsanov kernel** processes γ and φ (we view γ as a column-vector process, hence the transpose γ_t^*) are predictable and suitably integrable, and where φ must satisfy the condition

$$\varphi_t(z) > -1, \quad \text{for all } t, z \quad P\text{-a.s.}, \tag{19.18}$$

in order to ensure the positivity of the measure Q. From the Girsanov theorem 9.2 we also recall the following facts.

- We can write

$$dW = \gamma_t dt + dW^Q, \tag{19.19}$$

  where W^Q is a Q-Wiener process.
- The point process μ will under Q have an intensity λ^Q, given by

$$\lambda_t^Q(dz) = [\varphi_t(z) + 1] \lambda_t(dz). \tag{19.20}$$

The immediate problem is to find out how the kernel processes γ and φ above must be chosen in order to guarantee that Q actually is a martingale measure for S. To this end we apply the Girsanov theorem to obtain the Q-dynamics of S as

$$dS_t = S_t \{\alpha_t + \sigma_t \gamma_t\} dt + S_t \sigma_t dW_t^Q + S_{t-} \int_E \beta_t(z) \Psi(dt, dz).$$

We then compensate the point process Ψ under Q to obtain the Q-semimartingale representation of S as

$$dS_t = S_t \left\{ \alpha_t + \sigma_t \gamma_t + \int_E \beta_t(z) \lambda_t^Q(dz) \right\} dt + S_t \sigma_t dW_t^Q$$
$$+ S_{t-} \int_E \beta_t(z) \left\{ \Psi(dt, dz) - \lambda_t^Q(dz) dt \right\}. \tag{19.21}$$

Recalling from Proposition 16.7 that the measure Q is a martingale measure if and only if the local rate of return of S under Q equals the short rate r, we thus obtain the following **martingale condition**.

Proposition 19.7 *Assume that the measure Q is generated by the Girsanov kernels h, φ through (19.17) Then Q is a martingale measure if and only if the following conditions are satisfied:*

$$\alpha_t + \sigma_t \gamma_t + \int_E \beta_t(z) [1 + \varphi_t(z)] \lambda_t(dz) = r_t. \tag{19.22}$$

$$\varphi_t(z) \geq -1. \tag{19.23}$$

A Girsanov kernel process (γ, φ) for which the induced measure Q, is a martingale measure, i.e. a kernel process satisfying the martingale condition (19.22)–(19.23) will be referred to as an **admissible** Girsanov kernel.

19.3   Hansen–Jagannathan Bounds

In this section we will derive an extension of the so called Hansen–Jagannathan bounds, which were first discussed in Hansen & Jagannathan (1991). The HJ bound is a rather simple but very fundamental inequality relating the Sharpe ratio of an asset to the Girsanov kernels which are used to go from the objective measure P to the risk-neutral measure Q.

We start by noting that can rewrite the martingale condition (19.22) as

$$\alpha_t + \int_E \beta_t(z) \lambda_t(dz) - r_t = -\sigma_t \gamma_t - \int_E \varphi_t(z) \beta_t(z) \lambda_t(dz). \tag{19.24}$$

From (19.8) we recall that

$$\alpha_t + \int_E \beta_t(z)\lambda_t(dz) = \mu_t,$$

where μ is the local mean rate of return of the asset S. We can thus write (19.22) as

$$\mu_t - r_t = -\sigma_t \gamma_t - \int_E \varphi_t(z)\beta_t(z)\lambda_t(dz), \qquad (19.25)$$

and from (19.13) we recognize the risk premium RP in the left-hand side of this equation ,so we can write RP as

$$RP_t = -\sigma_t \gamma_t - \int_X \varphi_t(x)\beta_t(x)\lambda_t(dx) \qquad (19.26)$$

or in more detail as

$$RP_t = -\sum_{i=1}^{d} \sigma_t^i \gamma_t^i - \int_X \varphi_t(x)\beta_t(x)\lambda_t(dx). \qquad (19.27)$$

The reader acquainted with the Capital Asset Pricing Model (CAPM) will feel that (19.25) looks familiar. Now is the time then to introduce some concepts from financial economics.

Definition 19.8   We define the following terms

- The **market price vector of Wiener risk** γ_m is given by $\gamma_{mt} = -\gamma_t$.
- The **market price of jump risk** φ_m is given by $\varphi_{mt} = -\varphi_t$.

Note that the market prices of risk are not at all *prices* in the standard sense. You do *not* pay the market price of risk (in what units?) to get something (what would that be?). Instead, the market prices of risk tells you how the risk premium is affected by Wiener risk and jump risk according to (19.25). We can of course rewrite (19.27) in terms of market prices of risk to obtain

$$RP_t = \sum_{i=1}^{d} \sigma_t^i \gamma_{mt}^i + \int_X \varphi_{mt}(x)\beta_t(x)\lambda_t(dx), \qquad (19.28)$$

and we see that with increasing market prices of risk we will have a higher risk premium.
There is another concept, related to the market prices of risk, namely the jump-risk premium JRP.

Definition 19.9   The **jump-risk premium** is defined by

$$JRP = \frac{\lambda_t^Q(E)}{\lambda_t(E)} \qquad (19.29)$$

The jump-risk premium is thus the ratio between the intensity of the underlying counting process under Q and the intensity under P. If we think of jumps as "rare events" and if JRP > 1, then the rare events are more frequent under Q than under P. From Girsanov we have

$$\lambda_t^Q(dz) = [\varphi_t(z) + 1]\,\lambda_t(dz).$$

Integrating this over E and recalling the factorization $\lambda_t(dz) = \lambda^E \Gamma_t(dz)$ we obtain

$$\lambda_t^Q(E) = \lambda_t(E) \int_E [\varphi_t(z) + 1] \Gamma_t(dz),$$

so

$$\mathrm{JRP} = 1 + \int_E \varphi_t(z)\Gamma_t(dz).$$

We can now collect our results.

Proposition 19.10 *With notation as above we have*

- *For any asset price process π, derivative or underlying, with dynamics of the form*

$$d\pi_t = \pi_t \mu_t^\pi dt + \pi_t \sigma_t^\pi dW_t + \pi_{t-} \int_E \beta_t^\pi(z) \{\psi(dt, dz) - \lambda_t(dz)dt\}, \qquad (19.30)$$

  the risk premium is given by

$$\mu_t^\pi - r_t = -\gamma_t^* \sigma_t^\pi - \int_E \varphi_t(z)\beta_t(z)\lambda_t(dz), \qquad (19.31)$$

  which can be written as

$$\mu_t^\pi - r_t = -\gamma_t^* \sigma_t^\pi - \lambda^E E^{\Gamma_t} [\varphi_t(Z)\beta_t(Z)], \qquad (19.32)$$

  where Γ is the jump measure under P and Z is a generic mark.
- *The jump-risk premium is given by*

$$\mathrm{JRP} = 1 + \int_E \varphi_t(z)\Gamma_t(dz). \qquad (19.33)$$

- *Since Γ_t is a probability measure we can also write*

$$\mathrm{JRP} = 1 + E^{\Gamma_t} [Z]. \qquad (19.34)$$

Proof   The two last points follow directly from Girsanov. The other points have already been proved.   □

Note that since, in general, Γ_t is a (measure-valued) random process, this implies that the jump-risk premium is a random process. However, if Γ is deterministic and time-invariant, then we have the simple formula

$$\mathrm{JRP} = 1 + E^P [\varphi(Z)], \qquad (19.35)$$

where Z is a generic mark variable.

We now go on to the main result of this section, which is to derive an extension of the classical Hansen–Jagannathan bounds to a jump-diffusion framework.

Theorem 19.11 (Extended Hansen–Jagannathan bounds)   *For any arbitrage-free price process S as above, and for every admissible Girsanov kernel process (γ, φ) generating a martingale measure through*

$$\begin{cases} dL_t &= L_t \gamma_t^* dW_t + L_{t-} \int_E \varphi_t(z) \{\Psi(dt, dz) - \lambda_t(dz)dt\}, \\ L_0 &= 1, \end{cases} \qquad (19.36)$$

the following inequality holds for the Sharpe ratio:

$$|SR_t| \leq \|(\gamma_t, \varphi_t)\|_{\mathcal{H}}. \tag{19.37}$$

In more detail this inequality can be written as

$$|SR_t|^2 \leq \|\gamma_t\|_{R^d}^2 + \int_X \varphi_t^2(x)\lambda_t(dx). \tag{19.38}$$

In terms of market prices of Wiener and jump risk it looks the same:

$$|SR_t|^2 \leq \|\gamma_{mt}\|_{R^d}^2 + \int_X \varphi_{mt}^2(x)\lambda_t(dx). \tag{19.39}$$

Proof   A closer look at (19.26) reveals that the right-hand side can be viewed as an inner product in the Hilbert space $\mathcal{H} = R^d \times L^2[E, \lambda_t(dz)]$. Denoting this inner product by $\langle, \rangle_{\mathcal{H}}$ we can thus write

$$RP_t = \langle (\gamma_t, \varphi_t), (\sigma_t, \beta_t) \rangle_{\mathcal{H}} \tag{19.40}$$

and from the Schwartz inequality we obtain

$$|RP_t| \leq \|(\gamma_t, \varphi_t)\|_{\mathcal{H}} \cdot \|(\sigma_t, \beta_t)\|_{\mathcal{H}}. \tag{19.41}$$

The inequality (19.37) now follows immediately from (19.12), (19.14) and (19.41).   □

It is important to note that the HJ inequality not only holds for the given single underlying asset-price process S. To be more precise: suppose that we have an underlying market with several assets driven by W and Ψ, and suppose also that we have chosen a fixed pair of Girsanov kernels (γ, φ) generating an equivalent martingale measure for all the underlying assets. Suppose still further that we use this martingale measure to price various derivatives. Then the HJ inequality holds, not only for all underlying assets, but also for all derivatives and for all self-financing portfolios based on the underlying and the derivatives. In other words, for a given admissible choice of (γ, φ), the HJ inequality gives us a uniform upper bound of Sharpe ratios for the entire economy.

This argument can of course also be reversed. If, for example we model a concrete existing (incomplete) market where we empirically have observed a certain Sharpe ratio, then we must choose a martingale measure, i.e. choose (γ, φ), such that the HJ bounds are satisfied. In other words, for an given Sharpe ratio, the HJ bounds gives restrictions on the class of martingale measures.

19.4   Introducing Dividends

In this section we introduce dividends. The calculations are a very similar, but a little bit more messy than for the non-dividend case. WE will thus study a market with a price–dividend pair (S, D) and the usual money market account B. We recall that the D process is the cumulative dividend process, with the interpretation that over a small time interval $[t - dt, t]$ we obtain

$$dD_t = D_t - D_{t-dt}$$

dollars (paid at t).

19.4.1 The Model

Assumption 19.12 *We assume that the P-dynamics of the price–dividend process are given by*

$$dS_t = \alpha_t S_t dt + S_t \sigma_t dW_t + S_{t-} \int_E \beta_t(z) \Psi(dt, dz), \tag{19.42}$$

$$dD_t = \alpha_t^d S_t dt + S_t \sigma_t^d dW_t + S_{t-} \int_E \beta_t^d(z) \Psi(dt, dz), \tag{19.43}$$

$$dB_t = r_t B_t dt, \tag{19.44}$$

where α, α^d and r are scalar optional process, σ and σ^d are optional d-dimensional row-vector processes, whereas $\beta \geq -1$ and $\beta^d \geq -1$ are $\mathcal{P}$-predictable.

Note the structure of the D-dynamics. The appearance of S in front of every term makes the formulas below much easier, but it is just a normalizing factor so you can derive the theory without it. We now need to define the return process for S. Over a small time-interval we get a return due to the price changes of S but we also have a direct return in terms of the dividends paid during the time interval.

Definition 19.13 We make the following definitions.

1. The **return** process R of the stock is defined by

$$dR_t = \frac{dS_t + dD_t}{S_{t-}} \tag{19.45}$$

2. The **mean rate of return** μ is defined by

$$\mu_t dt = E^P \left[dR_t | \mathcal{F}_{t-} \right]. \tag{19.46}$$

3. The total (squared) **volatility** v_t^2 is defined by the relation

$$v_t^2 = \frac{d\langle R, R \rangle_t}{dt}, \tag{19.47}$$

   where the angular bracket $\langle R, R \rangle_t$ was defined in Section 8.3.
4. The **Sharpe ratio** is defined by

$$SR = \frac{\mu_t - r_r}{v}. \tag{19.48}$$

As in the non-dividend case, the squared volatility admits the informal interpretation

$$v_t^2 dt = \text{Var} \left[dR_t | \mathcal{F}_{t-} \right],$$

so v_t^2 is the conditional variance of return per unit time.
We now have

$$dR_t = \left[\alpha_t + \alpha_t^d \right] dt + \left[\sigma_t + \sigma_t^d \right] dW_t + \int_E \left[\beta_t(z) + \beta_t^d(z) \right] \Psi(dt, dz). \tag{19.49}$$

Compensating the point process we obtain the following result.

Proposition 19.14 *The mean rate of return (under P) is given by*

$$\mu_t = \left[\alpha_t + \alpha_t^d\right] dt + \int_E \left[\beta_t(z) + \beta_t^d(z)\right] \lambda_t(dz).$$ (19.50)

The volatility is given by

$$v_t^2 = \|\sigma_t + \sigma_t^d\|_{R^d}^2 + \int_E \left[\beta_t(z) + \beta_t^d(z)\right]^2 \lambda_t(dz)$$ (19.51)

19.4.2   Martingale Measures

We now look for a martingale measure for our model, and to this end we recall the normalized gain process G^z from Section 16.8:

$$G_t^Z = \frac{S_t}{B_t} + \int_0^t \frac{1}{B_s} dD_s.$$ (19.52)

From Section 16.8 we also recall that Q is a martingale measure if and only if G^Z is a Q-martingale. With $Z = S/B$ we obtain

$$dG_t^Z = Z_t \left[\alpha_t + \alpha_t^d - r_t\right] + Z_t \left[\sigma_t + \sigma_t^d\right] dW_t$$

$$+ Z_{t-} \int_E \left[\beta_t(z) + \beta_t^d(z)\right] \Psi(dt, dz).$$ (19.53)

We now perform a Girsanov transformation with likelihood dynamics

$$dL_t = L_t \gamma_t^* dW_t + L_{t-} \int_E \varphi_t(dz) \{\Psi(dt, dz) - \lambda_t(dz)dt\}.$$ (19.54)

Under Q we will have

$$dW_t = \gamma_t dt + dW_t^Q,$$

where W^Q is Q-Wiener, and the Q-intensity is

$$\lambda_t^Q(dz) = [1 + \varphi_t(z)] \lambda_t(dz).$$

Applying this to (19.53) we obtain the semimartingale dynamics for G^Z under Q as

$$dG_t^Z = Z_t \left[\alpha_t + \alpha_t^d - r_t + (\sigma_t + \sigma_t^d)\gamma_t + \int_E [1 + \varphi_t(z)] \{\beta_t(z) + \beta_t^d(z)\}\right] dt$$

$$+ Z_t \left[\sigma_t + \sigma_t^d\right] dW_t^Q$$

$$+ Z_{t-} \int_E \left[\beta_t(z) + \beta_t^d(z)\right] \{\Psi(dt, dz) - (1 + \varphi_t(z)\lambda_t(dz)dt\}.$$ (19.55)

We have thus proved the following.

Proposition 19.15 *The measure Q is a martingale measure for the market if and only the **martingale condition***

$$r = \alpha_t + \alpha_t^d + (\sigma_t + \sigma_t^d)\gamma_t + \int_E [1 + \varphi_t(z)] \{\beta_t(z) + \beta_t^d(z)\} \lambda_t(dz)$$ (19.56)

holds for all t, P-a.s.

The diligent reader can easily check that the martingale condition also can be expressed as

$$\mu_t^Q = r_t, \tag{19.57}$$

where μ_t^Q is the mean rate of return under Q, defined by

$$\mu_t^Q = E^Q[R_t | \mathcal{F}_{t-}]. \tag{19.58}$$

19.4.3 Hansen–Jagannathan

Using Proposition 19.14 we can write the martingale condition (19.56) as

$$\mu_t - r_t = -(\sigma_t + \sigma_t^d)\gamma_t - \int_E \varphi_t(z)\left\{\beta_t(z) + \beta_t^d(z)\right\}\lambda_t(dz) \tag{19.59}$$

As before we may introduce the **market prices of risk**

$$\gamma_{mt} = -\gamma_t, \tag{19.60}$$
$$\varphi_{mt} = -\varphi_t, \tag{19.61}$$

and write

$$\mu_t - r_t = (\sigma_t + \sigma_t^d)\gamma_{mt} + \int_E \varphi_{mt}(z)\left\{\beta_t(z) + \beta_t^d(z)\right\}\lambda_t(dz). \tag{19.62}$$

Using (19.51) we can now proceed as in the non-dividend case to obtain the following result for assets with dividends.

Theorem 19.16 (Hansen–Jagannathan Bounds for Dividends)   *For any arbitrage-free price–dividend pair (S, D) as above, and for every admissible Girsanov kernel process (γ, φ) generating a martingale measure through*

$$\begin{cases} dL_t &= L_t\gamma_t^* dW_t + L_{t-}\int_E \varphi_t(z)\left\{\Psi(dt, dz) - \lambda_t(dz)dt\right\}, \\ L_0 &= 1, \end{cases} \tag{19.63}$$

the following inequality holds for the Sharpe ratio:

$$|SR_t| \leq \|(\gamma_t, \varphi_t)\|_{\mathcal{H}}. \tag{19.64}$$

In more detail this inequality can be written as

$$|SR_t|^2 \leq \|\gamma_t\|_{R^d}^2 + \int_X \varphi_t^2(x)\lambda_t(dx). \tag{19.65}$$

In terms of market prices of Wiener and jump risk it looks the same:

$$|SR_t|^2 \leq \|\gamma_{mt}\|_{R^d}^2 + \int_X \varphi_{mt}^2(x)\lambda_t(dx). \tag{19.66}$$

19.5   Completeness

We have already seen in Chapter 18, that the jump-diffusion market (B, S) discussed above is, in the generic case, incomplete. To see this more formally within the present setting we look for an equivalent risk-neutral martingale measure Q and, with a likelihood process given by (19.17), Proposition 19.7 says that the Girsanov kernel (γ, φ) generates a martingale measure if and only if $\varphi_t(z) \geq -1$, for all $z \in E$, and

$$\alpha_t + \sigma_t \gamma_t + \int_E \beta_t(z)\,[1 + \varphi_t(z)]\,\lambda_t(dz) = r_t. \tag{19.67}$$

This is, for any fixed t, a scalar equation for the determination of $\gamma_t^1, \ldots, \gamma_t^d$ and $\varphi_t(z)$ for all $z \in E$. If we denote the cardinality of E by k we thus see that (19.67) is a scalar equation for the determination of $d + k$ unknowns. Unless $d + k = 1$ we thus have a underdetermined equation with infinitely many solutions. The martingale measure is thus not unique and the market is not complete. The only possibility for a complete market model is thus that $d + k = 1$, which leaves us with the following two possibilities.

- The Wiener process W is scalar and there is no jump part, so $\Psi = 0$
- There is no Wiener process and the mark space consists of a single point.

An obvious idea is then to complete the market by adding a sufficiently large number of underlying assets, but this is also problematic. Suppose that, instead of having only one underlying asset S, we model a market with N risky assets $S^1, \ldots, S^N$ while, of course, keeping the filtered space $(\Omega, \mathcal{F}, P, \mathbf{F})$, the d-dimensional Wiener process W and the marked point process Ψ. For simplicity we also assume that the filtration is the internal one, so $\mathbf{F} = \mathbf{F}^{\Psi, W}$, and that the short rate is constant. The stock price dynamics are then as follows:

$$dS_t^i = \alpha_t^i S_t^i\,dt + S_t^i \sigma_t^i\,dW_t + S_{t-}^i \int_E \beta_t^i(z)\Psi(dt, dz) \quad i = 1, \ldots, N, \tag{19.68}$$

and we have a bank account B with dynamics

$$dB_t = r B_t\,dt. \tag{19.69}$$

We now look for an equivalent martingale measure. With a likelihood process given by (19.17), Proposition 19.7 says that the Girsanvo kernel (γ, φ) generates a martingale measure if and only if $\varphi_t(z) \geq -1$, for all $z \in E$, and

$$\alpha_t^i + \sigma_t^i \gamma_t^i + \int_E \beta_t^i(z)\,[1 + \varphi_t(z)]\,\lambda_t(dz) = r, \quad i = 1, \ldots, N. \tag{19.70}$$

This is obviously an N-dimensional equation system, and the unknowns are, for each t, the following:

- the Wiener coefficients $\gamma_t^1, \ldots, \gamma_t^d$;
- for each $z \in E$ the value $\varphi_t(z)$.

If we denote the cardinality of E by k we thus see that (19.70) is a linear system of N equations for $d + k$ unknowns. In the generic case such a system admits infinitely many solutions if $N < d + k$, a unique solution if $N = d + k$ and no solution if $N > d + k$. To see what this means more concretely in our case, suppose for simplicity that we only have a scalar Wiener process, i.e. $d = 1$. If k is finite this implies that in order to have a complete market we need $k + 1$ risky assets. This implies, however, that all the risky assets will be driven by the same scalar Wiener process. Between jumps, the assets will thus have perfectly correlated price processes and this is extremely unrealistic from an economic point of view.

The only realistic scenario in this case is that we model only one underlying risky asset and complete the market with k derivatives on the underlying. If k is infinite it is clear that we need an infinite number of assets in order to have a complete market: this idea will in fact be pursued in Section 25.2. See (Björk, Kabanov & Runggaldier 1995) for the full story.

We can also refer to the Rule-of-Thumb 16.14. It tells us that if we have N risky assets and R "sources of randomness", then generically: the model is arbitrage-free if and only if $N \leq R$; it is complete if and only if $N \geq R$; and it is complete and arbitrage-free if and only if $N = R$.

In our case, each Wiener process and each point in E counts a a source of randomness, so we conclude that (always in the generic case) the following hold.

Intuition 19.17 *If W is d-dimensional, the cardinality of the mark space E equals k and we have a model with N risky assets, then we can in the generic case expect the following.*

1. *The model is arbitrage-free if and only if $N \leq d + k$.*
2. *The model is complete if and only if $N \geq d + k$.*
3. *The model is complete and arbitrage-free if and only if $N = d + k$.*
4. *In particular, if E is an infinite set, then the model is (very) incomplete.*

19.6   Methods for Handling Market Incompleteness

The long and short of the previous section is that all non-trivial jump-diffusion asset-price models are incomplete. This implies that there is no unique martingale measure, which in turn implies that, for a given derivative asset, there will not exist a unique arbitrage-free price process. What, then are we to do?

There are in fact a number of techniques and arguments that allow us to price derivatives in an incomplete market, and a partial list is as follows, where for simplicity we assume that there is only one risky underlying asset.

1. Disregard the objective measure P and (ruthlessly) specify the asset dynamics directly under Q. We can then price derivatives using standard risk-neutral valuation. This is quite often done in interest-rate theory. See Chapter 25.
2. Instead of looking at all possible Girsanov transformations, we restrict ourselves to a family of Girsanov transformations parameterized by a scalar parameter and hope that this will give us a unique Q. See Chapter 21.

3. Introduce a distance $d(p, Q)$ between probability measures and then choose the equivalent martingale measure Q which (within the class of EMMs) minimizes $d(P, Q)$. See Chapter 21.
4. Add a sufficiently large number of derivatives to complete the market. See Chapter 26.
5. Embed the market model within a full-fledged equilibrium model. The market equilibrium will then produce a unique martingale measure. See Chapter 26.
6. Give up the project of finding a unique martingale-pricing measure. Instead we try to find "reasonable" (in some sense) arbitrage-free *bounds* for derivative asset prices. See Chapter 22
7. Use a diversification argument. This will (perhaps) allow you to put zero market price of risk on the diversifiable risk factors, thus reducing the class of EMMs. See Chapter 23.
8. Use utility indifference techniques, inspired by economic theory. See Section 21.5.

19.7   Change of Numeraire

In this section we will study the effect of a change of numeraire on the asset-price dynamics. More precisely we will go from the risk-neutral measure Q to the measure Q^S where S is the numeraire. We recall the pricing formula (16.10):

$$\Pi_t[\mathcal{Z}] = S_t E^S \left[\frac{\mathcal{Z}}{S_T} \middle| \mathcal{F}_t \right]. \tag{19.71}$$

This is a nice formula but in order to use it we need to know the asset-price dynamics under Q^S, so we now investigate this. We formalize the setup as an assumption.

Assumption 19.18   *For the rest of this section we assume that the market uses a fixed risk-neutral measure Q for pricing.*

We know from general theory that the local rate of return under Q must equal the short rate r, so the Q-dynamics must have the form

$$dS_t = r_t S_t dt + S_t \sigma_t dW_t^Q + S_{t-} \int_R \beta_t(z) \left\{ \Psi(dt, dz) - \lambda_t^Q(dz) dt \right\}, \tag{19.72}$$

where W^Q is Q-Wiener and $\lambda_t(dz)$ is the Q-intensity measure for Ψ. We now change the measure from Q to Q^S; we want to find the Girsanov kernels associated with this change of measure. We therefore define the likelihood process L^S by

$$L_t^S = \frac{dQ^S}{dQ}, \quad \text{on } \mathcal{F}_t,$$

and from Proposition 16.15 we know that, apart from a normalizing constant, we have $L_t^S = S_t/B_t$. Using (19.72), an easy application of the Itô formula gives the following result, where we also have used the Girsanov theorem.

Proposition 19.19   *If S has Q-dynamics as in (19.72), then the following hold.*

1. *The likelihood process L^S is given by*

$$dL_t^S = L_t^S \sigma_t dW_t^Q + L_{t-}^S \int_R \beta_t(z) \left\{ \Psi(dt, dz) - \lambda_t^Q(dz)dt \right\}. \tag{19.73}$$

2. *We can write*

$$dW_t^Q = \sigma_t^* dt + dW_t^S, \tag{19.74}$$

*where * denotes transpose and W^S is Q^S-Wiener.*

3. *The Q^S-intensity measure of Ψ is given by*

$$\lambda_t^S(dz) = [1 + \beta_t(z)] \lambda_t^Q(dz). \tag{19.75}$$

4. *The Q^S-semimartingale dynamics of S are*

$$dS_t = S_t \left\{ r_t + \|\sigma_t\|^2 + \int_E \beta_t^2(z) \lambda_t^Q(dz) \right\} dt + S_t \sigma_t dW_t^S \tag{19.76}$$

$$+ S_{t-} \int_R \beta_t(z) \left\{ \Psi(dt, dz) - \lambda_t^S(dz)dt \right\}. \tag{19.77}$$

Proof The proof is left as an exercise. □

20 The Merton Model

In this chapter we introduce a simple point-process version of the standard Black–Scholes model where we specify the asset price process S under P as a geometric Lévy process. As a special case we study the famous *Merton model* from Merton (1976).

20.1 Setup

We consider a filtered space $(\Omega, \mathcal{F}, P, \mathbf{F})$ where P is the objective measure and the space carries the following objects.

- A scalar P-Wiener process W.
- A marked point process Ψ, with constant deterministic P-intensity measure $\lambda(dz)$.

We assume that the filtration is the internal one, so $\mathbf{F} = \mathbf{F}^{W, \Psi}$. The price semimartingale dynamics are

$$dS_t = \mu S_t dt + S_t \sigma dW_t + S_{t-} \int_E \beta(z) \left\{ \Psi(dt, dz) - \lambda(dz)dt \right\}, \qquad (20.1)$$

$$dB_t = r B_t dt, \qquad (20.2)$$

where μ, σ and r are constants and β a deterministic function. We recall that the constant μ is the conditional mean rate of return, and if the point process has a mark z at time t then S will have a jump with relative jump size $\beta(z)$. In order to avoid negative stock prices we also demand that $\beta(z) \geq -1$ for all $z \in E$.

We see that Ψ is compound Poisson. We also recall from Definition 7.12 that we can decompose the intensity measure into the local characteristics (λ^E, Γ) as $\lambda(dz) = \lambda^E \Gamma(dz)$, where

$$\lambda^E = \lambda(E),$$

$$\Gamma(dz) = \frac{\lambda(dz)}{\lambda(E)}.$$

Thus λ^E is the intensity for the event process and Γ the probability distribution of marks given an event. This type of model is the simplest non-trivial extension of the classical Black–Scholes model, and we recall from Section 8.6.2 that S is in fact a geometric Lévy process. In Merton (1976), it is assumed that the jump distribution $\Gamma(dz)$ is log-normal but we will start with a slightly more general model.

For a model like this with non-empty mark space E, the market is incomplete, so there is no unique martingale measure Q. Merton then had the following interesting idea.

Idea   Let us assume that the price process S above is just one of many price processes on the market. Let us furthermore assume that the jump risk Ψ is *idiosyncratic*, i.e it is specific to the process S, whereas the other price processes on the market have their own independent jump processes. Then, by forming well-diversified portfolios, we can, by the law of large numbers, get rid of the jump risk in S (and in every other price process). Standard economic reasoning would then imply that the jump risk is not priced by the market, which in mathematical terms says that $\lambda^Q(dz) = \lambda(dz)$, i.e. that we have the same intensity process under Q as under P.

If we accept Merton's diversification idea then the likelihood process from P to Q would have to be of the form

$$dL_t = L_t \gamma_t^* dW_t,$$

meaning that only the distribution of the Wiener part is changed under the measure transformation. From Girsanov we then get the Q-dynamics as

$$dS_t = S_t \{\mu + \sigma\gamma\}\, dt + \sigma S_t dW_t^Q + S_{t-} \int_E \beta(z) \{\Psi(dt, dz) - \lambda(dz)dt\},$$

where, by the Merton assumption, the point-process term is indeed also a Q-martingale. The condition for Q to be a martingale measure is, as always, that the local rate of return must equal the short rate under Q so we have our martingale condition

$$\mu + \sigma\gamma = r,$$

which has the unique solution

$$\gamma = -\frac{\mu - r}{\sigma},$$

which the reader will recognize from the Black–Scholes model. We have thus found our unique martingale measure Q so the Q-dynamics of S are

$$dS_t = rS_t dt + \sigma S_t dW_t^Q + S_{t-} \int_E \beta(z) \{\Psi(dt, dz) - \lambda(dz)dt\}. \tag{20.3}$$

Given this we may now start pricing options and other derivatives by the usual risk-neutral evaluation formula

$$\Pi_t[X] = e^{-r(T-t)} E^Q [X | \mathcal{F}_t]. \tag{20.4}$$

where X is an arbitrary T-claim.

20.2   The Merton Model

In order to use (20.3)–(20.4) in a concrete case, we must make more specific assumptions about λ and β. Here we again follow Merton. We know that in the absence of the MPP Ψ, we have a standard Black–Scholes model with log-normal prices. It is then natural to model also the relative jump size as a displaced log-normal distribution; this can be done in two natural ways.

1. We set $\beta(z) = (z - 1)$ and define E by $E = R_+$. We then define the mark distribution $\Gamma(dz)$ as a log-normal distribution on R. This means that we identify the mark z with the outcome of a log-normal variable.
2. We define β as $\beta(z) = e^z - 1$ and $\Gamma(dz)$ as a Gaussian distribution on R. This means that we identify z with the outcome of a normal variable.

Both choices are possible, but we choose the second so we have the following assumption, which specifies the Merton model.

Assumption   *We assume that the* $\beta(z) = e^z - 1$ *and that*

$$\lambda(dz) = \lambda^E f(z)dz.$$

where f is the density of a Gaussian distribution $N\left[\mu_z, \sigma_z^2\right]$. With a slight abuse of language we will, to simplify notation, write λ instead of λ^E.

Our Q-model is now as follows:

$$dS_t = rS_t dt + \sigma S_t dW_t^Q$$

$$+S_{t-}\int_R (e^z - 1)\left\{\Psi(dt, dz) - \lambda f(z)dzdt\right\}, \tag{20.5}$$

$$dB_t = rB_t dt. \tag{20.6}$$

We will thus have jumps in the price process according to a Poisson process with intensity λ, and the relative jump size will be (displaced) log-normal. We now turn to the pricing of derivatives and, in order to have any sort of analytical tractability, we restrict ourselves to the case of *simple* derivatives, i.e. we want to price a T-claim X of the form

$$X = \Phi(S_T)$$

for some deterministic function Φ. For a claim of this sort, the price process will be given by a pricing function and we now introduce some notation which will be convenient below.

Definition 20.1   We introduce two functions F^Ψ and F as follows.

- For the jump-diffusion model above we define the function $F^\Psi(t, s)$ by

$$F^\Psi(t, s) = e^{-r(T-t)}E_{t,s}^Q\left[\Phi(S_T)\right]. \tag{20.7}$$

- For the case when there is no driving point process, so $\Psi(dt, dz) = 0$ and $\lambda(dz) = 0$, we define the function $F(t, s; \sigma)$ by

$$F(t, s : \sigma) = e^{-r(T-t)}E_{t,s}^Q\left[\Phi(S_T)\right]. \tag{20.8}$$

In more concrete terms, $F^\Psi(t, s)$ is the pricing function for the full jump-diffusion model, whereas $F(t, s; \sigma)$ is the pricing function for the same claim but in a standard, purely Wiener-driven, Black–Scholes model with volatility σ. Our project is now to investigate if we, in some way, can express the pricing function F^Ψ in terms of the Black–Scholes pricing function F.

Under the assumption above we will have jump times $\{T_n\}_{n=1}^{\infty}$ according to a Poisson process N with intensity λ, and the i.i.d. marks $\{Z_n\}_{n=1}^{\infty}$ will have a normal distribution. Suppose now that we have a jump at time t. It then follows that

$$S_t = e^Z S_{t-},$$

where Z is the generic normal variable above. Suppose now that t is the first jump time of S. Then S_{t-} is log-normal (why?) and so is e^Z. Since W and Ψ are independent, this implies that also S_t will be log-normal, since it is the product of two independent log-normal variables, and this gives us some hope to be able to connect F^Ψ and F.

More specifically we solve the price equation (20.5) to obtain

$$S_t = S_0 \exp\left\{ \left(r - \frac{1}{2}\sigma^2 - \lambda m \right) t + \sigma W_t^Q + \sum_{k=0}^{N_t} Z_k \right\}. \tag{20.9}$$

Here

$$m = \int_R (e^z - 1) f(z) dz = E^Q\left[e^Z - 1 \right] = e^{\mu_z + \frac{1}{2}\sigma_z^2} - 1, \tag{20.10}$$

where Z is the generic $N[\mu_z, \sigma_z^2]$ mark variable, and N_t is the Poisson process with intensity λ counting the jump events. For notational simplicity we restrict ourselves to finding the price $F^\Psi(0, s)$ at time $t = 0$. We then condition on N_T to write the price as

$$F^\Psi(0, s) = e^{-rT} \sum_{n=0}^{\infty} E_{0,s}^Q\left[\Phi(S_T) | N_T = n \right] Q(N_T = n),$$

where

$$Q(N_T = n) = e^{-\lambda T} \frac{(\lambda T)^n}{n!}.$$

It now remains to compute $E_{0,s}^Q\left[\Phi(S_T) | N_T = n \right]$. From (20.9) we see that, conditional on $N_T = n$, we have

$$S_T = s \cdot \exp\left\{ \left(r - \frac{1}{2}\sigma^2 - \lambda m \right) T + \sigma W_T^Q + \sum_{k=0}^{n} Z_k \right\}. \tag{20.11}$$

Given the definition of the Z variables as $N[\mu_z, \sigma_z^2]$ we can write

$$S_T = se^\eta,$$

where

$$\eta \sim N\left[\left(r - \frac{1}{2}\sigma^2 - \lambda m \right) T + n\mu_z, \sigma^2 T + n\sigma_z^2 \right].$$

We can then rewrite this as

$$\eta \sim N\left[\left(r - \frac{1}{2}\sigma^2 - \lambda m \right) T + n\mu_z, \sigma_{n,T}^2 T \right],$$

where

$$\sigma_{n,T}^2 = \sigma^2 + \frac{n}{T}\sigma_z^2, \tag{20.12}$$

and as

$$\eta \sim N\left[\left(r - \sigma_{n,T}^2 T\right) + \left(\left[\sigma_{n,T}^2 T - \frac{1}{2}\sigma^2 - \lambda m\right]T + n\mu_z\right), \sigma_{n,T}^2 T\right].$$

We can thus write S_T as

$$S_T = \left[se^{\left(\sigma_{n,T}^2 T - \frac{1}{2}\sigma^2 - \lambda m\right)T + n\mu_z}\right]e^\nu, \tag{20.13}$$

where $\nu \sim N[r - \frac{1}{2}\sigma_{n,T}^2 T, \sigma_{n,T}^2 T]$. From this we see that we can write

$$e^{-rT}E_{0,s}^Q\left[\Phi(S_T)|N_T = n\right] = e^{-rT}E^Q\left[\Phi\left(se^\kappa \cdot e^{\left(r - \frac{1}{2}\sigma_{n,T}^2\right)T + \sigma_{n,T}W_T}\right)\right], \tag{20.14}$$

where

$$\kappa_T = \left(\sigma_{n,T}^2 - \frac{1}{2}\sigma^2 - \lambda m\right)T + n\mu_z = \left(\sigma^2 + \frac{n}{T}\sigma_z^2 - \lambda m\right)T + n\mu_z. \tag{20.15}$$

In the final term in (20.14) we recognize the arbitrage-free price for the claim $\Phi(S_T)$ in a standard Black–Scholes model with no point process present, with a volatility given by $\sigma_{n,T}$ and with initial stock price se^κ.

We thus have

$$e^{-rT}E_{0,s}^Q\left[\Phi(S_T)|N_T = n\right] = F(0, se^{\kappa_T}; \sigma_{n,T})$$

and by summing and an easy change of time variable we have the following classic result from Merton (1976).

Proposition 20.2 *Denote the pricing function for the derivative $\Phi(S_T)$ for the jump model by $F^\Psi(t,s)$ and the pricing function for the same derivative in a standard Black–Scholes model with Wiener volatility σ by $F(t,s;\sigma)$. Then we have*

$$F^\Psi(t,s) = \sum_{n=0}^\infty F\left(t, se^{\kappa_T - t}; \sigma_{n,T-t}\right) \cdot e^{-\lambda(T-t)}\frac{\lambda^n(T-t)^n}{n!}. \tag{20.16}$$

20.3   Pricing an Asset-or-Nothing Put Option

In this section we will exemplify the results from Section 19.7 on change of numeraire. We assume that we have the standard Merton model of the previous section. The T-claim X to be priced is given by

$$X = S_T \cdot I\{S_T \le K\}. \tag{20.17}$$

In everyday language this means that if the stock price at T is below the strike price K then you get one unit of the stock, otherwise you get nothing. We could of course also study the asset-or-nothing call (which perhaps seems more natural). We chose the put because we get less messy calculations.

In order to compute the price we will use the stock price S itself as the numeraire. From general theory in Section 16.6 we recall that for any T-claim X we have

$$\Pi_t[X] = S_t E^S \left[\frac{X}{S_T} \middle| \mathcal{F}_t \right], \tag{20.18}$$

where E^S denotes expectations under Q^S and where Q^S is the martingale measure for the numeraire S. Plugging in our expression for X we thus obtain

$$\Pi_t[X] = S_t \cdot Q_t^S (S_T \leq K), \tag{20.19}$$

where Q_t^S denotes conditional probability under Q^S. We thus have to compute the probability $Q_t^S (S_T \leq K)$. To do this we need the Q^S-dynamics of S and we will get those by starting with the Q-dynamics and then performing a Girsanov transformation from Q to Q^S.

We now recall the Q-dynamics of S as

$$dS_t = rS_t dt + \sigma S_t dW_t^Q$$
$$+ S_{t-} \int_R (e^z - 1) \{\Psi(dt, dz) - \lambda f(z) dz dt\}. \tag{20.20}$$

We also recall that the intensity process was not changed from P to Q, so λ is the Poisson intensity driving the jumps and the generic Z is $N[\mu_z, \sigma_z^2]$. From Proposition 19.19 we conclude that:

1. we can write

$$dW_t^Q = \sigma dt + dW_t^S, \tag{20.21}$$

   where W^S is Q^S-Wiener;
2. the Q^S-intensity measure of Ψ is given by

$$\lambda_t^S(dz) = e^z \lambda^Q(dz) = \lambda e^z f(z) dz. \tag{20.22}$$

Recalling the definition of f as $N[\mu_z, \sigma_z^2]$ we obtain (after some calculations)

$$\lambda_t^S(dz) = \lambda^{S,E} g(z) dz, \tag{20.23}$$

where

$$\lambda^{S,E} = \lambda e^{\mu_z + \frac{1}{2}\sigma_z^2} \tag{20.24}$$

and g is the density for $N[\mu_z + \sigma_z^2, \sigma_z^2]$. We can write the Q^S-dynamics of S as

$$dS_t = S_t \left\{ r + \sigma^2 - \int_R (e^z - 1)\lambda f(z) dz \right\} dt + \sigma S_t dW_t^S$$
$$+ S_{t-} \int_R (e^z - 1) \Psi(dt, dz). \tag{20.25}$$

We obviously have

$$\int_R 1 \lambda f(z) dz = \lambda$$

and

$$\int_R e^z \lambda f(z) dz = \int_R \lambda^{S,E} g(z) dz = \lambda^{S,E} = \lambda e^{\mu_z + \frac{1}{2}\sigma_z^2}.$$

It thus follows that

$$dS_t = S_t \{r + \sigma^2 - \lambda m\} dt + \sigma S_t dW_t^S$$
$$+ S_{t-} \int_R (e^z - 1) \Psi(dt, dz), \tag{20.26}$$

with m given by (20.10), so

$$m = e^{\mu_z + \frac{1}{2}\sigma_z^2} - 1. \tag{20.27}$$

Denoting the underlying counting process by N_t, we obtain

$$S_t = s \exp\left\{\left(r + \frac{1}{2}\sigma^2 - \lambda m\right)t + \sigma W_t^S + \sum_{k=0}^{N_t} Z_k\right\}, \tag{20.28}$$

where, under the measure Q^S, we have $Z_k \sim N[\mu_z + \sigma_z^2, \sigma_z^2]$ and N_t has intensity $\lambda^{S,E}$ given above. To compute $Q_t^S(S_T \leq K)$ we now condition on N_T and, for simplicity of notation, we set $t = 0$:

$$Q_0^S(S_T \leq K) = \sum_{n=0}^{\infty} Q^S(S_T \leq K | N_T = n) Q^S(N_T = n). \tag{20.29}$$

Conditional on $N_T = n$, we have

$$S_T = s \exp\left\{\left(r + \frac{1}{2}\sigma^2 - \lambda m\right)T + \sigma W_T^S + \sum_{k=0}^{n} Z_k\right\} \tag{20.30}$$

so we can write

$$S_T = s e^{X_{n,T}},$$

where $X_{n,T} \sim N\left[\mu_{n,T}, \sigma_{n,T}^2\right]$ and where

$$\mu_{n,T} = \left(r + \frac{1}{2}\sigma^2 - \lambda m\right)T + (\mu_z + \sigma_z^2)n \tag{20.31}$$

$$\sigma_{n,T}^2 = \sigma^2 T + n\sigma_z^2. \tag{20.32}$$

We now normalize to obtain our final result.

Proposition 20.3   *The price at $t = 0$ with $S_0 = s$, of the asset-or-nothing put is given by*

$$s \cdot \sum_{n=0}^{\infty} \Phi\left(\frac{\ln(\frac{K}{s}) - \mu_{n,T}}{\sigma_{n,T}}\right) e^{-\lambda^{S,E}T} \frac{(\lambda^{S,E}T)^n}{n!}, \tag{20.33}$$

where Φ is the cumulative distribution function for $N[0,1]$.

21 Determining a Unique Q

In this chapter we will present some of the methods that have been used in the literature in order to find a unique risk-neutral martingale measure in an incomplete market model. As a laboratory example we will mostly use a simple geometric Lévy model under P. We consider a filtered space $(\Omega, \mathcal{F}, P, \mathbf{F})$ carrying the following objects.

- A scalar Q-Wiener process W.
- A marked point process Ψ, with constant deterministic P-intensity measure $\lambda(dz)$ so we have compound Poisson.

We assume that the filtration is the internal one, so $\mathbf{F} = \mathbf{F}^{W,\Psi}$. The price dynamics are

$$dS_t = \mu S_t dt + S_t \sigma dW_t + S_{t-} \int_E \beta(z) \{\Psi(dt, dz) - \lambda(dz)dt\}, \qquad (21.1)$$

$$dB_t = r B_t dt, \qquad (21.2)$$

where μ, σ and r are constants, and β a deterministic function. In order to avoid negative stock prices we also demand that $\beta(z) \geq -1$ for all $z \in E$.

We know from Section 19.5 that, except for degenerate cases, this market is generically incomplete, and consequently there is no unique martingale measure. Nevertheless we need a unique risk-neutral measure Q if we want to price derivatives and the question is on which grounds can we choose such a Q.

Before we start searching for Q, we note, as before, that with likelihood dynamics of the form

$$dL_t = L_t \gamma_t^* dW_t + L_{t-} \int_E \varphi_t(z) \{\Psi(dt, dz) - \lambda(dz)dt\},$$

the Girsanov kernel (γ, φ) generates a martingale measure if and only if we have $\varphi(z) \geq -1$, for all $z \in E$, and

$$\mu + \sigma \gamma_t + \int_E \beta(z) \varphi_t(z) \lambda(dz) = r. \qquad (21.3)$$

21.1 A Simple Diversification Argument

The simplest way to find a unique Q is perhaps by a diversification argument. In the present context, the argument goes back to Merton (1976) and runs as follows.

- We assume that the the asset price above is just one price process in a big market with many other risky assets.
- We furthermore assume that, although the Wiener risk factor is correlated to the entire market, the jump risk is *idiosyncratic*, meaning that only the asset price S above is affected by Ψ. Other assets will be driven by other, independent, point processes.
- This would seem to imply that the *jump* risk is *diversifiable*, meaning that in a well-diversified portfolio, the jump risk will vanish. This would then seem to imply that the market price of jump risk φ_m should be zero, i.e. the Girsanov kernel φ should be zero.

If we accept this argument, the martingale equation (21.3) reduces to

$$\alpha + \sigma \gamma_t = r,$$

with the unique solution

$$\gamma_t = -\frac{\alpha - r}{\sigma}.$$

We have thus found a unique Q, and from the Girsanov theorem we deduce that

$$dW_t = \gamma dt + dW_t^Q,$$

where W^Q is Q-Wiener, and that Ψ has the same distribution under Q as under P, i.e. that $\lambda^Q(dz) = \lambda(dz)$. We thus obtain the Q-dynamics on semimartingale form as

$$dS_t = rS_t dt + S_t \sigma dW_t^Q + S_{t-} \int_E \beta(z) \{ \Psi(dt, dz) - \lambda(dz)dt \}, \qquad (21.4)$$

$$dB_t = rB_t dt. \qquad (21.5)$$

We started by specifying S as geometric Lévy process under P, and we note that, by using the diversification argument, S is also a geometric Lévy process under Q. We can thus go on, as in Section 20.2, to price derivatives by analytical or numerical methods.

We thus see that the diversification argument above leads to a comparatively tractable model under Q. From an economic point of view the argument also sounds reasonably convincing, but it leads to the following interesting question.

Problem *Is the argument above only heuristic, or is it possible to prove formally that absence of arbitrage implies that diversifiable risk factors should command zero market price of risk?*

This is a non-trivial and very interesting problem. We will provide a partial answer in Chapter 23.

21.2   The Esscher Transform

In this section we will try to determine a unique equivalent martingale measure by restricting the admissible Girsanov transformations to small subclass of transformations, parameterized by a single scalar parameter. We will do this by using the so-called *Esscher transform*, which is a particular type of absolutely continuous measure transformation.

21.2.1 Basics

The Esscher transform has a long history in insurance mathematics (see the Notes) and the general picture is that we consider a filtered space $(\Omega, \mathcal{F}, P, \mathbf{F})$ carrying a random process X.

Assumption 21.1 *We assume that $X_0 = 0$ and that X has independent stationary increments. We also assume that X has exponential moments of all orders.*

Given this assumption we define the mapping $h : R_+ \times R \rightarrow R$ by

$$h(t, a) = E\left[e^{aX_t}\right]. \tag{21.6}$$

From the independent stationary increment assumption it follows easily that

$$h(t + s, a) = h(t, a) \cdot h(s, a)$$

so we have in fact

$$h(t, a) = h^t(a), \tag{21.7}$$

where $h(a) = h(1, a)$. From this we see (why?) that, if we define the process L^a by

$$L_t^a = \frac{e^{aX_t}}{h^t(a)}, \tag{21.8}$$

then L^a is a positive $(P, \mathbf{F})$-martingale with $L_0^a = 1$. This allows us to define the Esscher transform as follows.

Definition 21.2 For any real number a, the probability measure Q^a is defined by

$$L_t^a = \frac{dQ^a}{dP}, \quad \text{on } \mathcal{F}_t. \tag{21.9}$$

The measure change from P to Q^a is referred to as an **Esscher transform**.

Suppose now, that X is a process of the form

$$dX_t = \alpha dt + \sigma dW_t + \int_E \beta(z)\Psi(dt, dz), \tag{21.10}$$

where W is Wiener and Ψ has constant deterministic intensity measure $\lambda(dz)$, so Ψ is compound Poisson.

Remark This is a very simple special case of a **Lévy process**. The difference between the process X above and a general Lévy process is that in the latter you allow for infinitely many (small) jumps in finite time. Lévy processes are very interesting and they are widely used, both in finance and other areas, but they are outside the scope of this text. See Protter (2004) for more information.

Defining X by $X_t = e^{aX_t}$, an easy application of the Itô formula shows that

$$dX_t = X_t \left[a\alpha + \frac{1}{2}\sigma^2 \right] dt + X_t a\sigma dW_t + X_{t-} \int_E \left[e^{a\beta(z)} - 1 \right] \Psi(dt, dz).$$

Since $h(t, a)$ is deterministic we can write $dh(t, a) = g(t, a)h(t, a)dt$ for some $g(t, a)$ which we do not bother to compute. Applying Itô to the formula $L_t = X_t/h$ gives us

$$dL_t = L_t \left[a\alpha - g(t, a) + \frac{1}{2}\sigma^2 \right] dt + L_t a\sigma dW_t + L_{t-} \int_E \left[e^{a\beta(z)} - 1 \right] \Psi(dt, dz).$$

Compensating the point process gives us

$$dL_t = L_t \left[a\alpha - g(t, a) + \frac{1}{2}\sigma^2 + \int_E \left[e^{a\beta(z)} - 1 \right] \lambda(dz) \right] dt + L_t a\sigma dW_t$$

$$+ L_{t-} \int_E \left[e^{a\beta(z)} - 1 \right] \{\Psi(dt, dz) - \lambda(dz)dt\}.$$

We know, however, *a priori* that L is a P-martingale, so the drift must vanish, leaving us with

$$dL_t^a = L_t^a a\sigma dW_t + L_{t-}^a \int_E \left(e^{a\beta(z)} - 1 \right) \{\Psi(dt, dz) - \lambda(dz)dt\}. \qquad (21.11)$$

From the Girsanov theorem we have the following result.

Proposition 21.3   *With X-dynamics as in (21.10) the following hold.*

- *We can write*

$$dW_t = a\sigma dt + dW_t^a, \qquad (21.12)$$

  where W^a is Q^a-Wiener.
- *The Q^a-intensity $\lambda^a(dz)$ of Ψ is given by*

$$\lambda^a(dz) = e^{a\beta(z)}\lambda(dz). \qquad (21.13)$$

- *In particular we see that Ψ is compound Poisson also under Q^a.*

Remark 21.4   Note that if we instead model X in semimartingale form as

$$dX_t = \mu dt + \sigma dW_t + \int_E \beta(z) \{\Psi(dt, dz) - \lambda(dz)dt\},$$

then we we still get exactly the same likelihood process as in (21.11) above.

21.2.2   Application to a Geometric Lévy Price Process

We now consider a geometric Lévy price process S and a bank account B with the usual dynamics

$$dS_t = \alpha S_t dt + S_t \sigma dW_t + S_{t-} \int_E \beta(z)\Psi(dt, dz), \qquad (21.14)$$

$$dB_t = rB_t dt. \qquad (21.15)$$

As we have seen earlier, this market model is incomplete so there no unique risk-neutral martingale measure. We now restrict the class of measure transformations to the subclass of Esscher transformations, using the return process as our X-process, so $dX_t = dS_t/S_{t-}$. This implies that

$$dX_t = \alpha dt + \sigma dW_t + \int_E \beta(z)\Psi(dt, dz),$$

which gives us exactly the transform studied above. We may thus use Proposition 21.3 to obtain the Q^a-dynamics of S as

$$dS_t = \left[\alpha + a\sigma^2 + \int_E \beta(z)e^{a\beta(z)}\lambda(dz)\right] S_t dt + S_t \sigma dW_t^a$$
$$+ S_{t-} \int_E \beta(z)\left\{\Psi(dt, dz) - \lambda^a(dz)dt\right\}.$$

The condition for Q^a to be a martingale measure is then

$$\alpha + a\sigma^2 + \int_E \beta(z)e^{a\beta(z)}\lambda(dz) = r \qquad (21.16)$$

and the question is now whether this equation can be solved. To see this we define the function $g(a)$ by

$$g(a) = \alpha + a\sigma^2 + \int_E \beta(z)e^{a\beta(z)}\lambda(dz).$$

We then have

$$g'(a) = \sigma^2 + \int_E \beta^2(z)e^{a\beta(z)}\lambda(dz) \geq \sigma^2 > 0,$$

which shows the following.

Proposition 21.5 *Consider the model (21.14)–(21.15). Then the following hold.*

- *There is a unique $\widehat{a}$, which solves the martingale equation (21.16).*
- *The corresponding measure $Q^{\widehat{a}}$ is the unique martingale measure generated by an Esscher transform. It is referred to as the **Esscher martingale measure** and denoted by Q^E.*
- *The semimartingale dynamics of the prices process S under the Q^E are given by*

$$dS_t = rS_t dt + \sigma S_t dW_t^E + S_{t-} \int_E \beta(z)\left\{\Psi(dt, dz) - \lambda^E(dz)dt\right\}, \qquad (21.17)$$

where W^E is Q^E-Wiener and

$$\lambda^E(z) = e^{\widehat{a}\beta(z)}\lambda(z); \qquad (21.18)$$

so Ψ is compound Poisson also under Q^E.

21.2.3   A Generalization of the Esscher Transform

Suppose now that we are in a more general situation where the S-dynamics has the form

$$dS_t = \alpha_t S_t dt + S_t \sigma_t dW_t + S_{t-} \int_E \beta_t(z) \Psi(dt, dz), \qquad (21.19)$$

$$dB_t = r B_t dt, \qquad (21.20)$$

where μ, σ, β and λ are processes instead of constants. The natural candidate for an Esscher transform would be as before; namely, the return process X with dynamics

$$dX_t = \alpha_t dt + \sigma_t dW_t + \int_E \beta_t(z) \Psi(dt, dz). \qquad (21.21)$$

Unfortunately, since this process does not have independent and stationary increments, it cannot be used directly for an Esscher transform. Thus formula (21.8) cannot be directly generalized to our new model. On the other hand, formula (21.11) can easily be extended to our present case.

Definition 21.6   Assume that X is given by (21.21). For any predictable process a_t, we then define the **generalized Esscher transform** by the likelihood process

$$dL_t^a = L_t^a a_t \sigma_t dW_t + L_{t-}^a \int_E \left(e^{a_t \beta_t(z)} - 1 \right) \{ \Psi(dt, dz) - \lambda_t(dz) dt \}. \qquad (21.22)$$

We may now proceed exactly as before and obtain a unique martingale measure by solving the equation

$$\alpha_t + a_t \sigma_t^2 + \int_E \beta_t(z) \left(e^{a_t \beta_t(z)} - 1 \right) \lambda_t(dz) = r \qquad (21.23)$$

for the process a.

21.3   The Minimal Martingale Measure

If we look for a unique martingale measure, then a very natural idea is to choose that measure Q which, in some sense, is closest to P. Suppose that we have a rather general model of the form

$$dS_t = \alpha_t S_t dt + S_t \sigma_t dW_t + S_{t-} \int_E \beta_t(z) \{ \Psi(dt, dz) - \lambda_t(dz) dt \}, \qquad (21.24)$$

$$dB_t = r B_t dt, \qquad (21.25)$$

with a scalar Wiener process and and a general MPP Ψ with intensity process $\lambda_t(dz)$. As usual we go from P to Q via a Girsanov transformation of the form

$$dL_t = L_t \gamma_t dW_t + L_{t-} \int_E \varphi_t(z) \{ \Psi(dt, dz) - \lambda_t(dz) dt \},$$

with $\varphi \geq -1$, and, also as usual, we have

$$\lambda_t^Q(dz) = [\varphi_t(z) + 1] \lambda_t(dz).$$

It is now natural to look at $d_t^2(P, Q)$, defined by

$$d_t^2(P, Q) = \gamma_t^2 + \int_E \varphi_t^2(Z)\lambda(dz),$$

as a measure of how far Q is from P (at time t). We see that this is really a (squared) norm in a product Hilbert space, so in fact we have

$$d_t^2(P, Q) = \|\gamma_t\|_{R^d}^2 + \|\varphi_t\|_{\lambda_t}^2.$$

The martingale condition for the Girsanov transformation is, as always, given by the equation

$$r = \mu_t + \sigma_t \gamma_t + \int_E \beta_t(z)\varphi_t(z)\lambda_t(dz), \tag{21.26}$$

Definition 21.7   Consider the problem of minimizing

$$\|\gamma_t\|_R^2 + \|\varphi_t\|_{\lambda_t}^2, \tag{21.27}$$

subject to the constraints

$$r = \mu_t + \sigma_t \gamma_t + \int_E \beta_t(z)\varphi_t(z)\lambda_t(dz). \tag{21.28}$$

We denote by γ^M and φ^M the optimal Girsanov kernels for the problem above, and define the **minimal martingale measure** Q^M as the measure generated by γ^M and φ^M.

We note that the problem above is a minimum norm problem with a single scalar constraint. A standard result in functional analysis is that the optimal vector $(\widehat{\gamma}_t, \widehat{\varphi}_t)$ must be parallel to the vector (σ, β_t). We thus have

$$\widehat{\gamma}_t = \eta\sigma_t,$$
$$\widehat{\varphi}_t(z) = \eta\beta_t(z),$$

where the constant η is determined by the linear constraint above, so we have in fact

$$\eta = \frac{\mu_t - r}{\sigma_t^2 + \int_E \beta_t^2(z)\lambda_t(dz)}.$$

21.4   Minimizing f-Divergence

We now return to the problem of defining a concept of "closeness" between the two measures P and Q. This will be done using the concept of f-divergence.

21.4.1   Definition and Basic Properties

One possible way of defining how far apart two measures are, would of course be to define a metric on the convex space of probability measures, but it turns out that another construction is very useful in statistics, finance, physics and in many other applications.

Definition 21.8 Let $f : R_+ \to R$ be a strictly convex function with $f(1) = 0$. Let furthermore P and Q be two measures such that $Q \ll P$ on $(\Omega, \mathcal{F})$, and let L denote the Radon–Nikodym derivative

$$L = \frac{dQ}{dP}, \quad \text{on } \mathcal{F}.$$

The f-**divergence** $f(Q, P)$ between P and Q is defined by

$$f(Q, P) = \begin{cases} E^P [f(L)] & \text{if the expectation above is well defined} \\ f(Q, P) = +\infty & \text{otherwise.} \end{cases} \quad (21.29)$$

We start by stating two simple properties of the f-divergence.

Proposition 21.9 *The following hold.*

(1) *We have* $f(Q, P) \geq 0$ *for all* $Q \ll P$.
(2) $f(Q, P) = 0$ *if and only if* $Q = P$.

Proof By the Jensen inequality we have

$$f(Q, P) = E^P [f(L)] \geq f \left(E^P [L] \right) = f(1) = 0,$$

which proves (1). Since f is strictly convex, the Jensen inequality is an equality if and only if L is deterministic, and this happens if and only if $L = 1$, which is equivalent to saying that $Q = P$. $\square$

It follows from this result that it is natural to interpret $f(Q, P)$ as a measure of the "distance" between Q and P. Note, however, that in general $f(Q, P)$ is not a metric.

Definition 21.10 Let $\mathcal{K}$ be a convex set of probability measures, dominated by P, and consider the problem of minimizing $f(Q, P)$, over $Q \in \mathcal{K}$. If there exists an optimal Q^*, i.e.

$$f(Q^*, P) = \inf_{Q \in \mathcal{K}} F(Q, P),$$

then we say that Q^* is the f-**projection** of P onto $\mathcal{K}$: we denote it by $f(\mathcal{K}, P)$.

The interpretation is of course that Q^* is the measure $Q \in \mathcal{K}$ which is closest to P in the sense of f-divergence.

Below is a list of some of the most commonly used f-divergences and their trade names.

$$\begin{array}{rcll} f(x) & = & x \ln(x) & \text{Relative entropy} \\ f(x) & = & -\ln(x) & \text{Reverse relative entropy} \\ f(x) & = & x^2 - 1 & \text{Variance} \\ f(x) & = & \frac{1}{2}|x - 1| & \text{Total variation} \\ f(x) & = & (\sqrt{x} - 1)^2 & \text{Hellinger distance} \end{array}$$

The way the f-divergence is typically used in finance is as follows, where we are given a model of an incomplete market.

1. Set $\mathcal{K} = \mathcal{M}$, where $\mathcal{M}$ is the class of equivalent martingale measures.

2. Choose a concrete f-divergence from the list above.
3. Minimize the divergence $f(Q, P)$ over $Q \in \mathcal{M}$, thus projecting P on $\mathcal{M}$.
4. Use the f-projection Q^* to price all derivatives in the market.

Depending on the choice of f we then obtain such objects as the minimal entropy measure, the minimal reverse entropy measure, the minimal variance measure, etc.

It is not an easy task to determine the f-projection for a concrete model, the only exception being the reverse entropy $f(x) = -\ln(x)$, which is treated in Section 21.4.2 below. These problems have been intensively studied in the literature and there is also an interesting duality theory for f-divergence. See the Notes. The most frequently used divergence in finance seems to be the relative entropy.

21.4.2   Minimal Reverse Entropy

We consider our usual scalar model

$$dS_t = \mu_t S_t dt + S_t \sigma_t dW_t + S_{t-} \int_E \beta_t(z) \{\Psi(dt, dz) - \lambda_t(dz)dt\}, \quad (21.30)$$

$$dB_t = r B_t dt, \quad (21.31)$$

and we want to minimize reverse relative entropy over the class of martingale measures. As usual we go from P to Q via a Girsanov transformation of the form

$$dL_t = L_t \gamma_t dW_t + L_{t-} \int_E \varphi_t(z) \{\Psi(dt, dz) - \lambda_t(dz)dt\},$$

with $\varphi \geq -1$, and, also as usual, we have

$$\lambda_t^Q(dz) = [\varphi_t(z) + 1] \lambda_t(dz).$$

The mathematical problem is then to minimize $-E^P[\ln(L_T)]$, which implies that we have the following problem.

$$\text{maximize} \quad E^P[\ln(L_T)]$$

over (γ, φ), subject to the martingale constraint

$$r = \mu_t + \sigma_t \gamma_t + \int_E \beta_t(z) \varphi_t(z) \lambda_t(dz). \quad (21.32)$$

From Itô we get

$$d \ln(L_t) = -\left\{ \int_E \varphi_t(z) \lambda(dz) + \frac{1}{2} \gamma_t^2 \right\} dt + \gamma_t dW_t + \int_E \ln[1 + \varphi_t(z)] \Psi(dz, dt).$$

Compensating the jump part gives us

$$d \ln(L_t) = -\left\{ \int_E [\varphi_t(z) - \ln(1 + \varphi_t(z))] \lambda(dz) + \frac{1}{2} \gamma_t^2 \right\} dt$$

$$+ \gamma_t dW_t + \int_E \ln[1 + \varphi_t(z)] \{\Psi(dz, dt) - \lambda_t(z)dt\},$$

and we obtain

$$
\ln(L_T) = - \int_0^t \left\{ \int_E [\varphi_t(z) - \ln(1 + \varphi_t(z))] \, \lambda(dz) + \frac{1}{2} \gamma_t^2 \right\} dt
$$

$$
+ \int_0^T \gamma_t \, dW_t + \int_0^T \int_E \ln[1 + \varphi_t(z)] \, \{\Psi(dz, dt) - \lambda_t(z)dt\} \, .(21.33)
$$

The two last terms are P-martingale terms so, taking expectations, we obtain

$$
E^P [\ln(L_T)] = -E^P \left[\int_0^t \left\{ \int_E [\varphi_t(z) - \ln(1 + \varphi_t(z))] \, \lambda(dz) + \frac{1}{2} \gamma_t^2 \right\} dt \right]. \qquad (21.34)
$$

Our problem is thus the following:

$$
\text{minimize} \quad E^P \left[\int_0^t \left\{ \int_E [\varphi_t(z) - \ln(1 + \varphi_t(z))] \, \lambda(dz) + \frac{1}{2} \gamma_t^2 \right\} dt \right]
$$

subject to the constraints

$$
r = \mu_t + \sigma_t \gamma_t + \int_E \beta_t(z) \varphi_t(z) \lambda_t(dz). \qquad (21.35)
$$

This is a decoupled problem in the t-variable so we can, for every t and ω, solve the standard optimization problem to minimize

$$
\int_E [\varphi_t(z) - \ln(1 + \varphi_t(z))] \, \lambda(dz) + \frac{1}{2} \gamma_t^2
$$

subject to the constraints above. The relevant Lagrangian $\mathcal{L}$ is given by

$$
\int_E [\varphi_t(z) - \ln(1 + \varphi_t(z))] \, \lambda(dz) + \frac{1}{2} \gamma_t^2 - \eta \left(\mu_t + \sigma_t \gamma_t + \int_E \beta_t(z) \varphi_t(z) \lambda_t(dz) \right).
$$

The problem of minimizing $\mathcal{L}$ is decoupled in the z-variable, and $\lambda_t(z)$ is positive, so we are left with the problem of minimizing

$$
\varphi_t(z) - \ln(1 + \varphi_t(z)) + \frac{1}{2} \gamma_t^2 - \eta \gamma_t \sigma - \eta \beta_t(z) \varphi_t(z)
$$

over $\varphi_t(z)$ and γ_t for every fixed (t, z). The first-order conditions give us, after some manipulations, the following result.

Proposition 21.11 *The minimal reverse entropy measure, denoted by Q^{RE}, is generated by the Girsanov kernel*

$$
\varphi_t(z) = \frac{\eta \beta_t(z)}{1 - \eta \beta_t(z)}, \qquad (21.36)
$$

$$
\gamma_t = \eta \sigma, \qquad (21.37)
$$

where the Lagrange multiplier η is determined by the constraint (21.35).

21.5 Utility Indifference Pricing

We end the chapter by a very brief discussion of utility indifference pricing.

21.5.1   Global Indifference Pricing

Consider an incomplete market and a fixed T-claim Y. In this market we have an agent with utility function U, so the utility, at time $t = 0$, of receiving the claim at time $t = T$ is given by $U(Y)$. The agent has initial capital x. Let us now define the function $V(x, z)$ by

$$V(x, z) = \sup_{X_T \in \mathcal{K}(x)} E^P \left[U(X_T + zY) \right], \qquad (21.38)$$

where $\mathcal{K}(x)$ denotes the wealth profiles at time T that can be reached by a self-financing portfolio with initial capital x. The entity $V(x, z)$ is thus the maximal utility that you can obtain by starting with initial capital x, trading in the market and at time $t = T$, in addition to your portfolio value X_T, getting z units of Y for free. Now, in real life you seldom get anything for free, so the question is how much you are willing to pay, at time $t = 0$, for these z units of Y. A reasonable answer to this question is the *utility indifference buy price* $p(z)$ which is defined as follows.

Definition 21.12   The **utility indifference buy price** $p(z)$ is defined by

$$V(x - p(z), z) = V(x, 0). \qquad (21.39)$$

In other words, you are indifferent between the following two alternatives:

- Paying $p(z)$ at time $t = 0$ and receiving z units of Y above your portfolio value X_T at time $t = T$.
- Paying zero at time $t = 0$ and receiving nothing extra above your portfolio value X_T at $t = T$.

Note that $p(z)$ will typically be non-linear in z. There is a substantial literature on global indifference pricing but the problems are hard, since in order to solve (21.39) we need to find the function V from (21.38). However, to compute V we need to solve a utility maximization problem in an incomplete market and such problems are notoriously very difficult.

21.5.2   Marginal Indifference Pricing

The reader with a background in microeconomics may feel uncomfortable with global utility indifference pricing, the reason being that in microeconomics "all" prices are determined by *marginal* objects, such as marginal utility, marginal productivity, marginal rates of substitution etc. Therefore it seems reasonable to develop a theory of marginal utility indifference pricing and in doing so we follow Davis (1997).

Consider the incomplete market and the claim X of the previous section and define the functions $V(x)$ and $V(x, p, \delta)$ by

$$V(x) = \sup_{X_T \in \mathcal{K}_T(x)} E^P \left[U(X_T) \right], \qquad (21.40)$$

$$V(x, p, \delta) = \sup_{X_T \in \mathcal{K}_T(x)} E^P \left[U \left(X_T + \frac{\delta}{p} X \right) \right]; \qquad (21.41)$$

so we have in fact $V(x) = V(x, p, 0)$ for all (x, p). The function V is the usual indirect utility function of wealth and the interpretation of $V(x, p, \delta)$ is as follows.

- The price (at $t = 0$) per unit of the derivative X is given by p.
- At time $t = 0$ you get (for free) δ dollars to invest in the derivative X. This will allow you to buy δ/p units of X.
- Your initial wealth to invest in the stock market is still equal to x.
- At time T you will have your portfolio value X_T plus the value of δ/p units of X.
- The value $V(x, p, \delta)$ is your utility after investing optimally in the stock market.

We can now define the marginal indifference price.

Definition 21.13 Assume that $V(x, p, \delta)$ is continuously differentiable, and that there is a unique solution $p(x)$ to the equation

$$\frac{\partial V}{\partial \delta}(x, p, 0) = \frac{\partial V}{\partial x}(x, p, 0). \tag{21.42}$$

Then we refer to $p(x)$ as the **marginal indifference price**.

The intuition of this is that, at the indifference price, you are indifferent between investing a small amount in the stock market or investing the same amount in the derivative X. From this argument it seems that the marginal indifference pricing is, in some sense, closely connected to equilibrium pricing and we will see that this is indeed the case.

It follows from an envelope theorem that

$$\frac{\partial V}{\partial \delta}(x, p, 0) = E^P \left[U' \left(\widehat{X}_T \right) \frac{X}{p} \right],$$

where prime denotes derivative, and one can prove (see Björk, 2020) that

$$\frac{\partial V}{\partial x}(x, p, 0) = E^P \left[U' (X_T) \right].$$

We thus have the equation

$$E^P \left[U' \left(\widehat{X}_T \right) \frac{X}{p} \right] = E^P \left[U' \left(\widehat{X}_T \right) \right],$$

which gives us our main result.

Proposition 21.14 *Suppose that $V(x)$ is differentiable and that $V'(x) > 0$ for all x. The the marginal indifference price is given by*

$$p(x) = \frac{E^P \left[U'(\widehat{X}_T)X \right]}{V'(x)} \tag{21.43}$$

or, equivalently, by

$$p(x) = \frac{E^P \left[U'(\widehat{X}_T)X \right]}{E^P \left[U' \left(\widehat{X}_T \right) \right]}.$$

From a microeconomic point of view, this is a very pleasing result. We can interpret the object

$$\frac{U'(\widehat{X}_T)}{E^P\left[U'\left(\widehat{X}_T\right)\right]} \tag{21.44}$$

as an equilibrium stochastic discount factor, and formulas of this type will turn up again in the context of dynamic equilibrium theory in Chapter 26.

21.6   Notes

The Esscher transformation was introduced in Esscher (1932) and further developed in Gerber & Shiu (1994), Kallsen & Shiryayev (2002) and many other papers. The minimal martingale measure turns up as a natural object in many areas of finance, such as quadratic hedging theory and the theory of local risk minimization. It is a canonical object of study in mathematical finance and it has been the object of intensive research. See Föllmer & Sondermann (1986), Schweizer (1991) and Schweizer (2001). Goll & Rüschendorff (2001) is a basic reference for the general theory of f-divergence and the connections to finance. Minimal entropy was introduced in Miyahara (1976) and developed by, among others, Frittelli in several papers such as Frittelli (2000). For a detailed study of entropy in Lévy-driven markets, including an extensive bibliography, see Miyahara (2011). For a collection of papers on indifference pricing see Carmona (2009). The basic reference on marginal indifference pricing is Davis (1997). For a textbook treatment of hedging in incomplete markets, see Dana & Jeanblanc (2003).

22 Good-Deal Bounds

In Chapter 21 we discussed incomplete markets and we considered several ways of choosing one "canonical" martingale measure from the (infinite) class of equivalent martingale measures. The approach of the present chapter is different in the sense that we do *not* look for a unique EMM. Instead we try to derive reasonable (in some sense) pricing *bounds* for financial derivatives.

22.1 General Ideas

As an example, let us consider a T-claim X on some incomplete market. One possible way of obtaining pricing bounds would be to compute the upper and lower no-arbitrage bounds. This means that, for every t, we compute the upper bound $\Pi_t^s[X]$ and the lower bound $\Pi_t^i[X]$ as

$$\Pi_t^s[X] = \sup_{Q \in M} e^{-r(T-t)} E^Q[X|\mathcal{F}_t],$$

$$\Pi_t^i[X] = \inf_{Q \in M} e^{-r(T-t)} E^Q[X|\mathcal{F}_t],$$

where s and i stand for sup and inf respectively. It is then clear that for any arbitrage-free price process $\Pi_t[\Phi]$ we must have

$$\Pi_t^i[X] \leq \Pi_t[X] \leq \Pi_t^s[X].$$

The drawback with this approach is that the bounds obtained in this way are typically extremely wide and completely useless from a practical point of view. As an example of this consider the following lottery:

- The lottery will take place once and only once in history (so it is not repeated) and only one lottery ticket is sold.
- The outcome of the lottery is decided by the flip of a fair coin
  - If the result is "heads", then the holder of the ticket receives one million dollars.
  - If the result is "tails", then the holder of the ticket receives nothing.
- The price of the lottery ticket is one dollar.

It is clear that, from a commonsense point of view, pricing the lottery ticket by setting the price to one dollar, as we did above, amounts to an almost astronomical mispricing and you would of course never expect a price system like this in any "reasonable" market. Note, however, that the pricing of the ticket *is* arbitrage-free.

This example indicates that we should perhaps rule out not only those prices which are violating the no-arbitrage restriction, but also those prices which in some sense would represent "deals which are too good".

We then have to formalize the idea of a "good deal" and this is exactly what was done in Cochrane & Saá Requejo (2000). They defined a "good deal" as an asset price process with a high Sharpe ratio (where the term "high" obviously has to be defined). To see more concretely what is gong on, we recall the definition of the Sharpe ratio SR of an asset as

$$SR_t = \frac{\mu_t - r_t}{v_t},$$

where μ is the local mean rate of return, r is the short rate and v is the volatility. The way we think about this is that the Sharpe ratio is the risk premium per unit of volatility, so it tells us something about the aggregate risk-aversion on the market. To be more concrete, let us assume that the underlying asset is a common stock. Given empirical data, a reasonable volatility would then be anything in the range 20–60% per annum, so let us assume that $v = 40\%$ per annum. In percentage terms we would then have

$$\mu - r = SR \cdot 40.$$

This implies that a Sharpe ratio of SR $= 5$ implies that the asset would have an excess rate of return of 200% per annum, which of course is extremely high. Empirical studies do in fact tell us that typical values of the Sharpe ratio are somewhere in the interval $[0.5, 1]$.

As a first attempt, a mathematical formalization of the pricing problem would then be to find the maximum (minimum) arbitrage-free price process for the derivative, subject to an upper bound on the Sharpe ratio. However, this way of formalizing the problem turns out have two major drawbacks.

1. The optimization problem turns out to be mathematically intractable.
2. A much more serious problem is the following: suppose that we have found upper and lower pricing bounds on a derivative, subject to a bound on the Sharpe ratio of the derivative. Then it may in principle still be possible to form a self-financing portfolio, based on the underlying assets and the newly introduced derivative, such that the portfolio has a very high Sharpe ratio.

What we need is thus a formalization of the pricing problem which gives us a mathematically tractable problem and which at the same time allows us to have complete control over the Sharpe ratios of *all portfolios* based on the underlying assets and the derivative.

This is precisely where the Hansen–Jagannathan bounds from Section 19.3 comes in useful, and the idea in Cochrane & Saá Requejo (2000) was that instead of putting a bound on the Sharpe ratio of the derivative under study, we put a bound on the right-hand side of the Hansen–Jagannathan inequality (i.e. the norm of the market price of risk vector). In the final formulation, the pricing problem is thus that of finding the maximum (minimum) arbitrage-free price process, given a bound on the right-hand side of the HJ inequality.

In the original paper (Cochrane & Saá Requejo 2000) this was done under the objective measure P, using the formalism of stochastic discount factors, but we will instead follow Björk & Slinko (2006) where the analysis is done under Q using the formalism of martingale measures rather than SDFs, the reason being that the analysis becomes much more streamlined in this way.

22.2   The Model

We now move on to a concrete model and consider a simple factor model of the form

$$dS_t = \alpha(X_t)S_t dt + S_t \sigma(X_t)dW_t + S_{t-} \int_E \beta(X_{t-}, z)\Psi(dt, dz), \tag{22.1}$$

$$dX_t = a(X_t)dt + b(X_t)dW_t + S_{t-} \int_E c(X_{t-}, z)\Psi(dt, dz), \tag{22.2}$$

$$dB_t = r(X_t)B_t dt, \tag{22.3}$$

where S is the price process of a traded asset and X is some underlying non-financial factor process. The model can easily be extended to a multidimensional setting and the coefficients can be allowed to depend on S as well as on X, but the present model is complex enough to illustrate the main ideas.

The precise probabilistic specification of the model is given by the following standing assumption.

Assumption 22.1

1. *We assume that $\alpha(y)$ and $a(y)$ are deterministic scalar functions, $\sigma(y)$ and $b(y)$ are deterministic two-dimensional row-vector functions and $\beta(y, z)$ and $c(y, z)$ are deterministic scalar functions. In order to avoid negative asset prices we also assume that $\beta(y, z) \geq -1$ for all i and all (y, z).*
2. *All functions above are assumed to be regular enough to allow for the existence of a unique strong solution for the system of SDEs.*
3. *The point process Ψ has a predictable P-intensity measure λ. More precisely we assume that the P-compensator $\nu(dt, dx)$ has the form*

$$\nu(dt, dz) = \lambda(X_{t-}, dz)dt, \tag{22.4}$$

   where $\lambda(y, dz)$ is deterministic. For brevity of notation we will often denote $\lambda(X_{t-}, dz)$ by $\lambda_t(dz)$.
4. *We assume the existence of a short rate r of the form*

$$r_t = r(X_t).$$

5. *We assume that the model is free of arbitrage in the sense that there exists a (not necessarily unique) risk-neutral martingale measure Q.*

22.3   The Good-Deal Bounds

Given the setup above, our task is to price a T-claim X of the form

$$X = \Phi(S_T, X_T).$$

We start by searching for a martingale measure Q so, as usual, we introduce a likelihood process $L = dQ/dP$ by

$$\begin{cases} dL_t & = & L_t\gamma_t^* dW_t + L_{t-} \int_E \varphi_t(z) \left\{ \Psi(dt, dz) - \lambda_t(dz)dt \right\}, \\ L_0 & = & 1 \end{cases}$$

(22.5)

where $\varphi \geq -1$. The Girsanov theorem implies that $dW_t = \gamma_t dt + dW_t^Q$ and that that the Q-intensity of Ψ is given by $\lambda_t^Q(dz) = [\varphi_t(z) + 1]\lambda(X_{t-}, dz)$. Using this, the condition for Q to be a martingale measure is easily seen to be

$$\alpha(y) + \gamma_t\sigma(y) + \int_E \beta(y, z)\left\{1 + \varphi_t(z)\right\}\lambda(y, dz) = r(y),$$

and from Theorem 19.11 we recall the Hansen–Jagannathan bounds as

$$(SR_t)^2 \leq \|\gamma_t\|_{R^d} + \int_E \varphi_t^2(z)\lambda(y, dz).$$

We are now ready to define the upper and lower good-deal pricing bounds.

Definition 22.2   The **upper good-deal pricing bound** process is defined as the optimal value process for the following optimal control problem:

$$\max_{\gamma, \varphi} \quad e^{-r(T-t)} E^Q\left[\Phi(S_T)| \mathcal{F}_t\right],$$

(22.6)

where (S, X) have Q-dynamics

$$dS_t = S_t \left\{r(X_t) - \int_E \beta(z)\left\{1 + \varphi_t(z)\right\}\lambda(dz)\right\} dt + S_t\sigma(X_t)dW_t^Q$$

$$+ S_{t-}\int_X \beta(z)\Psi(dt, dz),$$

(22.7)

$$dX_t = \left\{a(X_t) + b(X_t)\gamma_t\right\} dt + b(X_t)dW_t^Q$$

$$+ S_{t-}\int_E c(X_{t-}, z)\Psi(dt, dz).$$

(22.8)

The process W^Q is Q-Wiener, and Ψ has Q-intensity $\lambda_t^Q(dz)$ given by

$$\lambda_t^Q(dz) = [\varphi_t(z) + 1]\lambda(X_{t-}, dz).$$

(22.9)

The predictable processes γ and φ are subject to the constraints

$$\alpha(y) + \sigma(y)\gamma_t + \int_E \beta(y, z)\left\{1 + \varphi_t(z)\right\}\lambda(y, dz) = r(y),$$

(22.10)

$$\|\gamma_t\|_{R^d}^2 + \int_E \varphi_t^2(z)\lambda(y, dz) \leq B^2,$$

(22.11)

$$\varphi_t(z) \geq -1, \quad \text{for all } t, z. \quad (22.12)$$

Some comments are perhaps in order.

- The expected value in (22.6) is the standard risk-neutral valuation formula for contingent claims.

- In (22.10) we have the conditions on h and φ, guaranteeing that the induced measure Q is indeed a martingale measure for S.

- The induced Q-dynamics of S and X are given in (22.7)–(22.8).

- The constraint (22.11) is the constraint to rule out "good deals", not only for the underlying asset price S, but also for all derivatives and all portfolios based on those underlying assets and derivatives.

- The constraint (22.12) is needed to ensure that Q is a positive measure.

- In order to obtain the lower pricing bound, we solve the corresponding minimum problem.

Remark   As usual in a Markovian optimal control problem, we restrict ourselves to feedback control laws, so φ and h are of the form

$$\gamma_t = \gamma_t(t, S_{t-}, X_{t-}),$$
$$\varphi_t(z) = \varphi(t, S_{t-}, X_{t-}, z),$$

where, with a slight abuse of notation, γ and φ in the right-hand side of the equations denote deterministic functions.

Since we are in a standard setting for dynamic programming (DP), we know from general DP-theory that the optimal value function $V(t, s, y)$ will satisfy the Hamilton–Jacobi–Bellman equation on the time interval $[0, T]$. After some standard calculations we thus have the following result.

Theorem 22.3   *The upper good-deal bound function is the solution V to the following boundary value problem:*

$$\frac{\partial V}{\partial t}(t, s, y) + \sup_{\gamma_t, \varphi} \mathbf{A}^{\gamma_t, \varphi} V(t, s, y) - r(y)V(t, s, y) = 0, \tag{22.13}$$

$$V(T, s, y) = \Phi(s, y), \tag{22.14}$$

subject to the constraints

$$\alpha(y) + \sigma(y)\gamma_t(t, s, y) + \int_E \beta(y, z) \{1 + \varphi_t(t, s, y, z)\} \lambda(y, dz) = r(y),$$

$$\|\gamma_t(t, s, y)\|^2 + \int_E \varphi^2(t, s, y, z)\lambda(y, dz) \leq B^2,$$

$$\varphi(t, s, y, z) \geq -1.$$

The infinitesimal operator $\mathbf{A}^{\gamma_t, \varphi}$ *is given by*

$$\mathbf{A}^{\gamma, \varphi} V(t, s, y) = s \left\{ r - \int_E \beta(y, z) \{1 + \varphi(t, y, z)\} \lambda(y, dz) \right\} \frac{\partial V}{\partial s}(t, s, y)$$

$$+ \frac{\partial V}{\partial y}(t, s, y) \{a(y) + b(y)\gamma_t^*(t, s, y)\}$$

$$+ \frac{1}{2} \frac{\partial^2 V}{\partial s^2}(t, s, y)s^2\sigma^2(y) + \frac{1}{2} \frac{\partial^2 V}{\partial y^2}(t, s, y)b^2(y)$$

$$+ \frac{\partial^2 V}{\partial s \partial y}(t, s, y)s\sigma(y)b(y).$$

$$+ \int_E V_\beta(t, s, y, z) \{1 + \varphi(t, s, y, z)\} \lambda_t(s, y, dz),$$

where V_β is defined by

$$V_\beta(t, s, y, z) = V(t, s[1 + \beta(y, z)], y + c(y, z)) - V(t, s, y).$$

As usual in dynamic programming we have a static mathematical programming problem embedded in the HJB equation; in other words, we need to find

$$\sup_{\gamma, \varphi} \mathbf{A}^{h, \varphi} V(t, s, y).$$

We must point out that this is generically a hard problem. The vector γ is a finite-dimensional vector, but $\varphi(z)$ has to be determined for each $z \in E$. Thus, if E contains infinitely many points then φ is a vector in an infinite-dimensional function space, implying that we have a fully-fledged variational problem sitting in the HJB equation. There is very little hope for explicit solutions, so one is forced to numerical methods.

22.4 An Example

Assumption 22.4 *We consider a financial market and a scalar price process S satisfying the SDE*

$$dS_t = S_t \alpha dt + S_t \sigma dW_t + S_{t-} \int_E \beta(x) \Psi(dt, dx). \tag{22.15}$$

For this model we furthermore assume that

1. *the Wiener process W is one-dimensional;*
2. *the drift α and diffusion volatility σ are deterministic constants;*
3. *the jump function β is a time-invariant deterministic function of x only, i.e. β is a mapping $\beta : E \to R$;*
4. *the point process Ψ has a P-intensity of the form $\lambda(dz)$, where λ is a time-invariant deterministic finite non-negative measure on $(E, \mathcal{E})$;*
5. *the short rate r is constant.*

Under these assumptions the model parameters α, σ, β and λ are thus deterministic objects which do not depend on the stock price S. In particular the assumption about λ implies that the point process Ψ is compound Poisson.

$$\frac{1}{\lambda(E)}\lambda(dx). \tag{22.16}$$

In order to get a feeling for the techniques used, we study a very simple example.

22.4.1   The Poisson–Wiener Model

The simplest special case in the jump-diffusion setting above is when we define the point process Ψ as a standard Poisson process with constant intensity. In terms of the notation above this means that the mark space E contains a single point denoted by z_0. Hence $E = \{z_0\}$, the measure $\lambda(dz)$ is just a point mass $\lambda(z_0)$ at z_0 and the jump function β is just a real number $\beta(z_0)$. For brevity we will denote $\lambda(z_0)$ by λ and $\beta(z_0)$ by β. We thus have the following P-dynamics of S:

$$dS_t = S_t\alpha dt + S_t\sigma dW_t + S_{t-}\beta dN_t, \tag{22.17}$$

where N is Poisson with constant intensity λ.

In this case the kernel function γ_t is scalar and the kernel φ_t does not depend upon z. The upper good-deal bound function $V(t, s)$ is the solution to the following boundary value problem:

$$\frac{\partial V}{\partial t}(t, s) + \sup_{\gamma,\varphi} \{\mathbf{A}^{\gamma,\varphi}V(t, s)\} - rV(t, s) = 0, \tag{22.18}$$

$$V(T, s) = \Phi(s), \tag{22.19}$$

where for the moment we suppress the constraints and where

$$\mathbf{A}^{\gamma,\varphi}V(t, s) = \frac{\partial V}{\partial s}s\{r - \beta\lambda(1 + \varphi)\} + \frac{1}{2}s^2\sigma^2\frac{\partial^2 V}{\partial s^2}$$
$$+ \{V(t, s(1 + \beta)) - V(t, s)\}\lambda(1 + \varphi). \tag{22.20}$$

The static optimization problem thus becomes

Problem 22.5

$$\max_{h,\varphi} \quad \{V(t, s(1 + \beta)) - V(t, s) - V_s(t, s)s\beta\}\varphi \tag{22.21}$$

subject to the constraints

$$\alpha + \sigma\gamma + \beta\lambda\{1 + \varphi\} = r, \tag{22.22}$$

$$\|\gamma\|^2 + \varphi^2\lambda \le B^2, \tag{22.23}$$

$$\varphi \ge -1. \tag{22.24}$$

To study the static problem in more detail we need some notation.

Definition 22.6 Define $(\gamma_{\max}, \varphi_{\max})$ as the optimal solution to the programming problem

$$\max_{\gamma, \varphi} \varphi, \qquad (22.25)$$

subject to the constraints (22.22)–(22.24); and $(\gamma_{\min}, \varphi_{\min})$ as the optimal solution to the problem

$$\min_{\gamma, \varphi} \varphi, \qquad (22.26)$$

subject to the same constraints.

We will need $\gamma_{\max}$, $\varphi_{\max}$, $\gamma_{\min}$, $\varphi_{\min}$ below, so we should describe these constants in terms of the given model parameters. This is a simple exercise in constrained optimization theory, but a bit messy, and the result is as follows.

Lemma 22.7 *Denote the excess return $\alpha + \beta\lambda - r$ by R. Then the following hold.*

- *The constants $\gamma_{\max}$ and $\varphi_{\max}$ are given by*

$$\gamma_{\max} = -\frac{\sigma R}{(\sigma^2 + \beta^2\lambda)\lambda} - \frac{\beta\sqrt{B^2(\sigma^2 + \beta^2\lambda) - R^2)}}{(\sigma^2 + \beta^2\lambda)\sqrt{\lambda}}, \qquad (22.27)$$

$$\varphi_{\max} = -\frac{\beta R}{\sigma^2 + \beta^2\lambda} + \frac{\sigma\sqrt{B^2(\sigma^2 + \beta^2\lambda) - R^2)}}{(\sigma^2 + \beta^2\lambda)\sqrt{\lambda}}. \qquad (22.28)$$

- *The constants $\gamma_{\min}$ and $\varphi_{\min}$ are given by the following expressions.*
  1. *If*

$$-\frac{\beta R}{\sigma^2 + \beta^2\lambda} - \frac{\sigma\sqrt{B^2(\sigma^2 + \beta^2\lambda) - R^2)}}{(\sigma^2 + \beta^2\lambda)\sqrt{\lambda}} > -1, \qquad (22.29)$$

  then

$$\gamma_{\min} = -\frac{\sigma R}{(\sigma^2 + \beta^2\lambda)\lambda} + \frac{\beta\sqrt{B^2(\sigma^2 + \beta^2\lambda) - R^2)}}{(\sigma^2 + \beta^2\lambda)\sqrt{\lambda}}, \qquad (22.30)$$

$$\varphi_{\min} = -\frac{\beta R}{\sigma^2 + \beta^2\lambda} - \frac{\sigma\sqrt{B^2(\sigma^2 + \beta^2\lambda) - R^2)}}{(\sigma^2 + \beta^2\lambda)\sqrt{\lambda}}. \qquad (22.31)$$

  2. *If*

$$-\frac{\beta R}{\sigma^2 + \beta^2\lambda} - \frac{\sigma\sqrt{B^2(\sigma^2 + \beta^2\lambda) - R^2)}}{(\sigma^2 + \beta^2\lambda)\sqrt{\lambda}} \leq -1, \qquad (22.32)$$

  then

$$\gamma_{\min} = \frac{r - \alpha}{\sigma}, \qquad (22.33)$$

$$\varphi_{\min} = -1. \qquad (22.34)$$

Proof   A direct application of Kuhn–Tucker.   □

We can now present a preliminary description of the optimal kernels.

Proposition 22.8   *The optimal kernels $(\widehat{\gamma}, \widehat{\varphi})$ for the static problem (22.21)–(22.24) have the following structure.*

1. *For all (t, s) such that*

$$V(t, s(1 + \beta)) - V(t, s) - V_s(t, s)s\beta \geq 0, \tag{22.35}$$

   the optimal kernels $(\widehat{\gamma}, \widehat{\varphi})$ are given by

$$\widehat{\gamma}(t, s) = \gamma_{\max}, \quad \widehat{\varphi}(t, s) = \varphi_{\max}. \tag{22.36}$$

2. *For all (t, s) such that*

$$V(t, s(1 + \beta)) - V(t, s) - V_s(t, s)s\beta < 0, \tag{22.37}$$

   the optimal kernels $(\widehat{\gamma}, \widehat{\varphi})$ are given by

$$\widehat{\gamma}(t, s) = \gamma_{\min}, \quad \widehat{\varphi}(t, s) = \varphi_{\min}. \tag{22.38}$$

Proof   Obvious from the arguments above.   □

We thus see that the optimal kernels have a so-called bang–bang structure, i.e. they switch between the extremal choices $(\gamma_{\max}, \varphi_{\max})$ and $(\gamma_{\min}, \varphi_{\min})$. For an arbitrarily chosen problem, switches will indeed occur, and the number of switches will of course depend upon the optimal value function V through the conditions (22.35) and (22.37).

22.5   Notes

The good-deal theory presented above is first investigated in Cochrane & Saá Requejo (2000) where also other examples are studied. The theory was extended and streamlined in Björk & Slinko (2006). Here we follow Björk & Slinko (2006), where more theory and examples can be found. For an interesting, but slightly different, view of good-deal bounds, see Černý (2003) and Černý & Hodges (2002). A related approach to obtaining asset price bounds, based on gain–loss ratios, is presented in Bernardo & Ledoit (2000). See Rodriguez (2000) for an interesting connection between the approach of Bernardo & Ledoit (2000) and linear programming.

23 Diversifiable Risk

In this chapter we will take a closer look at the concept of "diversifiable risk", which we encountered in Section 21.1. We recall that the argument in Section 21.1 was that "diversifiable risk should not be priced", and we now investigate whether this conjecture can be put on a more firm theoretical base. In order to do this we will below formalize the idea of a large market containing an infinite number of risky assets, where each asset is driven by two random sources: one Wiener source which is common to the entire market, and an idiosyncratic jump risk which is unique to the specific asset. The question to investigate is then whether absence of arbitrage will force the market price of the idiosyncratic jump risk to be zero. It should be noted that most of the arguments below are a bit heuristic, so they are not exactly stringent. At the end we will, however, be more precise, and we also give references to the literature in the Notes.

23.1   The Basic Model

We consider a financial market containing a countable set of traded assets, and in the sequel this market will be studied on a finite fixed time interval $[0, T]$. The price of asset number i, with $i = 1, 2, \ldots$ is denoted by S_i and we assume that the dynamics of S^i are given as follows, under an objective probability measure P:

$$dS_t^i = \alpha_i S_t^i dt + \beta_i S_{t-} dN_t^i + \sigma_i S_t^i dW_t. \tag{23.1}$$

Here α_i, β_i and σ_i are assumed to be known deterministic constants (see also the technical assumptions below). The process W, which is common to all assets, represents the systematic risk in the market: it is assumed to be a standard P-Wiener process. The process N^i, on the other hand, represents the asset-specific risk of asset number i and we will assume that N^i is Poisson. In this model, the Poisson process N^i represents the occurence of sudden asset-specific shocks. If $\beta_i > 0$ then the shocks will increase the value of the stock, which could be interpreted as, for example, a technological breakthrough or suddenly improved market conditions. When $\beta_i < 0$, on the other hand, the stock price will decrease at every jump time of the Poisson process, representing negative shocks. The most striking example is perhaps the case when $\beta_i = -1$, in which case the stock price will drop to zero at the fist jump time of N_i. This has a very natural interpretation as a model of (complete) *default* for asset number i, and in this context it would be natural to study various credit risk derivatives.

The basic assumptions are as follows.

Assumption 23.1 *We assume that*

- *The process W is a P-Wiener process.*
- *The process N^i is P-Poisson with intensity λ_i.*
- *The processes $W, N^1, N^2, \ldots$ are P-independent.*
- *The parameters $\alpha_i, \beta_i, \sigma_i$ and $\lambda_i, i = 1, 2, \ldots$ are uniformly bounded from below and above, in the sense that there exist constants A and B with $0 < A \le B$ such that*

$$A \le \alpha_i, \beta_i, \sigma_i, \lambda_i \le B, \quad \text{for all } i.$$

We also assume the existence of a risk-free asset with price process B, with constant short rate r, so the market model has the form

$$dS_t^i = \alpha_i S_t^i dt + \beta_i S_{t-} dN_t^i + \sigma_i S_t^i dW_t, \quad i = 1, 2, \ldots, \tag{23.2}$$

$$dB = rB dt. \tag{23.3}$$

We note that the mean rate of return including jumps is given by $\alpha_i + \beta_i \lambda_i$ so, defining $\widehat{beta_i}$ by $\widehat{beta_i} = \beta_i \lambda_i$, the risk premium RP can be written

$$\text{RP} = \alpha_i + \widehat{\beta}_i - r.$$

23.2   Well-Diversified Portfolios

Given the market model above, we are allowed to form finite portfolios in the various assets, and we will now see what happens when we form certain infinitely well-diversified portfolios. We need a small technical lemma.

Lemma 23.2 *Define $\widehat{\beta}_i$ by $\widehat{\beta}_i = \beta_i \lambda_i$. Then there exists at least one infinite increasing subsequence, $I = \{i_j\}_{j=1}^{\infty}$, of the integers and real numbers $\alpha_I, \widehat{\beta}_I$ and σ_I (depending on the choice of the subsequence), such that*

$$\lim_{i \in I} \alpha_i = \alpha_I, \qquad \lim_{i \in I} \widehat{\beta}_i = \widehat{\beta}_I, \qquad \lim_{i \in I} \sigma_i = \sigma_I.$$

Proof Choose a subsequence I_0 such that $\lim_{i \in I} \alpha_i = \lim \sup_i \alpha_i$. Then choose a subsequence $I_1 \subseteq I_0$ such that $\lim_{i \in I_1} \widehat{\beta}_i = \lim \sup_{i \in I_0} \widehat{\beta}_i$, and finally choose a subsequence $I_2 \subseteq I_1$ such that $\lim_{i \in I_2} \sigma_i = \lim \sup_{i \in I_1} \sigma_i$. The assumption of uniform boundedness guarantees that the limits are finite.   □

Definition An index set (subsequence) I as in the lemma above will be called a **converging index set**. We introduce the natural equivalence relation $\sim$ by writing $I \sim \mathcal{J}$ if and only if $\alpha_I = \alpha_{\mathcal{J}}, \widehat{\beta}_I = \widehat{\beta}_{\mathcal{J}}$ and $\sigma_I = \sigma_{\mathcal{J}}$. The family of (equivalence classes of) converging index sets is denoted by $\mathcal{C}$.

Observe that generically there will exist several converging index sets and that I is unique (i.e., $\mathcal{C}$ is a singleton) if and only if $\lim_i \alpha_i, \lim_i \widehat{\beta}_i$ and $\lim_i \sigma_i$ all exist, (where the limits are taken with respect to the set of natural numbers).

We now turn to the construction of well-diversified self-financing portfolio strategies. Let us therefore consider a fixed finite index set $I = \{i_1,\ldots,i_K\}$ and recall that a self-financed portfolio in the assets $\{S^i\}_{i \in I}$ is specified by:

1. the initial capital $V(0)$;

2. a sequence of predictable (and sufficiently integrable) predictable stochastic processses $\{w_i; i \in I\}$ such that $\sum_{i \in I} w_{it} = 1$.

Remark 23.3 Unless otherwise specified, the filtration $\{\mathcal{F}_t\}_{t \geq 0}$ under consideration is defined by

$$\mathcal{F}_t = \sigma \left\{ W_s, N_s^i; \ s \leq t, \ i = 1, 2, \ldots \right\}. \tag{23.4}$$

Thus the prefix "predictable" above, means $(\mathcal{F}_t)$-predictable. In some cases it will be of interest to consider portfolios that are adapted to a smaller filtration than $\{\mathcal{F}_t\}_{t \geq 0}$, and in such cases we will give separate remarks to clarify the situation.

As usual w_{it} above is interpreted as the relative share of the total portfolio value invested in asset number i at time t; we also recall that the evolution of the value process V is given by

$$dV_t = V_{t-} \sum_{i \in I} w_{it} \frac{dS_t^i}{S_{t-}^i}. \tag{23.5}$$

Definition 23.4 Consider a converging index set $I = \{i_j\}_{j=1}^{\infty}$ and define I_n by $I_n = \{i_j\}_{j=1}^n$. A **diversifying strategy along I** is defined a sequence $\{w^n\}_{n=0}^{\infty}$ of self-financing portfolios such that:

1. for all i, n and t we have $w_{it}^n \geq 0$;

2. the strategy w^n is based on the assets $\{S^i; \ i \in I_n\}$;

3. for all t, the following relation holds P-almost surely:

$$\lim_{n \to \infty} \sup_{i \in I_n, t \geq 0} w_{it}^n = 0.$$

This definition formalizes the idea of a well-diversified portfolio, in the sense that each asset asymptotically contributes an infinitesimal part to the entire portfolio. The simplest concrete example is of course given by $w_{it}^n(t) \equiv 1/n$ for $i = 1, \ldots, n$. Let us now consider a fixed diversifying strategy along I and denote the value process corresponding to w^n by V^n. We then have the following portfolio dynamics:

$$dV_t^n = V_t^n \left(\sum_{i \in I_n} w_{it}^n \alpha_i \right) dt + V_{t-}^n \sum_{i \in I_n} w_{it}^n \beta_i dN_t^i + V_t^n \left(\sum_{i \in I_n} w_{it}^n \sigma_i \right) dW_t. \tag{23.6}$$

The immediate project is to study the behavior of this equation as n tends to infinity.

23.3   The Asymptotic Jump Model

In order to facilitate the analysis it is convenient to write (23.6) in P-semimartingale form as

$$dV_t^n = V_t^n \left(\sum_{i \in I_n} w_{it}^n \left[\alpha_i + \widehat{\beta}_i \right] \right) dt + V_{t-}^n \sum_{i \in I_n} w_{it}^n \beta_i \left[dN_t^i - \lambda_i dt \right]$$

$$+ V_t^n \left(\sum_{i \in I_n} w_{it}^n \sigma_i \right) dW_t, \tag{23.7}$$

where as before $\widehat{\beta}_i$ is defined as

$$\widehat{\beta}_i = \beta_i \lambda_i. \tag{23.8}$$

We now let n tend to infinity. From the definition of a diversifying strategy it is easy to see that, in this limit, we have the following results:

$$\sum_{i \in I_n} w_i^n \left[\alpha_i + \widehat{\beta}_i \right] \to \alpha_I + \widehat{\beta}_I, \tag{23.9}$$

$$\sum_{i \in I_n} w_i^n \sigma_i \to \sigma_I. \tag{23.10}$$

For the middle term in (23.7) we see that the differential $dN_t^i - \lambda_i dt$ is a martingale increment and thus it has expected value zero. The infinitesimal variance is given by

$$\mathrm{Var}_P \left[\sum_{i \in I_n} w_{it}^n \beta_i \left[dN_t^i - \lambda_i dt \right] \right] = \sum_{i \in I_n} (w_{it}^n)^2 \beta_i^2 \lambda_i dt.$$

Defining B_n by

$$B_n = \sup_{i \in I_n, t \geq 0} |w_{it}^n|$$

we have

$$\sum_{i \in I_n} (w_{it}^n)^2 \beta_i^2 \lambda_i \leq B_n \sum_{i \in I_n} w_{it}^n \beta_i^2 \lambda_i dt.$$

As $n \to \infty$ we have $\sum_{i \in I_n} w_{it}^n \beta_i^2 \lambda_i \to \beta_I^2 \lambda_I$, whereas $B_n \to 0$, so we conclude that

$$\sum_{i \in I_n} w_{it}^n \beta_i \left[dN_t^i - \lambda_i dt \right] \to 0.$$

The asymptotic behavior of (23.7) is thus given by the equation

$$dV = (\alpha_I + \widehat{\beta}_I) V dt + V \sigma_I dW. \tag{23.11}$$

We now define the asymptotic jump model.

Definition 23.5   The **asymptotic jump model** consists of the following price processes.

1. For each i the process S^i with P-dynamics

$$dS_t^i = \alpha_i S_t^i dt + \beta_i S_{t-}^i dN_t^i + \sigma_i S_t^i dW_t, \quad i = 1, 2, \ldots \qquad (23.12)$$

2. For each converging index set $I \in C$, the process S^I with P-dynamics defined by

$$dS_t^I = (\alpha_I + \widehat{\beta}_I)S_t^I dt + \sigma_I S_t^I dW_t, \quad I \in C. \qquad (23.13)$$

3. The risk-free asset price B with

$$dB_t = r B_t dt. \qquad (23.14)$$

The asymptotic model is an idealized picture of a financial market where we are allowed to trade infinitely diversified portfolios. It is an idealization in the same way that "continuous trading" and "frictionless markets" are idealizations of real-world phenomena. The real-world counterparts to the S^I-processes are of course the mutual funds, which can contain hundreds of different assets.

Since we will work under the martingale measure Q, as well as under P, we will keep to the measure-invariant formulation (23.12).

23.4 No-Arbitrage Conditions

We now make an assumption concerning absence of arbitrage.

Assumption *We assume that the asymptotic model above is free of arbitrage in the sense that there exist no arbitrage possibilities when we are allowed to trade in finite portfolios.*

We have made the requirement of finite portfolios in order to be able to use the "classical" theory of no-arbitrage. Note however that since we now are working with the asymptotic model, the assumption of no-arbitrage is really an assumption of "no asymptotic arbitrage" in terms of the original model (23.2)–(23.3).

Given this assumption we now go on to find the market prices of risk for the diffusion part as well as for the various jump parts. We follow the following scheme.

• Carry out a Girsanov transformation for the Wiener process using a likelihood process of the form

$$dL_t^0 = L_t^0 \cdot h_t dW_t,$$
$$L_0^0 = 1.$$

• For each i we carry out a Girsanov transformation using a likelihood process of the form

$$dL_t^i = L_{t-}^i \varphi_i(t) \left[dN_t^i - \lambda_i dt \right],$$
$$L_0^i = 1.$$

In this way we obtain a measure Q with the property that

- Under Q the process W has the decomposition

$$dW_t = h_t dt + dW_t^Q,$$

where W^Q is Q-Wiener.
- The process N^i has Q-intensity λ_i^Q given by

$$\lambda_i^Q(t) = [1 + \varphi_i(t)] \lambda_i.$$

The asset-price dynamics under Q are thus given by

$$dS_t^i = S_t^i \left[\alpha_i + \sigma_i h + \widehat{\beta}_i(1 + \varphi_i) \right] dt + S_t^i \sigma_i dW_t^Q$$

$$+ S_{t-}^i \beta_i \left[dN_i - (1 + \varphi_i)\lambda_i \right], \tag{23.15}$$

$$dS_t^I = S_t^I \left\{ \alpha_I + \widehat{\beta}_I + \sigma_I h \right\} dt + S_t^I \sigma_I dW_t^Q, \quad I \in C. \tag{23.16}$$

Now we want to choose the Girsanov kernels in such a way that all asset prices have a local rate of return equal to the short rate of interest; i.e., we want to solve the following set of equations:

$$\alpha_i + \sigma_i h + \widehat{\beta}_i(1 + \varphi_i) = r, \tag{23.17}$$

$$\alpha_I + \widehat{\beta}_I + \sigma_I h = r. \tag{23.18}$$

This system is easily solved as

$$h = \frac{r - \alpha_I - \widehat{\beta}_I}{\sigma_I}, \tag{23.19}$$

$$\varphi_i = \frac{r - \alpha_i - \widehat{\beta}_i - \frac{\sigma_i}{\sigma_I}(r - \alpha_I - \widehat{\beta}_I)}{\widehat{\beta}_i}. \tag{23.20}$$

These are the no-arbitrage Girsanov kernels and we have the following first result.

Proposition 23.6   *The following hold.*

- *The model (23.12)–(23.14) is free of arbitrage if and only if there exist constants h and $\{\varphi_i\}_{i=1}^{\infty}$ such that*

$$h = \frac{r - \alpha_I - \widehat{\beta}_I}{\sigma_I}, \tag{23.21}$$

$$\varphi_i = \frac{r - \alpha_i - \widehat{\beta}_i - \frac{\sigma_i}{\sigma_I}(r - \alpha_I - \widehat{\beta}_I)}{\widehat{\beta}_i}. \tag{23.22}$$

$$\varphi_i > -1 \tag{23.23}$$

for all i and all converging index sets $I \in C$.
- *We have the asymptotic result*

$$\varphi_i \to 0.$$

Remark We stress again the fact that the arguments above are partly heuristic. We have performed a Girsanov transformation for each N^i-process, not just for the W-process. Each of these changes of measure has been an equivalent measure transformation, but since we have a countable number of measure transformations there is no guarantee that the prospective "martingale measure" Q obtained above is globally equivalent to the objective measure P. This question will be handled in Section 23.6 below.

From this we see that absence of arbitrage generically does *not* in general imply that the Girsanov kernel of diversifiable risk φ_i equals zero. We do however have an interesting asymptotic result.

Proposition 23.7

1. We have the asymptotic result

$$\lim_{i \to \infty} \varphi_i = 0.$$

2. The market price of jump risk of type i equals zero if and only if the following relation hold for all $I \in C$:

$$\frac{\alpha_i + \widehat{\beta}_i - r}{\sigma_i} = \frac{\alpha_I + \widehat{\beta}_I - r}{\sigma_I}.$$

Proof The second part of the proposition follows immediately from (23.20). To prove the first part we first note that, for any converging index set J, (23.20) and Proposition 23.6 imply that

$$\lim_{i \in J} \varphi_i = \frac{r - \alpha_J - \widehat{\beta}_J - \frac{\sigma_J}{\sigma_I}(r - \alpha_I - \widehat{\beta}_I)}{\widehat{\beta}_i} = 0.$$

Now take any infinite subsequence J of the natural numbers. It is easily seen that J will contain some converging set J_0. Thus, using the result above, we see that any subsequence of the full sequence $\{h_i\}_1^\infty$ will contain a further subsequence converging to zero. Hence the full sequence has to converge to zero. □

We thus see that the market price of jump risk of type i equals zero if and only if the risk premium per unit of systematic volatility of asset i equals the risk premium per unit of systematic volatility for any asymptotic asset. Furthermore, we see that if all assets are identical under P, i.e. the coefficients for asset i do not depend on the index i, then in fact all market prices of diversifiable risks equal zero.

Remark 23.8 We may rewrite (23.20) as

$$-\varphi_i = \frac{1}{\widehat{\beta}_i} \left\{ \alpha_i + \widehat{\beta}_i - r - \frac{1}{\sigma_I^2} \cdot \frac{\text{Cov}\left(\frac{dS_i}{S_i}, \frac{dS_I}{S_I}\right)}{dt} \left(\alpha_I + \widehat{\beta}_I - r \right) \right\}.$$

Using Proposition 23.7 this implies that that we are asymptotically on the CAPM line:

$$\alpha_i + \widehat{\beta}_i - r = \frac{1}{\sigma_I^2} \cdot \frac{\text{Cov}\left(\frac{dS_i}{S_i}, \frac{dS_I}{S_I}\right)}{dt} \left(\alpha_I + \widehat{\beta}_I - r \right)$$

Remark   As in the diffusion case we stress the fact that the arguments above are partly heuristic, since there is no guarantee that the prospective "martingale measure" Q obtained above is globally equivalent to the objective measure P. This question will be handled in Section 23.6 below.

23.5   Completeness for the Asymptotic Model

In order to study completeness we have to specify the relevant classes of contingent claims to be considered. To make things simple we assume that the class of convergent index sets is a singleton, i.e. that all model coefficients converge. Thus we only have one asymptotic asset, which will be denoted by S, and its P-dynamics will be written as

$$dS_t = S_t(\alpha + \widehat{\beta})dt + \sigma S_t dW_t.$$

We now define two natural classes of contingent claims.

Definition 23.9   Fix some point in time T. A (sufficiently integrable) stochastic variable X is said to be a

1. **Contingent claim** if $X \in \sigma \{W(t), N_i(t); t \le T, i = 1,\ldots\}$.
2. **Finite contingent claim** if there exist a number n such that

$$X \in \sigma \{W(t), N_i(t); t \le T, i = 1,\ldots,n\}.$$

The main result concerning the original model is not surprising. Remember that, both in the original and in the asymptotic model, we are by assumption only allowed to trade in finite portfolios.

Proposition 23.10   *The original model is not even complete with repect to finite (market observable) claims.*

Proof   It is immediately clear that no claim of the form $\Phi(N_1(T))$, where Φ is any non-constant function, can be replicated using a finite set of assets.   □

For the asymptotic model the situation is much brighter.

Proposition 23.11   *In the asymptotic model every (Q-square integrable) finite claim can be replicated.*

Proof   Suppose that the claim X is of the form

$$X \in \sigma \{W(t), N_i(t); t \le T, i = 1,\ldots,n\}.$$

Then we adjoin the asymptotic asset S to $S^1,\ldots,S^n$, and it is a standard exercise to see that X can be replicated using a portfolio based upon $S, S^1,\ldots,S^n$.   □

We thus see that the introduction of the asymptotic asset – i.e. the possibility of forming well-diversified portfolios – has the function of completing the model (see the end of the next section for a simple concrete example).

23.6   The Martingale Approach

The strategy in the previous sections has roughly been the following:

- we extended the original model to the asymptotic model by including the set of well-diversified portfolios as asymptotic assets;
- we assumed that the asymptotic model was free of arbitrage in the sense that there did not exist arbitrage opportunities for any finite submodel;
- in this way we claimed that absence of (finite) arbitrage in the asymptotic model could be interpreted as absence of "asymptotic arbitrage" in the original model.

A different approach would be to give a precise definition of "asymptotic arbitrage" for the original model and then to investigate the implications of absence of asymptotic arbitrage. A natural conjecture in this context is of course that absence of asymptotic arbitrage is connected to the existence of a "martingale measure" $Q \sim P$, such that all discounted asset prices are Q-martingales.

In this section we present a simple version of such a theory. We show that the existence of a global martingale measure implies absence of arbitrage and we also investigate the models above in the light of this theory.

Let us denote the finite market consisting only of the assets $S^1,\ldots,S^n$ by M_n. We consider a fixed time horizon T. We write h^n for a self-financing trading strategy on M_n and denote the corresponding value process by V^n. There are several reasonable definitions of asymptotic arbitrage, the simplest perhaps being the following.

Definition 23.12   An **asymptotic arbitrage** is a sequence of strategies $\{h^n\}_{n=1}^{\infty}$ such that, for some real number $c > 0$, we have:

1.
$$V^n(t) \geq -c, \quad \text{for all } t \leq T \text{ and for all } n$$

2.
$$V^n(0) = 0, \quad \text{for all } n.$$

3.
$$\liminf_{n \to \infty} V^n(T) \geq 0, \ P\text{-a.s.}$$

4.
$$P\left(\liminf_{n \to \infty} V^n(T) > 0\right) > 0.$$

We use a straightforward definition of martingale measures.

Definition 23.13   A measure Q is called a **martingale measure** if the following conditions are satisfied.

1. $P \sim Q$ on $\sigma\{W(s), N_i(s); \ s \leq T, \ i = 1,2,\ldots\}$
2. All normalized asset prices S_t^i/B_t are Q-martingales.

Exactly as in the finite case we now have the following central (and extremely easy) result.

Proposition 23.14 *Assume that there exists a martingale measure Q. Then there is no asymptotic arbitrage.*

Proof Without loss of generality we may assume that the short rate of interest equals zero. Suppose now that $\{h^n\}_1^\infty$ actually realizes an asymptotic arbitrage. Then every V^n is a local martingale under Q and, because of the requirement $V^n \geq -c$, we see that every V^n is in fact a Q-supermartingale. Thus we have $0 = V^n(0) \geq E^Q[V^n(T)]$, and Fatou's lemma gives us $E^Q[\liminf V^n(T)] \leq 0$. However, the definition of an asymptotic arbitrage, plus the equivalence between P and Q, implies the inequality $E^P[\liminf V^n(T)] > 0$ which leads to a contradiction. $\square$

We will now study our earlier models in the light of Proposition 23.14, and we start by viewing W as the coordinate process on the canonical space $C[0,T]$ equipped with Wiener measure. In the same way we view each Poisson process N^i as the coordinate process on $D[0,T]$ with the appropriate Poisson measure. We can thus write $\Omega = \bigotimes_{i=0}^\infty \Omega_i$, where $\Omega_0 = C[0,T]$ and $\Omega_i = D[0,T]$ for $n = 1,2,\ldots$ In the same spirit we can write $P = \prod_{i=0}^\infty P_n$, where P_0 is Wiener measure but where P_n is Poisson measure with intensity λ_i for $n \geq 1$.

On Ω_0 we now perform a Girsanov transformation of the form

$$dL_t = L_t h_t \, dW_t,$$

giving us a measure $Q_0 \sim P_0$. On each Ω_i with $i \geq 1$ we perform the transformation

$$dL_t = L_{t-}\varphi_i(t)\left[dN_t^i - \lambda_i dt\right],$$

giving us a measure $Q_i \sim P_i$.

The condition for S_t^i/B_t to be a Q-martingale is

$$r = \alpha_i + \sigma_i h + \widehat{\beta}_i(1 + \varphi_i). \tag{23.24}$$

To see if Q is a martingale measure according to the definition above it now only remains to check if $\prod_{i=0}^\infty P_i$ is equivalent to $\prod_{i=0}^\infty Q_i$. To this end we use the Kakutani dichotomy theorem (see, for example, Durrett, 1996, p. 244) which says that $\prod_{i=0}^\infty P_i \sim \prod_{i=0}^\infty Q_n$ if and only if the following condition holds:

$$\prod_{i=0}^\infty E^P\left[\sqrt{L_T^i}\right] > 0, \tag{23.25}$$

where $L_i = dQ_i/dP_i$. We have

$$L_T^0 = e^{-hW_T - \frac{1}{2}h^2 \cdot T},$$

and

$$L_T^i = (1 + \varphi_i)^{N_T^i} e^{-\varphi_i \lambda_i T}.$$

After some calculations, this gives us

$$E\left[\sqrt{L_T^i}\right] = e^{-T\lambda_i\left\{\frac{1}{2}\varphi_i+1-\sqrt{\varphi_i+1}\right\}}.$$

Since, by Assumption 23.1, $\lambda_i \geq A > 0$ for all i, the the Kakutani condition (23.25) implies that

$$\sum_{i=1}^{\infty}\left\{\frac{1}{2}\varphi_i + 1 - \sqrt{\varphi_i + 1}\right\} < \infty$$

and this in can easily be seen to imply that $\varphi_i \to 0$.

If we now let $i \to \infty$ along a converging index set, formula (23.24) gives us

$$h = \frac{r - \alpha_I - \widehat{\beta_I}}{\sigma_I}$$

which is exactly the formula (23.19). If we plug this into (23.24) we obtain

$$\varphi_i = \frac{r - \alpha_i - \widehat{\beta_i} - \frac{\sigma_i}{\sigma_I}(r - \alpha_I - \widehat{\beta_I})}{\widehat{\beta_i}}.$$

which is formula (23.20). We have thus proved the following result.

Proposition 23.15   *The original model (23.2)–(23.3) is free of arbitrage if and only if there exist constants h and $\{\varphi_i\}_{i=1}^{\infty}$ such that*

$$h = \frac{r - \alpha_I - \widehat{\beta_I}}{\sigma_I}, \tag{23.26}$$

$$\varphi_i = \frac{r - \alpha_i - \widehat{\beta_i} - \frac{\sigma_i}{\sigma_I}(r - \alpha_I - \widehat{\beta_I})}{\widehat{\beta_i}}, \tag{23.27}$$

$$\varphi_i > -1, \tag{23.28}$$

for all i and all converging index sets $I \in C$.

Thus we obtain exactly the same formulas as in Proposition 23.6. The difference is that the martingale approach is more general than the "asymptotic model approach". The point of studying the latter is of course that it gives an intuitive interpretation of the abstract martingale results in terms of the well-diversified portfolios.

23.7   Notes

Our model is a continuous-time extension of the classical "arbitrage pricing theory" (APT) which was introduced by Ross (1976) for a one-period model and then developed and clarified in Hubermann (1982) and other papers.

The one-period APT results were extended in Chamberlain (1988) and Reisman (1992) to continuous-time models. More advanced studies can be found in Kabanov & Kramkov (1994), Klein & Schachermayer (1996), Klein & Schachermayer (2000), de Donno (2004) and references therein.

24 Credit Risk and Cox Processes

As an application of Cox processes we will now study the pricing of a risky bond. We study a simple case and, with some small variations, we follow Lando (1998).

24.1 Defaultable Claims and Risky Bonds

The model consists of a filtered space $(\Omega, \mathcal{F}, P, \mathbf{F})$ carrying the following objects

- An optional cadlag process X.
- A Cox counting process N with $\mathbf{F}$-intensity

$$\lambda_t = \lambda(X_t),$$

  where λ in the right-hand side denotes a deterministic function.

In this setting we now consider a financial market consisting (at least) of the following objects.

- A short-rate process r of the form

$$r_t = r(X_t),$$

  where r in the right-hand side denotes a deterministic function.
- A bank account process B with dynamics

$$dB_t = r(X_t)B_t dt.$$

- A contingent T-claim Z such that

$$Z \in \mathcal{F}_T^X.$$

The claim Z is defined as the *nominal* claim, i.e. the claim which is paid if no default occurs. From general theory we recall that the price, at t, of the T-claim X in the case of no default is given by

$$\Pi_t[Z] = E^Q \left[e^{-\int_t^T r_s ds} Z \, \middle| \, \mathcal{F}_t \right],$$

and as a special case, when $Z = 1$, the price of a non-defaultable bond with unit face value is given by

$$p(t, T) = E^Q \left[e^{-\int_t^T r_s ds} \, \middle| \, \mathcal{F}_t \right].$$

We now want to study pricing when there is a possibility of default. We model default as follows.

Definition 24.1 The default time τ is the first jump time of the Cox process N. If default occurs before T, i.e. if $\tau \leq T$, then nothing is paid to the owner of Z.

The actual payment at time T, denoted by Z_R, is thus given by

$$Z_R = Z \cdot I\{\tau > T\},$$

so

$$Z_R = Z \cdot I\{N_T = 0\}.$$

We then have the following basic result.

Proposition 24.2 Denote the risky price of Z (i.e. the price of Z_R) at t by $\Pi_t^R[Z]$, and the price of a risky zero coupon bond by $q(t,T)$. Then we have

$$\Pi_t^R[Z] = I\{N_t = 0\} \cdot E^Q\left[e^{-\int_t^T R_s \, ds} \cdot Z \middle| \mathcal{F}_t\right], \tag{24.1}$$

$$q(t,T) = I\{N_t = 0\} \cdot E^Q\left[e^{-\int_t^T R_s \, ds} \middle| \mathcal{F}_t\right], \tag{24.2}$$

where the **risk adjusted** interest rate R is defined by

$$R_t = r_t + \lambda_t.$$

Proof It is obviously enough to prove (24.1), and it suffices to consider the case $t = 0$. Using the Cox property and the measurability assumptions we obtain

$$\Pi_0^R[Z] = E^Q\left[e^{-\int_0^T r_s \, ds} \cdot Z \cdot I\{N_T = 0\}\right]$$

$$= E^Q\left[E^Q\left[e^{-\int_0^T r_s \, ds} \cdot Z \cdot I\{N_T = 0\} \middle| \mathcal{F}_T^X\right]\right]$$

$$= E^Q\left[e^{-\int_0^T r_s \, ds} \cdot Z \cdot E^Q\left[I\{N_T = 0\} \middle| \mathcal{F}_T^X\right]\right]$$

$$= E^Q\left[e^{-\int_0^T r_s \, ds} \cdot Z \cdot e^{-\int_0^T \lambda_s \, ds}\right] = E^Q\left[e^{-\int_0^T R_s \, ds} \cdot Z\right]. \qquad \square$$

We see that we have the same structure in the valuation formula as in the standard non-defaultable case. The difference is that we have to discount using the risk-adjusted interest rate R. These results have also been generalized to much more complicated payoff structures: See Lando (2004) for an overview.

24.2 A Defaultable Payment Stream

We can extend the result above to the case of valuing a risky payment stream. The model, concerning X, r and N, is as in the previous section, but there is one change. Instead of the T-claim Z we now consider a random process C_t, with the interpretation that C_t is a payment rate, in the sense that over the time interval $[t, t + dt]$ the owner of C receives the amount $C_t \, dt$ in dollars. The financial dimension of C_t is thus in dollars per unit time and we assume that the payments stop at a predetermined time T. We also need an obvious measurability assumption.

Assumption   *The process C is $\mathbf{F}^X$-adapted.*

We now define $\Pi_t[C]$ as the arbitrage-free value of the payment stream generated by C over the time interval $[t,T]$, and from standard theory we have the usual risk-neutral valuation formula:

$$\Pi_t[C] = E^Q\left[\int_t^T e^{-\int_t^s r_u\,du}\cdot C_s\,ds\,\middle|\,\mathcal{F}_t\right]. \tag{24.3}$$

We now want to price the payment stream C in the presence of risk of default. We therefore define default by stipulating the the payments defined by C will continue until default, i.e. until the first jump of N. After the time of default nothing will be paid.

Definition 24.3   The **defaultable** payment stream C^R is defined by

$$C_t^R = C_t \cdot I\{\tau > t\}, \tag{24.4}$$

or equivalently

$$C_t^R = C_t \cdot I\{N_t = 0\}. \tag{24.5}$$

We define the price process $\Pi_t^R[C]$ as the arbitrage-free price process for the payment stream C^R over the time interval $[t,T]$.

Using exactly the same arguments as in the previous section we obtain the following result.

Proposition 24.4   *The price process $\Pi_t^R[C]$ of the risky payment stream C^R, defined as above, is given by*

$$\Pi_t^R[C] = I\{N_t = 0\}\cdot E^Q\left[\int_t^T e^{-\int_t^s R_u\,du}\cdot C_s\,ds\,\middle|\,\mathcal{F}_t\right] \tag{24.6}$$

where the risk-adjusted interest rate R is defined by

$$R_t = r_t + \lambda_t.$$

24.3   Multiple Defaults

We now move on to study pricing when there a possibility of multiple default times. This can be modeled in many ways but we only consider a very simple case.

24.3.1   The Model

The general setup with X, N, and λ is exactly as in Section 24, but instead of loosing the entire remaining promised value at the first jump-time of N, we allow for partial loss at all the jump-times of N, by introducing a new object.

Definition 24.5 A **recovery function** k, is any mapping $k : \mathbf{N} \to [0, 1]$.

The interpretation is that if $N_t = n$ then you get $k(n)$ cents on the dollar on any claim at time t. More formally this means that if we consider a contract promising you the payment rate C over the time interval $[0, T]$ and a final payment of Z dollars at time T, then you will in fact receive the payment rate C^R defined by

$$C_t^R = C_t \cdot k(N_t),$$

and the final payment Z^R defined by

$$Z^R = Z \cdot k(N_T).$$

The single default case studied earlier is of course a special case of this setup if we define k by $k(0) = 1$, and $k(n) = 0$ for $n = 1, 2, \ldots$

We now want to price the risky pair (C^R, Z^R) and to that end we have the following result.

Proposition 24.6 *The price process* $\Pi_t^R[C, Z]$ *for the risky pair* (C^R, Z^R) *is given by*

$$\Pi_t^R[C, Z] = E^Q \left[\int_t^T e^{-\int_t^s R_u \, du} \cdot C_s \cdot k_s^s \, ds + e^{-\int_t^T R_u \, du} \cdot Z \cdot k_t^T \middle| \mathcal{F}_t \right], \quad (24.7)$$

where

$$R_t = r_t + \lambda_t, \quad (24.8)$$

$$k_t^s = \sum_{n=0}^{\infty} k \, (N_t + n) \cdot \frac{\left(\int_t^s \lambda_u \, du \right)^n}{n!}. \quad (24.9)$$

Proof For notational simplicity we only consider the case when $C = 0$. We have

$$\Pi_t^R[Z] = E^Q \left[e^{-\int_t^T r_s \, ds} \cdot Z \cdot k(N_T) \middle| \mathcal{F}_t \right]$$

$$= E^Q \left[E^Q \left[e^{-\int_t^T r_s \, ds} \cdot Z \cdot k(N_T) \middle| \mathcal{F}_T^X \vee \mathcal{F}_t^N \right] \middle| \mathcal{F}_t \right]$$

$$= E^Q \left[e^{-\int_t^T r_s \, ds} \cdot Z \cdot E^Q \left[k(N_T) \middle| \mathcal{F}_T^X \vee \mathcal{F}_t^N \right] \middle| \mathcal{F}_t \right].$$

Using (5.22) from the definition of the Cox process we have

$$E^Q \left[k(N_T) \middle| \mathcal{F}_T^X \vee \mathcal{F}_t^N \right] = e^{-\int_t^T \lambda_s \, ds} \cdot \sum_{n=0}^{\infty} k \, (N_t + n) \cdot \frac{\left(\int_t^T \lambda_u \, du \right)^n}{n!},$$

so we obtain

$$\Pi_t^R[Z] = E^Q \left[e^{-\int_t^T r_s \, ds} \cdot Z \cdot e^{-\int_t^T \lambda_s \, ds} \cdot k_t^T \middle| \mathcal{F}_t \right] = E^Q \left[e^{-\int_t^T R_u \, du} \cdot Z \cdot k_t^T \middle| \mathcal{F}_t \right].$$

The proof for the payment stream C is almost identical. $\square$

24.3.2   A Special Case

We now specialize the model above to the case when the recovery function k is given by

$$k(n) = \alpha^n, \quad n = 0, 1, \ldots, \tag{24.10}$$

for some real number α with $0 \leq \alpha \leq 1$, with the convention $0^0 = 1$ if $\alpha = 0$. In concrete economic terms this means that at every default, i.e. at every jump of N, your claim is reduced to $100 \cdot \alpha$ cents on the dollar, so if $\alpha = 0.8$ you loose 20% at every default time. From (24.9) we have

$$k_t^s = \sum_{n=0}^{\infty} \alpha^{(N_t + n)} \cdot \frac{\left(\int_t^s \lambda_u du \right)^n}{n!} = \alpha^{N_t} \sum_{n=0}^{\infty} \alpha^n \cdot \frac{\left(\int_t^s \lambda_u du \right)^n}{n!}$$

$$= \alpha^{N_t} \cdot e^{\alpha \int_t^s \lambda_u du}.$$

Plugging this into (24.7) gives us the following result.

Proposition 24.7 *With assumptions and notation as above we have*

$$\Pi_t^R[C, Z] = \alpha^{N_t} E^Q \left[\int_t^T e^{-\int_t^s R_u^\alpha du} \cdot C_s ds + e^{-\int_t^T R_u^\alpha du} \cdot Z \bigg| \mathcal{F}_t \right], \tag{24.11}$$

where

$$R_t^\alpha = r_t + (1 - \alpha)\lambda_t, \tag{24.12}$$

and with the convention that $0^0 = 1$ for the case $\alpha = 0$.

24.4   Notes

There is a huge literature on credit risk. In the text above we have largely followed Lando (1998). For textbook treatments with extensive bibliographies see Hunt & Kennedy (2000), Lando (2004).

25   Interest-Rate Theory

In this chapter we will study some aspects of interest-rate theory for jump-diffusion models. We start by setting the scene.

25.1   Basics

Our main object of study is the zero-coupon bond market and the related risk-free interest rates. For this we need some formal definitions.

Definition 25.1   A **zero-coupon bond** with **maturity date** T, also called a T-bond, is a contract which guarantees the holder $1 to be paid on the date T. The price at time t of a bond with maturity date T is denoted by $p(t,T)$.

We now make an assumption to guarantee the existence of a sufficiently rich bond market.

Assumption 25.2   *We assume that*

1. *There exists a (frictionless) market for T-bonds for every $T > 0$.*
2. *For every fixed T, the process $\{p(t,T); \; 0 \leq t \leq T\}$ is an optional stochastic process with $p(t,t) = 1$ for all t.*
3. *For every fixed t, $p(t,T)$ is P-a.s. continuously differentiable in the T-variable. This partial derivative is often denoted by*

$$p_T(t,T) = \frac{\partial p(t,T)}{\partial T}.$$

Note that with the assumptions above we are studying a financial market with an *infinite* number of assets – one asset price process $t \mapsto p(t,T)$ for each fixed maturity date T.

Given the bond market above, one can define a (surprisingly large) number of **riskless interest rates**, and the basic construction is as follows. Suppose that we are standing at time t and fix two other points in time S and T with $t < S < T$. The immediate project is to write a contract at time t which allows us to make an investment of $1 at time S and to have a *deterministic* rate of return, determined at the contract time t, over the interval $[S,T]$. This can easily be achieved as follows.

1. At time t we sell one S-bond. This will give us $p(t,S)$.

2. For this money we can by exactly $p(t,S)/p(t,T)$ T-bonds. Thus our net investment at time t equals zero.
3. At time S the S-bond matures, so we are obliged to pay out $1.
4. At time T the T-bonds mature at $1 apiece so we will receive $p(t,S)/p(t,T) \cdot 1$.
5. The net effect of all this is that, based on a contract at t, an investment of $1 at time S has yielded $p(t,S)/p(t,T)$ at time T.

We now go on to compute the relevant interest rates implied by the construction above. We will use one way (out of many possible) of quoting forward rates: namely, as continuously compounded rates.

The **continuously compounded** forward rate R is the solution to the equation

$$e^{R(T-S)} = \frac{p(t,S)}{p(t,T)}.$$

Solving this simple equation we make the following definitions.

Definition 25.3

1. The continuously compounded **forward rate for** $[S,T]$ **contracted at** t is defined as

$$R(t; S,T) = -\frac{\log p(t,T) - \log p(t,S)}{T - S}. \tag{25.1}$$

2. The **instantaneous forward rate with maturity** T, **contracted at** t, is defined by

$$f(t,T) = -\frac{\partial \log p(t,T)}{\partial T}. \tag{25.2}$$

3. The instantaneous **short rate at time** t is defined by

$$r_t = f(t,t). \tag{25.3}$$

The instantaneous forward rate, which will be of great importance below, is the limit of the continuously compounded forward rate when $S \to T$. It can thus be interpreted as the risk-free interest rate, contracted at t, over the infinitesimal interval $[T,T + dT]$. We see that the short rate r_t can be interpreted as a risk-free interest rate, contracted at t, over the infinitesimal interval $[t,t + dt]$.

We now go on to define the money account process B.

Definition 25.4   The **money account** process is defined by

$$B_t = e^{\int_0^t r(s)ds},$$

i.e.

$$\begin{cases} dB_t & = & r_t B_t dt, \\ B_0 & = & 1. \end{cases}$$

The interpretation of the money account is that you may think of it as describing a bank with the stochastic short rate r. It can also be shown that investing in a money account is equivalent to investing in a self-financing "rolling over" trading strategy, which at each time t consists entirely of "just maturing" bonds, i.e. bonds which will mature at $t + dt$.

We will study bond prices and interest rates within a jump-diffusion framework and, in order to fix notation, we make the following assumption.

Assumption 25.5 *We consider a financial market model living in a filtered probability space (stochastic basis) $(\Omega, \mathcal{F}, \mathbf{F}, P)$ where $\mathbf{F} = \{\mathcal{F}_t\}_{t\geq0}$. The basis is assumed to carry a m-dimensional Wiener process W^P as well as a marked point process $\Psi(dt, dx)$ with mark space $(E, \mathcal{E})$ and predictable intensity $\lambda_t^P(dx)$. The processes W^P and Ψ are not assumed to be independent.*

Within this framework, we will study models where the short rates, bond prices and forward rates have dynamics of the following form.

Short-rate dynamics

$$dr_t = a_t dt + b_t dW_t^P + \int_E q(t, z)\Psi(dt, dz), \tag{25.4}$$

Bond-price dynamics

$$dp(t, T) = p(t-, T)m(t, T)dt + p(t-, T)v(t, T)dW_t^P$$
$$+ p(t-, T)\int_E n(t, z, T)\Psi(dt, dz), \tag{25.5}$$

Forward-rate dynamics

$$df(t, T) = \alpha(t, T)dt + \sigma(t, T)dW_t^P + \int_E \beta(t, z, T)\Psi(dt, dz). \tag{25.6}$$

In order for the objects above to be well defined we need a number of technical assumptions which we collect below in an "operational" manner.

Assumption 25.6

1. *The processes a and b are assumed to be optional. For each fixed T, we assume that $m(\cdot, T)$, $v(\cdot, T)$, $\alpha(\cdot, T)$ and $\sigma(\cdot, T)$ are optional processes, whereas $q(\cdot, z)$, $n(\cdot, z)$ and $\delta(\cdot, z, T)$ are assumed to be predictable processes for each z and T.*
2. *For each fixed ω, t and, (in appropriate cases) z, all the objects $m(t, T)$, $v(t, T)$, $n(t, z, T)$, $\alpha(t, T)$, $\sigma(t, T)$ and $\delta(t, z, T)$ are assumed to be continuously differentiable in the T-variable. This partial T-derivative is sometimes denoted by $m_T(t, T)$, etc.*
3. *All processes are assumed to be regular enough to allow us to differentiate under the integral sign as well as to interchange the order of integration.*

These assumptions are rather *ad hoc* and one would, of course, like to give conditions which *imply* the desired properties above. This can be done but at a fairly high price in terms of technical complexity. As for the point-process integrals, these are made pathwise, so the standard Fubini theorem can be applied. For the stochastic Fubini theorem for the interchange of integration with respect to dW and dt see Protter (2004).

Going on to pricing, we recall from arbitrage theory that if $X \in \mathcal{F}_T$ is a T-claim, i.e.

a contingent claim paid out at time T, and Q is the risk-neutral martingale measure with B as the numeraire, then the arbitrage-free price is given by

$$\Pi(t; X) = E^Q \left[e^{-\int_0^t r_s ds} \cdot X \middle| \mathcal{F}_t \right]. \tag{25.7}$$

In particular we have the following formula for bond prices:

$$p(t,T) = E^Q \left[e^{-\int_0^t r_s ds} \middle| \mathcal{F}_t \right]. \tag{25.8}$$

As an immediate consequence of the definition in (25.2) we have the following useful formulas.

Lemma 25.7   *For $t \leq s \leq T$ we have*

$$p(t,T) = p(t,s) \cdot e^{-\int_s^T f(t,u)du}, \tag{25.9}$$

and in particular

$$p(t,T) = e^{-\int_t^T f(t,s)ds}. \tag{25.10}$$

We note the following.

- From the definition in (25.2) it follows that, for any fixed t, the forward-rate curve $T \mapsto f(t,T)$ is deterministically determined by the bond-pricing curve $T \mapsto p(t,T)$.
- From formula (25.10) it follows that, for any fixed t, the bond-pricing curve $T \to p(t,T)$ is deterministically determined by the forward rate curve $T \mapsto f(t,T)$.
- From formula (25.3) it follows that, for any fixed t, the short rate r_t is deterministically determined by the forward-rate curve $T \mapsto f(t,T)$. It is in fact determined by $f(t,t)$.
- Note, however, that for a fixed t, knowledge of r_t alone neither determines the forward-rate curve, nor the bond-pricing curve. In order to derive the bond-pricing curve at t from the short rate we need the entire short-rate process distribution under a martingale measure Q. We can then use formula (25.8).

If we wish to make a model for the bond market, it is thus obvious that this can be done in many different ways.

- We may specify the dynamics of the short rate and then try to derive bond prices using arbitrage arguments and (25.8).
- We may specify the dynamics of a finite set of underlying factors and then try to derive bond prices using arbitrage arguments.
- We may directly specify the dynamics of all possible bonds.
- We may specify the dynamics of all possible forward rates and then use Lemma 25.7 in order to obtain bond prices.

We now move on to study some interest-rate models, i.e. models where we specify the dynamics of one or more of the risk-free interest rates defined above. The number of interest-rate models in the literature is very large indeed (see the Notes) and in this chapter we will confine ourselves to two classes of models, namely *Markovian short-rate models* and *forward-rate models*.

Before going on to study these models we will, however, first provide a small toolbox, in the form of an important result which shows how bond-price dynamics, short-rate dynamics and forward-rate dynamics are related to each other. This will later be used in Section 25.5. Below we use lower-case index T to denote partial derivative.

Proposition 25.8

1. If $p(t,T)$ satisfies (25.5), then for the forward-rate dynamics we have

$$df(t,T) = \alpha(t,T)dt + \sigma(t,T)dW_t + \int_E \delta(t,x,T)\Psi(dt,dx),$$

where α, σ and δ are given by

$$\begin{cases} \alpha(t,T) & = & v_T(t,T) \cdot v(t,T) - m_T(t,T), \\ \sigma(t,T) & = & -v_T(t,T), \\ \delta(t,x,T) & = & -n_T(t,x,T) \cdot [1 + n(t,x,T)]^{-1}, \end{cases} \qquad (25.11)$$

where $v_T(t,T)v(t,T)$ denotes inner product.

2. If $f(t,T)$ satisfies (25.6) then the short rate satisfies

$$dr_t = a_t dt + b_t dW_t + \int_E q(t,x)\Psi(dt,dx),$$

where

$$\begin{cases} a_t & = & f_T(t,t) + \alpha(t,t), \\ b_t & = & \sigma(t,t), \\ q(t,x) & = & \delta(t,x,t). \end{cases} \qquad (25.12)$$

3. If $f(t,T)$ satisfies (25.6) then $p(t,T)$ satisfies

$$dp(t,T) = p(t,T)\left\{r_t - A(t,T) + \frac{1}{2}\|S(t,T)\|^2\right\}dt - p(t,T)S(t,T)dW_t$$

$$+p(t-,T)\int_E \left\{e^{-D(t,x,T)} - 1\right\}\Psi(dt,dx), \qquad (25.13)$$

where

$$\begin{cases} A(t,T) & = & \int_t^T \alpha(t,s)ds, \\ S(t,T) & = & \int_t^T \sigma(t,s)ds, \\ D(t,x,T) & = & \int_t^T \delta(t,x,s)ds. \end{cases} \qquad (25.14)$$

Proof **1:** We recall that $f(t,T) = -\frac{\partial}{\partial T}\ln p(t,T)$. From the P-dynamics and Itô we obtain

$$d\ln p(t,T) = \frac{1}{p(t,T)}p(t,T)\{m(t,T)dt + v(t,T)dW_t\} - \frac{1}{2}\frac{1}{p^2(t,T)}p^2(t,T)\|v(t,T)\|^2 dt$$

$$+ \int_E \{\ln[p(t-,T) + p(t-T)n(t,z,T)] - \ln p(t-,T)\}\Psi(dt,dz)$$

$$= \left(m(t,T) - \frac{1}{2}\|v(t,T)\|^2\right)dt + v(t,T)dW_t$$

$$+ \int_E \ln[1 + n(t,z,T)]\Psi(dt,dz).$$

If we now (ruthlessly) apply $\frac{\partial}{\partial T}$ to this formula and recall $f(t,T) = -\frac{\partial}{\partial T}\ln p(t,T)$, we obtain (25.11).

2: By definition we have $r_t = f(t,t)$, so

$$dr_t = df(t,T)|_{t=T} + f_T(t,t)dt$$

$$= \{\alpha(t,t) + f_T(t,t)\}\,dt + \sigma(t,t)dW_t + \int_E \delta(t,x,t)\Psi(dt,dx)$$

which proves (25.12).

3: We give a slightly heuristic argument. The full formal proof (see Björk et al., 1995) is an integrated version of the proof given here, but the infinitesimal version below is (we hope) easier to understand. Using the definition of the forward rates we may write

$$p(t,T) = e^{-X(t,T)}, \tag{25.15}$$

where X is given by

$$X(t,T) = \int_t^T f(t,s)ds. \tag{25.16}$$

We now want to compute $dX(t,T)$ and then apply Itô to (25.15). We have

$$dX(t,T) = d\left(\int_t^T f(t,s)ds\right),$$

but there's a problem because in the integral the t-variable occurs in two places: as the lower limit of integration, and in the integrand $f(t,s)$. This is a situation that is not covered by the standard Itô formula, but it is easy to guess the answer. The t appearing as the lower limit of integration should give rise to the term

$$\frac{\partial}{\partial t}\left(\int_t^T f(t,s)ds\right)dt.$$

Furthermore, since the stochastic differential is a linear operation, we should be allowed to move it inside the integral, thus providing us with the term

$$\left(\int_t^T df(t,s)ds\right).$$

We have therefore arrived at

$$dX(t,T) = \frac{\partial}{\partial t}\left(\int_t^T f(t,s)ds\right)dt + \int_t^T df(t,s)ds,$$

which, using the fundamental theorem of integral calculus, as well as the forward-rate dynamics, gives us

$$dX(t,T) = f(t,t)dt - \int_t^T \alpha(t,s)dt\,ds - \int_t^T \sigma(t,s)dW_t\,ds$$

$$- \int_t^T \int_E \delta(t,x,s)\Psi(dt,dx)ds.$$

We now use a stochastic Fubini theorem, allowing us to change the order of integration in the three integral terms. This gives us

$$dX(t,T) = -f(t,t)dt + \left(\int_t^T \alpha(t,s)ds \right) dt + \left(\int_t^T \sigma(t,s)ds \right) dW_t$$
$$+ \int_E \left(\int_t^T \delta(t,x,s)ds \right) \Psi(dt,dx).$$

We also recognize $f(t,t)$ as the short rate r_t, thus obtaining

$$dX(t,T) = -r_t dt + A(t,T)dt + S(t,T)dW_t + \int_E D(t,x,T)ds\Psi(dt,dx),$$

with A, S and D as in (25.14). Applying the Itô formula to

$$p(t,T) = e^{-X(t,T)}$$

gives us

$$dp(t,T) = p(t,T)\left(r_t + A(t,T) - \frac{1}{2}\|S(t,T)\|^2 \right) dt - p(t,T)S(t,T)dW_t$$
$$+\text{jump term}.$$

The jump term is, by Itô, given by

$$\int_E \left\{ e^{-(X(t-,T)+D(t,x,T))} - e^{-X(t-,T)} \right\} \Psi(dt,dx)$$

and recalling (25.15) we obtain the jump term as

$$p(t-,T) \int_E \left\{ e^{-D(t,x,T)} - 1 \right\} \Psi(dt,dx),$$

which proves (25.14). □

25.2 An Alternative View of the Bank Account

The object of this subsection is to show (heuristically) that the risk-free asset B can in fact be replicated by a self-financing strategy, defined by "rolling over" just-maturing bonds. This is a "folklore" result, which is very easy to prove in discrete time, but surprisingly tricky in a continuous-time framework.

Let us consider a self-financing portfolio which at each time t consists entirely of bonds maturing x units of time later (where we think of x as a small number). At time t the portfolio thus consists only of bonds with maturity $t + x$, so the value dynamics for this portfolio is, by (16.2), given by

$$dV_t = V_t \cdot 1 \cdot \frac{dp(t-,t+x)}{p(t-,t+x)}, \tag{25.17}$$

where the constant 1 indicates that the weight of the $t + x$-bond in the portfolio equals unity. We now want to study the behavior of this equation as x tends to zero, and to this end we use Proposition 25.8 to obtain

$$\frac{dp(t, t + x)}{p(t-, t + x)} = \left\{ r(t) + A(t, t + x) + \frac{1}{2} \|S(t, t + x)\|^2 \right\} dt + S(t, t + x) dW(t)$$

$$+ \int_E \left\{ e^{-D(t, z, t+x)} - 1 \right\} \Psi(dt, dz).$$

Letting x tend to zero, (25.14) gives us

$$\lim_{x \to 0} A(t, t + x) = 0,$$

$$\lim_{x \to 0} S(t, t + x) = 0$$

$$\lim_{x \to 0} D(t, t + x) = 0$$

and, substituting all this into equation (25.17), we obtain the value dynamics

$$dV_t = r_t V_t dt, \tag{25.18}$$

which we recognize as the dynamics of the money account.

The argument thus presented is of course only heuristic, and hard work is required to make it precise. Note, for example, that the rolling-over portfolio above does not fall into the general framework of self-financing portfolios, developed earlier. The problem is that, although at each time t, the portfolio only consists of one particular bond (maturing at $t + x$), over an arbitrarily short time interval, the portfolio will use an infinite number of different bonds.

Since the bond market contains an infinite number of assets – one bond price process $p(t, T)$ for very maturity T, we need to extend the portfolio theory of Section 16.1. This can be done roughly as follows.

Definition 25.9   A portfolio h on the bond market is a pair of processes $h_t = (h_t^B, h_t^P(dT))$ where

$$h_t^B = \text{is a real-valued predictable process}$$

$$h_t^P = \text{is a measure-valued predictable process.}$$

The interpretation is hat h_t^B is the number of units of the risk-free asset B we are holding at time t, whereas $h_t^P(dT)$ is the "number" of units of bond with maturities in $(T, T + dT)$ we are holding at time t.

Definition 25.10   Given a portfolio as above we make the follwing definitions

1. The **value process** is given by

$$V_t = h_t^B B_t + \int_0^\infty p(t, T) h_t^P(dT) \tag{25.19}$$

2. The portfolio is **self-financing** if

$$dV_t = h_t^B dB_t + \int_0^\infty h_t^P(dT) dp(t, T). \tag{25.20}$$

The problem now is to make sense of (25.20). This can be done but it quickly becomes very technical. See the Notes for references. If we have a bond market driven by a general MPP, then we really need a theory of measure-valued portfolios. The reason is given by the Rule-of-Thumb 16.14. If we want a complete market then we need to have as many assets as the number of random sources. In our framework, the number of random sources is the dimension of the Wiener process plus the number of points in the mark space E. If we have an infinite number of points in E then we have an infinite number of random sources, so we need to include an infinite number of bonds in our portfolio. Given a proper theory of measure-valued portfolios one can prove the following version of the First and Second Fundamental Theorems 16.22, 16.13.

Theorem 25.11

- *If there exists a martingale measure then the bond market is arbitrage-free.*
- *Assume that there exists a martingale measure for the bond market. Then the martingale measure is unique if and only if the set of claims which can be replicated by a self-financing portfolio is dense in the space of contingent claims.*

25.3   Short-Rate Models

In this section we will model the short rate as a Markov process, and in order to do this we define r as the solution to a jump-diffusion SDE. This can be done under the objective measure P or under a martingale measure Q, and the choice is free. In this chapter we choose to model under Q, i.e. we assume that the market has chosen a fixed martingale measure, Q, and we define the r-dynamics directly under Q. This approach is known as *martingale modeling*. See (Björk 2020) for a detailed discussion on this point. We now specialize the general short-rate dynamics (25.4) to the Markovian case.

Assumption 25.12   *We assume that, under the martingale measure Q, the short rate process r is the solution of a stochastic differential equation of the form*

$$dr_t = a(t, r_t)dt + b(t, r_t)dW_t + \int_E q(t, r_t, z)\Psi(dt, dz), \qquad (25.21)$$

where $a(t, r)$, $b(t, r)$ and $q(t, r, x)$ are given deterministic functions. The process W is a scalar Q-Wiener process and Ψ has a predictable Q-intensity λ of the form

$$\lambda_t(dz) = \lambda_t(r_{t-}, dz), \qquad (25.22)$$

where $\lambda_t(r, dz)$ is a deterministic measure for each t and r.

We start by presenting the fundamental partial differential-difference equation in this context as it concerns the pricing of simple claims in this Markovian setting.

Proposition 25.13   *Suppose that the short rate is given by (25.21) and consider, for a fixed T, any sufficiently integrable contingent claim X, to be paid at T, of the form*

$$X = \Phi(r_T). \qquad (25.23)$$

Then the arbitrage-free price process $\Pi_t[X]$ of this asset is given by

$$\Pi_t[X] = F(t, r_t),$$ (25.24)

where F is the solution of the Cauchy problem

$$\begin{cases} \frac{\partial F}{\partial t}(t,r) + \mathcal{A}F(t,r) - rF(t,r) &= 0, \\ F(T,r) &= \Phi(r), \end{cases}$$ (25.25)

with

$$\mathcal{A}F(t,r) = a(t,r)\frac{\partial F}{\partial r}(t,r) + \frac{1}{2}b^2(t,r)\frac{\partial^2 F}{\partial r^2}(t,r)$$
$$+ \int_E \{F(t,r+q(t,r,z)) - F(t,r)\}\,\lambda_t(r,dz).$$ (25.26)

Proof From (25.7) we have

$$\Pi_t[X] = E^Q\left[e^{-\int_0^t r_s\,ds} \cdot \Phi(r_T)\Big|\mathcal{F}_t\right].$$ (25.27)

The result now follows directly from the fact that r is Markov and Theorem 9.6 – which says that (25.25) is the Kolmogorov backward equation for the expected value above. □

25.4 Affine Term Structures

As soon as one moves from abstract theory to practical applications, and in particular to algorithms that have to be executed in real time on a computer, the need emerges of easily manageable analytical formulas. In the case of interest-rate derivatives one is particularly fortunate if the model possesses a so-called *affine term structure*.

Definition 25.14 The short-rate model (25.21) is said to have an **affine term structure** if bond prices can be written as

$$p(t,T) = F(t, r_t, T),$$ (25.28)

where F is of the form

$$F(t,r,T) = e^{A(t,T)-B(t,T)r},$$ (25.29)

and where A and B are deterministic functions. We sometimes use the notation

$$F(t,r,T) = F^T(t,r).$$

A model exhibiting an affine term structure occurs naturally only in a Markovian environment and so we again assume that r satisfies the SDE

$$dr_t = a(t,r_t)dt + b(t,r_t)dW_t + \int_E q(t,r_t,z)\Psi(dt,dz).$$ (25.30)

We now turn to the existence of the affine term structure. At first sight this looks like a difficult problem, but it is in fact surprisingly easy to obtain sufficient conditions. To do this we proceed as follows.

- Make the affine Ansatz

$$F^T(t,r) = e^{A(t,r) - B(t,r)r}.$$ (25.31)

- Plug this Ansatz into the fundamental equation (25.25).
- Take a look at the result and identify assumptions on a, b, q and λ which will turn the equation into one that is linear in r.
- Now we have separation of variables, so we are more or less done.

To carry out this small program, we use (25.31) to obtain the following formulas, where we have suppressed all variables except r. We use lower-case indices to denote partial derivative:

$$F_t = (A_t - B_t r)F,$$

$$F_r = -BF,$$

$$F_{rr} = B^2 F,$$

$$\int_E \{F(r + q(r,z)) - F(r)\} \, \lambda(r, dz) = F \cdot \int_E \left\{ e^{-Bq(r,z)} - 1 \right\} \lambda(r, dz).$$

Plugging this into (25.25) gives us (after dividing by the common factor F)

$$A_t - B_t r - a(r)B + \frac{1}{2} b^2(r) B^2 + \int_E \left\{ e^{-Bq(r,z)} - 1 \right\} \lambda(r, dz) = 0.$$ (25.32)

The idea is now to make assumptions which will turn this into an equation which is linear in the r-variable. It is more-or-less obvious how to do this. We thus assume that the the model parameters a, b, q, and λ have the structure

$$a(t,r) = \alpha_1(t) + \alpha_2(t)r,$$

$$b(t,r) = \sqrt{\beta_1(t) + \beta_2(t)r},$$

$$q(t,r,x) = q(t,x),$$

$$\lambda_t(r, dx) = \lambda_1(t, dx) + \lambda_2(t, dx)r.$$

Under this assumption, equation (25.32) becomes

$$A_t - B_t r - \alpha_1 B - \alpha_2 Br + \frac{1}{2}\beta_1 B^2 + \frac{1}{2}\beta_2 B^2 r$$

$$+ \int_E \left\{ e^{-Bq(x)} - 1 \right\} \lambda_1(dx) + r \int_E \left\{ e^{-Bq(x)} - 1 \right\} \lambda_2(dx) = 0.$$

Collecting terms we obtain

$$\left[A_t - \alpha_1 B + \frac{1}{2}\beta_1 B^2 + \int_E \left\{ e^{-Bq(x)} - 1 \right\} \lambda_1(dx) \right]$$

$$-r \left[B_t + \alpha_2 B - \frac{1}{2}\beta_2 B^2 - \int_E \left\{ e^{-Bq(x)} - 1 \right\} \lambda_2(dx) \right] = 0.$$

Since this equality holds for all values of r, the two brackets must be identically equal to zero, and this gives us a coupled system of differential–integral equations for A and B. We thus have the following result.

Proposition 25.15   *Suppose that the r-dynamics under Q is given by (25.21), and that the model parameters a, b, q and λ have structure*

$$
\begin{array}{rcl}
a(t,r) & = & \alpha_1(t) + \alpha_2(t)r, \\
b(t,r) & = & \sqrt{\beta_1(t) + \beta_2(t)r}, \\
q(t,r,z) & = & q(t,z), \\
\lambda_t(r,dz) & = & \lambda_1(t,dz) + \lambda_2(t,dz)r.
\end{array}
\tag{25.33}
$$

Then the model has an affine term structure of the form (25.28)–(25.29), where the functions $A(\cdot,T)$ and $B(\cdot,T)$ for $t \in [0,T]$ solve the following system of ODEs:

$$
\frac{\partial B}{\partial t}(t,T) + \alpha_2(t)B(t,T) - \frac{1}{2}\beta_2(t)B^2(t,T) + K_2(t,B(t,T)) = -1,
\tag{25.34}
$$
$$
B(T,T) = 0,
$$

$$
\frac{\partial A}{\partial t}(t,T) - \alpha_1(t)B(t,T) + \frac{1}{2}\beta_1(t)B^2(t,T) - K_1(t,B(t,T)) = 0,
\tag{25.35}
$$
$$
A(T,T) = 0,
$$

where

$$
K_i(t,y) = \int_E \left\{ 1 - e^{-yq(t,x)} \right\} \lambda_i(t,dx), \quad i = 1,2.
\tag{25.36}
$$

We note that the class of models satisfying (25.33) generalizes some well-known classical short-rate models such as those of Cox, Ingersoll & Ross (1985), Ho & Lee (1986), Hull & White (1990) and Vašíček (1977). The affine theory above can in fact be enormously generalized. For more on that, see the Notes.

25.5   Forward-Rate Models

In the previous sections we studied interest models where the short rate r is the only explanatory variable. The main advantages with such models are as follows.

- Specifying r as the solution of an SDE allows us to use Markov process theory, so we may work within a PDE framework.
- In particular it is often possible to obtain analytical formulas for bond prices and derivatives.

The main drawbacks of short-rate models are as follows.

- From an economic point of view it seems unreasonable to assume that the entire money market is governed by only one explanatory variable.
- It is hard to obtain a realistic volatility structure for the forward rates without introducing a very complicated short-rate model.
- As the short-rate model becomes more realistic, fitting the model to market data ("inverting the yield curve") becomes increasingly more difficult.

It thus seems natural to study models that have more than one state variable. One obvious idea would be, for example, to present an *a priori* model for the short rate as well as for some long rate, and one could of course also model one or several intermediary interest rates. The method proposed by Heath, Jarrow & Morton (1992) is at the far end of this spectrum – they choose the entire forward-rate curve as their (infinite-dimensional) state variable.

25.5.1   The HJM Framework

We now turn to the specification of the Heath–Jarrow–Morton (HJM) framework. Given the usual stochastic basis from Assumption 25.5 we copy the forward-rate dynamics from (25.6).

Assumption   *We assume that, for every fixed $T > 0$, the forward rate $f(\cdot,T)$ has a stochastic differential which, under the objective measure P, is given by*

$$df(t,T) = \alpha(t,T)dt + \sigma(t,T)dW_t^P + \int_E \delta(t,z,T)\Psi(dt,dz), \qquad (25.37)$$

$$f(0,T) = f^*(0,T), \qquad (25.38)$$

where $f^*(0,T)$ is the observed forward rate with maturity T at t = 0.

Note again that in equation (25.37) we have a scalar stochastic differential in the t-variable for each fixed choice of T, implying that (25.37) describes the evolution of an infinite-dimensional system parameterized by T. The index T thus only serves as a parameter in order to indicate which maturity we are looking at. Also note that we use the observed forward-rated curve $\{f^*(0,T); T \geq 0\}$ as the initial condition. This will automatically give us a perfect fit between observed and theoretical bond prices at $t = 0$, thus relieving us of the task of calibrating the model to market data.

The main questions for us are the following:

- how must the processes α, σ, δ and λ be related in order that the induced system of bond prices is free of arbitrage?
- what are the Q-dynamics of the forward rates?

The answer to the first question is given by the famous *HJM drift condition*, first presented in Heath et al. (1992), which we now will extend to include a driving-point process. The answer to the second question then follows immediately.

25.5.2   Absence of Arbitrage

Suppose now that we have specified α, σ, δ, λ and $\{f^*(0,T); T \geq 0\}$. Then we have specified the entire forward-rate structure and thus, by the relation

$$p(t,T) = e^{-\int_t^T f(t,s)ds}, \qquad (25.39)$$

we have in fact specified the entire term structure of bond prices

$$\{p(t,T); T > 0, \ 0 \leq t \leq T\}.$$

The first question we pose is thus very natural: How must the processes α, σ, δ and λ be related so that the induced system of bond prices is arbitrage-free? The answer is given by the HJM drift condition which we derive below.

To start the investigation we know from Proposition 25.8 that the induced bond-price dynamics are given by

$$dp(t,T) = p(t,T) \left\{ r_t - A(t,T) + \frac{1}{2} \|S(t,T)\|^2 \right\} dt - p(t,T)S(t,T)dW_t$$

$$+ p(t-,T) \int_E \left\{ e^{-D(t,x,T)} - 1 \right\} \Psi(dt, dx), \qquad (25.40)$$

where

$$\begin{cases} A(t,T) &= \int_t^T \alpha(t,s)ds, \\ S(t,T) &= \int_t^T \sigma(t,s)ds, \\ D(t,x,T) &= \int_t^T \delta(t,x,s)ds. \end{cases} \qquad (25.41)$$

We now look for a martingale measure, i.e. a measure Q such that $\frac{p(t,T)}{B_t}$ is a Q-martingale for every fixed T. Equivalently, we look for a measure Q such that the local rate of return equals the short rate for every T-bond. The obvious way to handle this is to perform a Girsanov transformation. We fix therefore an adapted column-vector process γ_t and a predictable process $\gamma_t(x)$ to define a likelihood process L by

$$dL_t = L_{t-}\gamma_t^* dW_t + \int_E \varphi_t(x) \left\{ \Psi(dt, dx) - \lambda_t(dx)dt \right\}, \qquad (25.42)$$

$$L_0 = 1. \qquad (25.43)$$

Assuming that L is a true martingale (not just a local one) we then change measure from P to Q by $dQ = L_T dP$ on $\mathcal{F}_T$. It now follows from Girsanov that we can write

$$dW_t = \gamma_t dt + dW_t^Q, \qquad (25.44)$$

where W^Q is Q-Wiener, and that Ψ has predictable Q-intensity

$$\lambda_t^Q(x) = \varphi_t(x)\lambda_t(x). \qquad (25.45)$$

Plugging (25.44) into (25.40) and compensating the jump term under Q gives us the Q-dynamics of $p(t,T)$ as

$$dp(t,T) = p(t,T) \left\{ r_t - A(t,T) + \frac{1}{2} \|S(t,T)\|^2 - S(t,T)\gamma_t \right\} dt$$

$$+ p(t,T) \left\{ \int_E \left[e^{-D(t,x,T)} - 1 \right] \varphi_t(x)\lambda_t(x)dx \right\} dt$$

$$- p(t,T)S(t,T)dW_t^Q$$

$$+ p(t-,T) \int_E \left[e^{-D(t,x,T)} - 1 \right] \left\{ \Psi(dt, dx) - \varphi_t(x)\lambda_t(x)dxdt \right\}.$$

The two last terms are Q-martingales so the martingale condition for Q is that

$$A(t,T) - \frac{1}{2}\|S(t,T)\|^2 - S(t,T)\gamma_t - \int_E \left[e^{-D(t,x,T)} - 1\right]\varphi_t(x)\lambda_t(x)dx = 0$$

for all (t,T). Since this is an identity in T we can take the T-derivative to obtain

$$\alpha(t,T) = \sigma(t,T)\int_t^T \sigma^*(t,s)ds + \sigma(t,T)\gamma_t + \int_E \delta(t,x,T)e^{-D(t,x,T)}\varphi_t(x)\lambda_t(x)dx,$$

where * denotes transpose. We have thus proved the following result

Theorem 25.16 (HJM Drift Condition 1) *Assume that forward-rate dynamics are given by (25.37). Then the induced bond market is arbitrage-free if and only if there exists an adapted process γ_t and a predictable process $\varphi_t(x)$ such that*

$$\alpha(t,T) = \sigma(t,T)\int_t^T \sigma^*(t,s)ds + \sigma(t,T)\gamma_t + \int_E \delta(t,x,T)e^{-D(t,x,T)}\varphi_t(x)\lambda_t(x)dx$$
$$(25.46)$$

for all (t,T).

Suppose now that we want to undertake martingale modeling, i.e. we specify the forward-rate dynamics directly under an equivalent martingale measure Q. We then note that in the arguments above we have never used the fact that P is the *objective* probability measure. It could have been be any measure equivalent to the objective one. In particular, all arguments and result carry over to the case when $P = Q$. The condition $P = Q$ implies that $\varphi_t(x) \equiv 1$ and $\gamma_t \equiv 0$, so we have the following result.

Theorem 25.17 (HJM Drift Condition 2) *Assume that the forward-rate dynamics under an equivalent martingale measure Q are given by*

$$df(t,T) = \alpha(t,T)dt + \sigma(t,T)dW_t + \int_E \delta(t,x,T)\Psi(dt,dx), \qquad (25.47)$$

where W is Q-Wiener and Ψ has predictable intensity process $\lambda_t(dz)$. Then $\alpha(t,T)$ is uniquely determined by the formula

$$\alpha(t,T) = \sigma(t,T)\int_t^T \sigma^*(t,s)ds + \int_E \delta(t,x,T)e^{-D(t,x,T)}\lambda_t(x)dx. \qquad (25.48)$$

The moral of the HJM drift condition is thus that when we specify the forward-rate dynamics (under Q) we may freely specify the volatility structure. The drift parameters are then uniquely determined.

25.5.3 An Example

To see how the theory above works we now consider the simplest non-trivial example under a martingale measure Q. We thus assume that the Q-Wiener process W is scalar and that

$$\sigma(t,T) = \sigma, \quad \delta(t,x,T) = \delta(x).$$

We then have $D(t, x, T) = \delta(x)(T - t)$ so by the HJM drift condition (25.48) we have the Q-dynamics given by

$$df(t,T) = \left\{ \sigma^2(T-t) + \int_E \delta(x) e^{-\delta(x)(T-t)} \lambda f(x) dx \right\} dt + \sigma dW_t$$

$$+ \int_E \delta(x) \psi(dt, dx).$$

Integrating this over $[0, t]$ gives us

$$f(t,T) = f^*(0,t) + \sigma^2 Tt - \frac{\sigma^2}{2} t^2 + \int_0^t \int_E \delta(x) e^{-\delta(x)(T-s)} \lambda f(x) dx ds$$

$$+ \sigma W_t + \int_0^t \int_E \delta(x) \psi(ds, dx).$$

We can now use Fubini on the $dxds$-integral term to obtain

$$f(t,T) = f^*(0,t) + \sigma^2 Tt - \frac{\sigma^2}{2} t^2 + \int_E \left\{ e^{-\delta(x)(T-t)} - e^{-\delta(x)T} \right\} \lambda f(x) dx$$

$$+ \sigma W_t + \int_0^t \int_E \delta(x) \psi(ds, dx).$$

This is a fairly explicit representation of the forward rate. In particular we can get the short rate by using $r_t = f(t, t)$. We then have

$$r_t = f^*(0,t) + \frac{\sigma^2}{2} t^2 + \int_E \left\{ 1 - e^{-\delta(x)t} \right\} \lambda f(x) dx$$

$$+ \sigma W_t + \int_0^t \int_E \delta(x) \psi(ds, dx).$$

The short-rate dynamics are thus given by

$$dr_t = \left\{ f_T^*(0,t) + \sigma^2 t + \int_E \delta(x) e^{-\delta(x)t} \lambda f(x) dx \right\} dt$$

$$+ \sigma dW_t + \int_E \delta(x) \psi(dt, dx). \tag{25.49}$$

We recognize this as a point-process extension of the classic Ho–Lee short-rate model (see Ho & Lee, 1986), calibrated to the initial term structure $\{f^*(0, T) : t \geq 0\}$.

25.5.4   The Musiela Parameterization

In many practical applications it is more natural to use time *to* maturity, rather than time *of* maturity, to parameterize bonds and forward rates. If we denote running time by t, time of maturity by T and time to maturity by x, then we have $x = T - t$, and in terms of x, the forward rates are defined as follows.

Definition 25.18   For all $x \geq 0$ the forward rates $r_t(x)$ are defined by the relation

$$r_t(x) = f(t, t + x). \tag{25.50}$$

Note that we have a slight change of notation. The short rate is now denoted by $r_t(0)$.

Suppose now that we have a HJM-type model for the forward rates under a martingale measure Q:

$$df(t,T) = \alpha(t,T)dt + \sigma(t,T)dW_t + \int_E \delta(t,z,T)\Psi(dt,dz). \qquad (25.51)$$

Note that since we use x to denote time to maturity (this is standard in the literature), we use z as the integration variable for Ψ.

We want to find the Q-dynamics for $r_t(x)$. We give a heuristic argument which can be made precise (see the Notes). Since $r_t(x) = f(t,t+x)$ we have

$$dr_t(x) = df(t,t+x).$$

Here we have a slightly non-standard expression since the variable t occurs in two places. The obvious guess (which can be proved) is that

$$dr_t(x) = df(t,T)\,|_{T=t+x} + \frac{\partial f}{\partial T}(t,t+x)dt,$$

where the differential in the term $df(t,t+x)$ only operates on the first t. We thus obtain

$$dr_t(x) = \alpha(t,t+x)dt + \sigma(t,t+x)dW_t + \int_E \delta(t,z,t+x)\Psi(dt,dz)$$

$$+ \frac{\partial}{\partial x}r(t,x)dt.$$

Using the HJM drift condition we obtain our result.

Proposition 25.19 (The Musiela equation) *Assume that the forward-rate dynamics under Q are given by (25.51). Then*

$$dr_t(x) = \{\mathbf{F}r_t(x) + G_0(t,x) + H_0(t,x)\}\,dt + \sigma_0(t,x)dW_t + \int_E \delta_0(t,z,x)\psi(dt,dz),$$

$$(25.52)$$

where

$$\sigma_0(t,x) = \sigma(t,t+x),$$
$$\delta_0(t,z,x) = \delta(t,z,t+x),$$
$$G_0(t,x) = \sigma_0(t,x)\int_0^x \sigma_0^*(t,s)ds,$$
$$H_0(t,x) = \int_E \delta_0(t,z,x)e^{-D_0(t,z,x)}\lambda_t(z)dz$$
$$D_0(t,z,x) = D(t,z,t+x)$$
$$\mathbf{F} = \frac{\partial}{\partial x}$$

The point of the Musiela parameterization is that it highlights equation (25.52) as an infinite-dimensional SDE. It has become an indispensable tool of modern interest-rate theory.

Up to this moment we have viewed (25.52) as an *infinite* number of *scalar* equations (one for each x). A more exciting alternative is to view (25.52) as a *single* equation describing the evolution of an *infinite-dimensional* object r evolving in some infinite-dimensional space. The object r is the *forward-rate curve* $x \mapsto r(x)$, which can be viewed as point (or vector) in some infinite-dimensional function space $\mathcal{H}$, and $\mathbf{F}$ is a linear operator on that space. Taking this point of view, and to emphasize that all functions of the variable x should be viewed as vectors in $\mathbf{F}$, we suppress x in (25.52) and write it as

$$dr_t = \{\mathbf{F}r_t + G_t + H_t\} dt + \sigma_t dW_t + \int_E \delta_t(z)\psi(dt, dz), \qquad (25.53)$$

where, for readability, we have slightly changed the notation so we have the interpretation

$$G_t = G_0(t, \cdot)$$
$$H_t = H_0(t, \cdot)$$
$$\sigma_t = \sigma_0(t, \cdot)$$
$$\delta_t(z) = \delta_0(t, z, \cdot).$$

We furthermore lump G and H together as $G_t + H_t = K_t$ so we have

$$dr_t = \{\mathbf{F}r_t + K_t\} dt + \sigma_t dW_t + \int_E \delta_t(z)\psi(dt, dz). \qquad (25.54)$$

In the general case, when K_t, σ_t and $\delta_t(z)$ (for a fixed x) are allowed to be stochastic, this is a highly non-linear equation and there is no hope of solving it. The special case when K_t, σ_t and $\delta_t(z)$ are deterministic can however, be analyzed in some detail so we now do this. For simplicity of notation we also assume that K, σ and δ are time-independent.

Assumption   *For the rest of this section we assume (with slight abuse of notation) that $K : R_+ \to \mathcal{H}$, $\sigma : R_+ \to \mathcal{H}$ and (for each fixed $z \in E$) $\delta(z)$ are deterministic functions.*

Given this assumption, we can write

$$dr_t = \{\mathbf{F}r_t + K\} dt + \sigma dW_t + \int_E \delta(z)\psi(dt, dz), \qquad (25.55)$$

and we see that (25.55) is a *linear SDE* evolving in the infinite-dimensional space $\mathcal{H}$. This looks familiar: we recall from Section 8.6.1 that the linear equation

$$\begin{cases} dX_t &= \{aX_t + b\} dt + \sigma dW_t + \int_E \beta(z)\Psi(dz, dt), \\ X_0 &= x_0 \end{cases} \qquad (25.56)$$

has the solution

$$X_t = e^{at}x_0 + \int_0^t e^{a(t-s)}b\,ds + \int_0^t e^{-a(t-s)\sigma} dW_s + \int_0^t \int_E e^{a(t-s)}\beta(z)\Psi(dz, ds). \quad (25.57)$$

We are thus led to conjecture that the solution to (25.55) is given by the formal expression

$$r_t = e^{\mathbf{F}t}r_0 + \int_0^t e^{\mathbf{F}(t-s)}K\,ds + \int_0^t e^{\mathbf{F}(t-s)}\sigma dW_s + \int_0^t \int_E e^{\mathbf{F}(t-s)}\beta(z)\Psi(dz, ds).$$

The question is now how we should interpret the expression $e^{\mathbf{F}}t$. We recall that r_t (for fixed t), K, σ and $\beta(z)$ are real-valued functions from R_+ to R. The formal exponential $e^{\mathbf{F}t}$ should thus act on real-valued functions, and we have to figure out how it operates. From the standard series expansion of the exponential function we are led to write

$$\left[e^{\mathbf{F}t}f\right](x) = \sum_{n=0}^{\infty} \frac{t^n}{n!} [F^n f](x) \tag{25.58}$$

for a function $f : R_+ \to R$. In our case $F^n = \frac{\partial^n}{\partial x^n}$, so (assuming f to be analytic) we have

$$\left[e^{\mathbf{F}t}f\right](x) = \sum_{n=0}^{\infty} \frac{t^n}{n!} \frac{\partial^n f}{\partial x^n}(x). \tag{25.59}$$

This is a Taylor-series expansion of f around the point x, so for analytic f we have $\left[e^{\mathbf{F}t}f\right](x) = f(x+t)$. We have in fact the following precise result (which can be proved rigorously).

Proposition 25.20   *The operator $\mathbf{F}$ is the infinitesimal generator of the semigroup of left-translations, i.e. for any $f \in C[0,\infty)$ we have*

$$\left[e^{\mathbf{F}t}f\right](x) = f(t+x).$$

The solution of the forward-rate equation (25.55) is given by

$$r_t(x) = e^{\mathbf{F}t}r_0(x) + \int_0^t e^{\mathbf{F}(t-s)}K(x)ds + \int_0^t e^{\mathbf{F}(t-s)}\sigma(x)dW_s, \tag{25.60}$$

or equivalently by

$$r_t(x) = r_0(x+t) + \int_0^t K(x+t-s)ds + \int_0^t \sigma(x+t-s)dW_s. \tag{25.61}$$

Remark   In the arguments above we have never precisely defined the pace, $\mathcal{H}$, of forward-rate curves. This is in fact quite tricky. See the references in the Notes.

25.6   Notes

There are many textbooks on bond markets, such as for example Fabozzi (2004) and Sundaresan (2009). Classic papers on short-rate models are Vasiček (1977), Hull & White (1990), Ho & Lee (1986), Cox et al. (1985), Dothan (1978) and Black, Derman & Toy (1990). For a multivariate extension of the affine term structure theory, see Duffie & Kan (1996). A massive extension of the short-rate affine framework to general process theory is given in Duffie, Filipovic & Schachermayer (2003). For an almost encyclopedic monograph on interest-rate theory, see Brigo & Mercurio (2001). The theory of measure-valued portfolios was developed in Björk et al. (1995) and Björk, Di Masi, Kabanov & Runggaldier (1997). The forward-rate methodology was introduced in the seminal paper by Heath et al. (1992), and the extension to the MPP case was done in Björk et al. (1995). The Musiela parameterization was developed in an unpublished working paper by Musiela, and in Brace & Musiela (1994) and Musiela (1993).

26 Equilibrium Theory

In this chapter we will take a brief look at dynamic equilibrium theory within a jump-diffusion framework. In the previous models we have studied, all asset price processes, as well as the short rate, have been exogenously given. Now we are going to study models where some objects are exogenously given, but other objects such as the short rate, the martingale measure, and the stochastic discount factor, will be determined in economic equilibrium.

26.1   The Model

26.1.1   Exogenous Objects

We start with the exogenous objects.

Assumption   *The following objects are considered as given a priori.*

1. *The usual stochastic basis, carrying an m-dimensional Wiener process and a marked point process* Ψ *with mark space E and predictable intensity process* λ. *The filtration is assumed to be the internal one, generated by W and* Ψ.

2. *A scalar and strictly positive process e of the form*

$$de_t = a_t dt + b_t dW_t + \int_E \eta_t(z)\Psi(dt, dz), \tag{26.1}$$

where a and b are adapted and η *is predictable.*

We interpret e as a an **endowment process**, meaning that if you own e then you will have a consumption stream at the rate e_t units of the consumption good per unit time, so during the time interval $[t, t + dt]$ the owner will obtain $e_t dt$ units of the consumption good. A more concrete interpretation is that you have a orchard (like the cherry orchard in the Chekhov play) you do not invest anything in the orchard and you do not work: the cherries simply fall off the trees for you to eat. Following its first appearance in Lucas (1978), the endowment process is in fact often referred to as a "Lucas tree" or a "Lucas orchard".. We will, however, assume that the endowment is measured in dollars per unit time.

26.1.2 Endogenous Objects

The endogenous objects in the model are as follows.

1. A risk-free asset B, in zero net supply, with dynamics

$$dB_t = r_t B_t dt,$$

   where the risk-free rate r is determined in equilibrium.
2. A price–dividend pair (S, D) in unit net supply, where by assumption

$$dD_t = e_t dt.$$

   In other words, holding the asset S provides the owner with the dividend process e over the time interval $[0, T]$. Since S is defined in terms of e we can write the dynamics of S as

$$dS_t = \alpha_t S_t dt + \sigma_t S_t dW_t + \int_E \beta_t(z) \Psi(dt, dz),$$

   where α, σ and β will be determined in equilibrium.
3. The stochastic discount factor $\mathbf{M}$, which will be determined in equilibrium.

We note again that the asset S provides the owner only with the dividend process e over $[0, T]$: nothing else. From risk-neutral valuation it now follows that

$$S_t = \frac{1}{\mathbf{M}_t} E^P \left[\left. \int_t^T \mathbf{M}_s e_s ds \right| \mathcal{F}_t \right],$$

where $\mathbf{M}$ is the equilibrium stochastic discount factor. In particular we will have

$$S_T = 0.$$

Remark   That B is in zero net supply means that for anyone to sell a unit of B then someone else has to buy it. In particular you can shortsell as many units as you want as long as someone is willing to buy them. Unit net supply means that there is a fixed amount, namely one unit, of the asset S.

26.1.3 Economic Agents

In our economy we will have a single *representative agent* who wants to maximize expected utility of the form

$$E^P \left[\int_0^T U(t, c_t) dt \right].$$

where c is the consumption rate (measured in dollars per time unit) and U is the utility function.

Assumption   *We assume that the utility function U is smooth, increasing in c, strictly concave and satisfying $U_c(t, 0) = +\infty$. We also assume that the agent has initial wealth $X_0 = S_0$. In other words, the agent has enough money to buy the right to the dividend process e. Another interpretation is that the agent is the initial owner of the single* *existing unit of S (unit net supply). At time t = 0 she sells this unit at the price S_0 and then the game is on.*

26.1.4   Equilibrium Conditions

We now define the relevant equilibrium concept. In order to make the theory we we need, however, one more assumption.

Assumption   *We assume that, apart from (S, D), we can add the necessary number of financial derivatives in zero net supply so that the market is complete.*

   This is a very bold assumption. If the number of random sources is finite, which will happen if the number of points in E is finite, then we can complete the market by a finite number of derivatives. If, however, the number of points in E is infinite, then we must add an infinite number of derivatives in order to get a complete market. A natural objection to the assumption above is also that *a priori* one would guess that any results derived would depend on the particular choice of derivatives used for market completion. As we will see below, however, the specific choice of derivatives does not have any impact on our results. We will in fact get a unique martingale measure and it will not depend on our way of completing the market. We will use the notation

$$h_t^S = \text{the number of units held in the risky asset,}$$
$$h_t^B = \text{the number of units held in the risk-free asset,}$$
$$h_t^D = \text{the portfolio vector held in the added derivatives,}$$
$$c_t = \text{rate of consumption.}$$

Remark   Denote the number of points in E by $|E|$. If $|E|$ is finite or countably infinite, then h^d will be a vector with dimension equal to $|E| + m$, where m is the dimension of W. If $|E|$ is uncountable, then we must extend our portfolio theory to include measure-valued portfolios. This is technically much more complicated. See Section 25.2 for more comments and references on this topic.

   We now describe in more detail how we define equilibrium. The story goes a follows.

- The representative agent takes the short-rate process, r and the stochastic discount factor, **M**, as given by the market.
- She then solves the optimization problem to maximize

$$E^P \left[\int_0^T U(t, c_t) dt \right]$$

over self-financing portfolio consumption pairs (h, c).

The natural equilibrium conditions are as follows: the agent will hold one unit (i.e. all) of the risky asset, the market will clear for the zero net supply assets, and the agent will consume all dividends. Formally this reads as follows.

Definition 26.1   An **equilibrium** in the model is a triple of processes $(\widehat{h}, \widehat{c}, \widehat{r})$ satisfying the following conditions.

- Given the short rate process $\widehat{r}$, the processes $(\widehat{h}, \widehat{c})$ are optimal for the agent.
- The processes $(\widehat{h}, \widehat{c})$ satisfy the **equilibrium conditions** which states that supply equals demand in all markets and that all dividends are consumed:

$$\widehat{h}_t^S = 1, \quad \text{(the agent is holding one unit of } (S, D)),$$
$$\widehat{h}_t^B = 0, \quad \text{(no holdings in } B),$$
$$\widehat{h}_t^D = 0, \quad \text{(no holdings in the added derivative assets),}$$
$$\widehat{c}_t = e_t, \quad \text{(market clearing for consumption).}$$

Note that the conditions $\widehat{h}_t B = 0$ and $\widehat{h}_t^D = 0$ are in fact redundant. The reason for this is that all assets apart from S, including the risk-free asset B, are in zero net supply. This implies that in equilibrium (where supply must equal demand) there is no trade in these assets. If someone was buying then someone else must be selling, but since there is only one agent in the market this cannot happen unless the traded amount is zero.

Note also that the market clearing condition $c_t = e_t$ for consumption is by no means self-evident. In particular it implies that you cannot save or invest the endowment $e_t dt$ that you get at time t in order to consume more at a later date. One possible interpretation is that the consumption good is instantly perishable: the cherries you harvest at time t are rotten already at time $t + dt$.

26.2 The Control Problem

We assume again that the initial wealth of the agent is given by $X_0 = S_0$. Since we have a complete market we can rely on the so-called *martingale approach* to reformulate the control problem. We then note that, because of market completeness, a consumption stream c can be replicated by a self-financing portfolio if and only if it satisfies the *budget constraint*

$$E^P \left[\int_0^T \mathbf{M}_t c_t dt \right] \leq S_0.$$

We may thus reformulate our initial problem to the following, much simpler, one:

$$\text{maximize } E^P \left[\int_0^T U(t, c_t) dt \right]$$

over consumption processes $c \geq 0$. subject to the constraints

$$c_t \geq 0, \qquad E^P \left[\int_0^T \mathbf{M}_t c_t dt \right] \leq S_0.$$

As usual, $\mathbf{M}$ denotes the stochastic discount factor. The first constraint is obvious and the second one is the budget constraint.

Since the asset S provides the owner only with the income stream defined by e and nothing else we can apply arbitrage theory and risk-neutral valuation to deduce that

$$S_0 = E^P \left[\int_0^T \mathbf{M}_t e_t dt \right].$$

We can thus rewrite the budget constraint as

$$E^P\left[\int_0^T \mathbf{M}_t c_t dt\right] \le E^P\left[\int_0^T \mathbf{M}_t e_t dt\right],$$

so our problem can be formulated as follows.

Problem

$$maximize\ E^P\left[\int_0^T U(t,c_t)dt\right]$$

subject to the constraints

$$c_t \ge 0, \qquad E^P\left[\int_0^T \mathbf{M}_t c_t dt\right] \le E^P\left[\int_0^T \mathbf{M}_t e_t dt\right].$$

It follows from the assumptions on U that the optimal consumption will always be strictly positive. The relevant Lagrangian is thus given by

$$E^P\left[\int_0^T \{U(t,c_t) - v\mathbf{M}_t c_t dt\}\right] + vE^P\left[\int_0^T \mathbf{M}_t e_t dt\right],$$

where v is the Lagrange multiplier. It now follows from general optimization theory that there will exist a Lagrange multiplier v such that the optimal c for the original problem with the budget constraint is the same as the optimal c for L, without any constraints on c.

The point of all this is that we now have a problem which is very easy to solve; namely, maximizing the Lagrangian

$$E^P\left[\int_0^T \{U(t,c_t) - v\mathbf{M}_t c_t dt\}\right] + vE^P\left[\int_0^T \mathbf{M}_t e_t dt\right]$$

over c. This, however is a completely decoupled problem; we can, for every t and ω, maximize

$$U(t,c_t) - v\mathbf{M}_t c_t$$

over c_t. The market clearing condition for c is thus

$$U_c(t,c_t) = Z_t, \tag{26.2}$$

where

$$Z_t = v\mathbf{M}_t. \tag{26.3}$$

Thus, up to a multiplicative constant, the SDF is equal to the marginal utility of consumption along the equilibrium consumption path. Recalling the general result

$$\pi_s = E^P\left[\frac{\mathbf{M}_t}{\mathbf{M}_s}\pi_t \Big| \mathcal{F}_s\right], \tag{26.4}$$

for any non-dividend price process, we obtain

$$\pi_s = E^P\left[\frac{U_c(t,c_t)}{U_c(s,c_s)}\pi_t \Big| \mathcal{F}_s\right], \tag{26.5}$$

which identifies the SDF over the interval $[s,t]$ with the corresponding marginal utility of equilibrium consumption.

26.3   Equilibrium

It is now surprisingly easy to derive formulas for the equilibrium short rate, stochastic discount factor (SDF) and Girsanov kernel. The clearing condition $\widehat{c}_t = e_t$ and the optimality condition (26.2) gives us

$$Z_t = U_c(t, e_t), \tag{26.6}$$

so we can use the Itô formula to obtain

$$
dZ_t = \left\{ U_{ct}(t-, e_{t-}) + a(e_{t-})U_{cc}(t-, e_{t-}) + \frac{1}{2}\|b(e_{t-})\|^2 U_{ccc}(t-, e_{t-}) \right\} dt
$$
$$
+ b(e_t)U_{cc}(t-, e_{t-})dW_t
$$
$$
+ \int_E \{U_c(t-, e_{t-} + \eta_t(z)) - U_c(t, e_{t-})\} \, \Psi(dt, dz).
$$

We now compensate the point process and recall that evaluation at t or $t-$ does not matter for the dt and dW terms and, since U and its derivatives are continuous, we have $U_{ct}(t-, e) = U_{ct}(t, e)$ and similarly for all other derivatives of U. In the point-process term, however, we must have evaluation at $t-$ in order to keep the integrand predictable. We thus obtain

$$
dZ_t =
$$
$$
\left\{ U_{ct}(t, e_t) + a(e_{t-})U_{cc}(t, e_t) + \frac{1}{2}\|b\|^2(e_t)U_{ccc}(t, e_{t-}) + \int_E U_{c,\eta}(t, e_{t-}, z)\lambda_t(dz) \right\} dt
$$
$$
+ b(e_t)U_{cc}(t, e_t)dW_t
$$
$$
+ \int_E U_{c,\eta}(t, e_{t-}, z) \{\Psi(dt, dz) - \lambda_t(dz)dt\}, \tag{26.7}
$$

where

$$U_{c,\eta}(t, e, z) = U_c(t, e + \eta_t(z)) - U_c(t, e). \tag{26.8}$$

From (26.7), (26.3) and (26.6) we then obtain

$$
dL_t =
$$
$$
L_t \cdot \frac{U_{ct}(t, e_t) + a(e_{t-})U_{cc}(t, e_t) + \frac{1}{2}\|b\|^2(e_t)U_{ccc}(t, e_{t-}) + \int_E U_{c,\eta}(t, e_{t-}, z)\lambda_t(dz)}{U_c(t, e_t)} dt
$$
$$
+ L_t \cdot \frac{b(e_t)U_{cc}(t, e_t)}{U_c(t, e_t)} dW_t
$$
$$
+ L_{t-} \cdot \int_E \frac{U_{c,\eta}(t, e_{t-}, z)}{U_c(t, e_{t-})} \{\Psi(dt, dz) - \lambda_t(dz)dt\}. \tag{26.9}
$$

We now recall from (16.25) that the P-dynamics of the stochastic discount factor $\mathbf{M}$ are given by

$$dL_t = -r_t L_t dt + L_t \gamma^* dW_t + L_{t-} \int_E \varphi_t(z) \{\Psi(dt, dz) - \lambda_t(dz)dt\}, \tag{26.10}$$

where L is the likelihood process taking us from P to the (unique) martingale measure Q, with Girsanov kernels γ and φ. Comparing (26.7) with (26.10), we can now directly identify the equilibrium short rate and the Girsanov kernel as follows.

Proposition 26.2 *The equilibrium stochastic discount factor is given by*

$$\mathbf{M}_t = \frac{U_c(t, e_t)}{U_c(0, e_0)}. \tag{26.11}$$

The equilibrium short rate is given by

$$r_t = r(t-, e_{t-}),$$

where the deterministic function $r_t(e)$ is given by

$$r_t(e) = -\frac{U_{ct}(t, e) + a(e)U_{cc}(t, e) + \frac{1}{2}\|b\|^2(e)U_{ccc}(t, e) + \int_E U_{c,\eta}(t, e, z)\lambda_t(dz)}{U_c(t, e)}. \tag{26.12}$$

The equilibrium Girsanov kernels are given by

$$\gamma_t = \frac{U_{cc}(t, e_{t-})}{U_c(t, e_{t-})} \cdot b(e_{t-}), \tag{26.13}$$

$$\varphi_t(z) = \frac{U_{c,\eta}(t, e_{t-}, z)}{U_c(t, e_{t-})} \tag{26.14}$$

and

$$U_{c,\eta}(t, e, z) = U_c(t, e + \eta_t(z)) - U_c(t, e). \tag{26.15}$$

This is a rather remarkable result, since it gives us the equilibrium SDF, short rate and Girsanov kernels in explicit formulas involving only exogenously given data. Note that in the formula for γ we can replace e_{t-} by e_t.

26.4   Risk Premia

Since have the Girsanov kernels on explicit form we can easily determine the risk premia according to Proposition 19.10.

Proposition 26.3 *Assume that some asset-price process π, derivative or underlying, has the following structure under P.*

$$d\pi_t = \mu_{\pi t} dt + \sigma_{\pi t} dW_t + \pi_{t-} \int_E \beta_{\pi t}(z) \{\Psi(dt, dz) - \lambda_t(dz)dt\}. \tag{26.16}$$

Then the risk premium is given by the formula

$$\mu_t - r = -\frac{U_{cc}(t, e_{t-})}{U_c(t, e_{t-})} \cdot b(e_{t-})\sigma_{\pi t} - \int_E \beta_{\pi t}(z)\frac{U_{c,\eta}(t, e_{t-}, z)}{U_c(t, e_{t-})}\lambda_t(dz) \tag{26.17}$$

and the jump-risk premium is given by

$$\frac{\lambda_t^{E,Q}}{\lambda_t^{E,P}} = 1 + \int_E \frac{U_{c,\eta}(t, e_{t-}, z)}{U_c(t, e_{t-})} \Gamma_t(dz), \tag{26.18}$$

where

$$U_{c,\eta}(t, e, z) = U_c(t, e + \eta_t(z)) - U_c(t, e). \tag{26.19}$$

We can also write

$$\frac{\lambda_t^{E,Q}}{\lambda_t^{E,P}} = 1 + E^{\Gamma_t}\left[\frac{U_{c,\eta}(t, e_{t-}, z)}{U_c(t, e_{t-})}\right], \tag{26.20}$$

where E^{Γ_t} denotes expectation under the probability measure Γ_t.

26.5 Log Utility

To exemplify we now specialize to a simple Markovian setting.

Assumption
- We have log utility, i.e. the local utility function has the form

$$U(t, c) = e^{-\delta t} \ln(c),$$

  where δ is the discount factor of the agent.
- The endowment process satisfies an SDE of the form

$$de_t = a(e_t)dt + b(e_t)dW_t + \int_E \eta(e_{t-}, z)\Psi(dt, dz),$$

  where a and b are deterministic real-valued functions of e, whereas η is a deterministic function of (e, z).
- The Wiener process W is m-dimensional.
- The point process Ψ is compound Poisson so λ has the form

$$\lambda_t(dz) = \lambda^E \cdot \Gamma(dz),$$

  where λ^E in the right-hand side is a constant and μ is a deterministic probability measure on E.

In this case we have

$$U_c(t, e) = \frac{1}{e}e^{-\delta t}, \quad U_{tc}(t, e) = -\frac{\delta}{e}e^{-\delta t}, \tag{26.21}$$

$$U_{cc}(t, e) = -\frac{1}{e^2}e^{-\delta t}, \quad U_{ccc}(t, e) = \frac{2}{e^3}e^{-\delta t}, \tag{26.22}$$

$$U_{c,\eta}(t, e, z) = -e^{-\delta t}\frac{\eta(e, z)}{e\,[e + \eta(e, z)]}. \tag{26.23}$$

Plugging this into (26.12) gives us the short rate as $r_t = r_{t-}(e_{t-})$, where

$$r(t, e) = \delta + \frac{a(e)}{e} - \frac{\|b(e)\|^2}{e^2} + \int_E \frac{\eta(e, z)}{e + \eta(e, z)}\lambda(dz) \tag{26.24}$$

and the Girsanov kernels are

$$\gamma_t = \frac{b(e_t)}{e_t},$$

$$\varphi_t(z) = \frac{\eta(e_{t-}, z)}{e_{t-} + \eta(e_{t-}, z)}.$$

Given the expression for (26.24), it is natural to specialize further to the case when the e-dynamics are of the form

$$de_t = ae_t\,dt + e_t b\,dW_t + e_{t-}\int_E \eta(z)\Psi(dt, dz), \qquad (26.25)$$

where (with a slight abuse of notation) a is a real constant and b is an m-dimensional row vector, whereas η is a deterministic function of z, so that

$$a(e) = a \cdot e, \quad b(e) = b \cdot e, \quad \eta(e, z) = e \cdot \eta(z). \qquad (26.26)$$

We are thus studying a geometric Levy process with a compound Poisson jump process. We then obtain a constant short rate of the form

$$r = \delta + a - \|b\|^2 + \int_E \frac{\eta(z)}{1 + \eta(z)}\lambda(dz) \qquad (26.27)$$

with deterministic Girsanov kernels

$$\gamma^* = -b \qquad (26.28)$$

$$\varphi_t(z) = \frac{\eta(z)}{1 + \eta(z)}. \qquad (26.29)$$

26.6   A Factor Model

In the previous sections we have assumed that the endowment process e satisfies a rather general SDE of the form

$$de_t = a(e_t)dt + b(e_t)dW_t + \int_E \eta(e_{t-}, z)\Psi(dt, dz).$$

A natural specification of this setup is to consider a factor model of the form

$$de_t = a(e_t, X_t)dt + b(e_t, X_t)dW_t + \int_E \eta(e_{t-}, X_t, z)\Psi(dt, dz),$$

$$dX_t = \mu(X_t)dt + \sigma(X_t)dW_t + \int_E \kappa(X_{t-}, z)\Psi(dt, dz),$$

where X is an underlying factor process, W is a two-dimensional Wiener process and Ψ is compound Poisson as in the previous section. For simplicity we again assume log utility, so

$$U(t, c) = e^{-\delta t}\ln(c).$$

In this case the equilibrium rate and the Girsanov kernels will be of the form $r_t = r(e_{t-}, X_{t-})$, $\gamma_t = \gamma(e_{t-}, X_{t-})$ and $\varphi_t = \varphi(e_{t-}, X_{t-})$, and from Proposition 26.2 we obtain

$$r(e, y) = \delta + \frac{a(e, y)}{e} - \frac{\|b(e, y)\|^2}{e^2} + \int_E \frac{\eta(e, y, z)}{e + \eta(e, y, z)} \lambda(dz),$$

$$\gamma(e, y) = -\frac{b(e, y)}{e},$$

$$\varphi(e, y) = \frac{\eta(e, y, z)}{e + \eta(e, y, z)}.$$

Given these expressions it is natural to make the further assumption that a and b are of the form

$$a(e, y) = e \cdot a(y),$$
$$b(e, y) = e \cdot b(y),$$
$$\eta(e, y, z) = e \cdot \eta(y, z),$$

which implies

$$r(y) = \delta + a(y) - \|b(y)\|^2 + \int_E \frac{\eta(y, z)}{1 + \eta(y, z)} \lambda(dz),$$

$$\gamma(y) = -b(y),$$

$$\varphi(y, z) = \frac{\eta(y, z)}{1 + \eta(y, z)}.$$

An obvious idea is now to specialize further to the case when

$$a(y) = a \cdot y,$$
$$b(y) = \sqrt{y} \cdot b,$$
$$\eta(y, z) = \eta(z),$$

and in order to guarantee positivity of X we assume

$$\mu(y) = \eta + \mu \cdot y,$$
$$\sigma(y) = \sigma \cdot \sqrt{y},$$
$$\kappa(y, z) = y \cdot \kappa(z),$$

where $2\eta \geq \|\sigma\|^2$ and $\kappa > -1$. We then have the following result.

Proposition 26.4 *Assume that the model has the structure*

$$de_t = ae_t X_t dt + e_t b\sqrt{X_t} dW_t + e_{t-} \int_E \eta(z) \Psi(dt, dz), \tag{26.30}$$

$$dX_t = \{\eta + \mu X_t\} dt + \sigma \sqrt{X_t} dW_t + X_{t-} \int_E \kappa(z) \Psi(dt, dz), \tag{26.31}$$

with $2\eta \geq \|\sigma\|^2$ and $\kappa > -1$. Then the equilibrium short rate and the Girsanov kernel *are given by*

$$r_t = \delta + \left(a - \|b\|^2\right) X_t + \int_E \frac{\eta(z)}{1 + \eta(z)} \lambda(dz), \tag{26.32}$$

$$\gamma_t = -\sqrt{X_t} \cdot b, \tag{26.33}$$

$$\varphi = \frac{\eta(z)}{1 + \eta(z)}. \tag{26.34}$$

The short rate is thus an affine transformation of the X process, so we have essentially constructed a point-process extension of the famous Cox–Ingersoll–Ross short rate model (see Cox et al., 1985).

26.7   Endowment Equilibrium under Partial Information

In this section we study a simple partially observable version of the endowment model of Section 26.1. In order to understand this section you should first have read Chapter 12.

26.7.1   The Model

The main assumptions are as follows.

Assumption   *We assume the existence of an endowment process e of the form*

$$de_t = a_t dt + b_t dW_t. \tag{26.35}$$

Furthermore, we assume the following.

- *The observable filtration is given by $\mathbf{F}^e$, i.e. all observations are generated by the endowment process e.*
- *The process a is not assumed to be observable, so it is not adapted to $\mathbf{F}^e$.*
- *The process b is adapted to $\mathbf{F}^e$.*
- *The process b is assumed to satisfy the non-degeneracy condition*

$$b_t > 0, \quad P\text{-a.s. for all } t. \tag{26.36}$$

Apart from these assumptions, the setup is that of Section 26.1, so we assume that there exists a risky asset S in unit net supply, giving the holder the right to the endowment e. We also assume the existence of a risk-free asset in zero net supply. The initial wealth of the representative agent is assumed to equal S_0 so the agent can afford to buy the right to the endowment e. The representative agent is as usual assumed to maximize utility of the form

$$E\left[\int_0^T U(t, c_t)dt\right].$$

26.7.2   Projecting the e-Dynamics

Since we only have a partially observed model we cannot immediately apply the methods and results from Sections 26.1–26.6. We therefore start by projecting the relevant process dynamics onto the observable filtration. We thus define the process Z by

$$dZ_t = \frac{de_t}{b_t} \qquad (26.37)$$

so that

$$dZ_t = \frac{a_t}{b_t} dt + dW_t,$$

and define the innovation process v as usual by

$$dv_t = dZ_t - \frac{\widehat{a}_t}{b_t} dt, .$$

where

$$\widehat{a}_t = E\left[a_t \mid \mathcal{F}_t^e\right].$$

This gives us the Z-dynamics on the $\mathbf{F}^e$-filtration as

$$dZ_t = \frac{\widehat{a}_t}{b_t} dt + dv_t,$$

and plugging this into (26.37) gives us the e-dynamics projected onto the $\mathbf{F}^e$-filtration as

$$de_t = \widehat{a}_t dt + b_t dv_t. \qquad (26.38)$$

Note that the e-process in (26.35) is, for each (t, ω), *exactly* the same process as the one in (26.38). The difference is that the dynamics in (26.35) gives us the semimartingale decomposition on the $\mathbf{F}$-filtration, whereas (26.38) gives us the decomposition on the $\mathbf{F}^e$-filtration.

26.7.3   Equilibrium

Given the formula (26.38) we are now back in a completely observable model, so we can quote Proposition 26.2 to obtain the main result.

Proposition 26.5   *For the partially observed model above, the following hold.*
The equilibrium stochastic discount factor is given by

$$\mathbf{M}_t = \frac{U_c(t, e_t)}{U_c(0, e_0)}. \qquad (26.39)$$

- *The equilibrium short-rate process is given by*

$$r_t = -\frac{U_{ct}(t, e) + \widehat{a}_t U_{cc}(t, e_t) + \frac{1}{2}\|b_t\|^2 U_{ccc}(t, e_t)}{U_c(t, e_t)}. \qquad (26.40)$$

- *The Girsanov kernel is given by*

$$\varphi_t = \frac{U_{cc}(t, e_t)}{U_c(t, e_t)} \cdot b_t. \qquad (26.41)$$

In an abstract sense we have thus completely solved the problem of equilibrium within a partially observable endowment model, but we see that if we want to use the model we have to compute the filter estimate $\widehat{a}_t$. In order to obtain more concrete results, we therefore need to impose some extra structure which will be done in the next section.

26.7.4   A Factor Model

In this section we specialize the model above to a factor model of the form

$$de_t = a(e_t, Y_t)dt + b(e_t)dW_t^e,$$
$$dY_t = \mu(Y_t)dt + \sigma(Y_t)dW_t^y,$$

where, for simplicity, we assume that W^e and W^y are independent. Note that we cannot allow b to depend on the factor Y. We also assume log utility, so that

$$U(t, c) = e^{-\delta t} \ln(c).$$

As in Section 26.6 we easily obtain

$$r_t = \delta + \frac{\widehat{a}_t}{e_t} - \frac{b^2(e_t)}{e_t^2},$$

$$\varphi_t = -\frac{b(e_t)}{e_t},$$

where

$$\widehat{a}_t = E\left[a(e_t, Y_t)| \mathcal{F}_t^e\right].$$

Given these expressions it is natural to specialize to the case when

$$a(e, y) = e \cdot a(y),$$
$$b(e) = b \cdot e,$$

where b is a constant. This gives us

$$r_t = \delta + \widehat{a}_t - b^2,$$
$$\varphi_t = -b.$$

In order to obtain a finite filter for $\widehat{a} = E\left[a(Y_t)| \mathcal{F}_t^e\right]$ it is now natural to look for a Kalman model and our main result is as follows.

Proposition 26.6   *Assume a model of the form*

$$de_t = ae_t Y_t dt + be_t dW_t^e,$$
$$dY_t = BY_t dt + CdW_t^y.$$

The risk-free rate and the Girsanov kernel are then given by

$$r_t = \delta - b^2 + a\widehat{y}_t, \tag{26.42}$$
$$\varphi_t = -b, \tag{26.43}$$

where $\hat{y}$ is given by the Kalman filter

$$d\hat{y} = B\hat{y}_t + H_t dv_t.$$

26.8   Notes

Basic references for endowment models are Huang (1987) and Karatzas, Lehoczky & Shreve (1990). For textbook treatments see Back (2017), Björk (2020), Duffie (2001), Dana & Jeanblanc (2003) and Karatzas & Shreve (1998).

A problem with additive utility of the form (11.1) is that it ties time preferences to risk aversion in a restrictive way. The theory of *recursive preferences* allows you separate time preferences from risk aversion while retaining the Bellman optimality principle. This approach is now more or less standard; some basic references are Kreps & Porteus (1978) and Epstein & Zin (1989) in discrete time and Duffie & Epstein (1992) in continuous time.

Another problem with time-additive utility is that it does not allow consumption at different points in time to be substitutes or complements. One way of dealing with this problem is to allow for *habit formation* models. In *internal* models you let the utility of consumption depend on a smoothed average of previous consumption (if you bought a new car yesterday your utility of a new one today is small). See Detemple & Zapatero (1991).

In *external* models you allow the utility of consumption to depend on some external factor, like for example average consumption in the economy ("catching up with the Joneses"). See Abel (1990). For point processes in equilibrium models, see Tsai & Wachter (2015), Barro (2006), Barro (2009), Gabaix (2012) and Rietz (1988). For more on partially observed equilibrium models see Brandt, Zeng & Zhang (2004), David (1997), David & Veronesi (2013) and Moore & Schaller (1996).

References

Abel, A. (1990), 'Asset prices under habit formation and catching up with the joneses', *American Economic Review* pp. 38–42.

Back, K. (2017), *Asset Pricing and Portfolio Choice Theory*, 2 edn, Oxford University press, Oxford.

Bain, A. & Crisan, D. (2009), *Fundamentals of Stochastic Filtering*, Springer Verlag, Berlin.

Barro, R. (2006), 'Rare disasters and asset markets in the twentieth century', *The Quarterly Journal of Economics* **121**, 823–866.

Barro, R. (2009), 'Rare disasters, asset prices, and welfare costs', *American Economic Review* **99**, 243–264.

Bernardo, A. & Ledoit, O. (2000), 'Gain. loss, and asset pricing', *Journal of Political Economy* **108**(1), 144–172.

Björk, T. (2020), *Arbitrage Theory in Continuous Time*, 4th edn, Oxford University Press, Oxford.

Björk, T., Di Masi, G., Kabanov, Y. & Runggaldier, W. (1997), 'Towards a general theory of bond markets', *Finance and Stochastics* **1**, 141–174.

Björk, T., Kabanov, Y. & Runggaldier, W. (1995), 'Bond market structure in the presence of a marked point process', *Mathematical Finance* **7**(2), 211–239.

Björk, T. & Slinko, I. (2006), 'Towards a general theory of good deal bounds', *Review of Finance* **10**, 221–260.

Black, F., Derman, E. & Toy, W. (1990), 'A one-factor model of interest rates and its application to treasury bond options', *Financial Analysts Journal* **33**, 33–39.

Black, F. & Scholes, M. (1973), 'The pricing of options and corporate liabilities', *Journal of Political Economy* **81**, 659–683.

Brace, A. & Musiela, M. (1994), 'A multifactor Gauss Markov implementation of Heath, Jarrow, and Morton', *Mathematical Finance* **4**, 259–283.

Brandt, M., Zeng, Q. & Zhang, L. (2004), 'Equilibrium stock return dynamics under alternative rules of learning about hidden states', *Journal of Economic Dynamics and Control* **28**, 1925–1954.

Brémaud, P. (1981), *Point Processes and Queues: Martingale Dynamics*, Springer-Verlag, Berlin.

Brigo, D. & Mercurio, F. (2001), *Interest Rate Models*, Springer, Berlin.

Carmona, R. E. (2009), *Indifference Pricing: Theory and Applications.*, Princeton University Press, Princeton.

Chamberlain, G. (1988), 'Asset pricing in multiperiod securities markets.', *Econometrica* **56**, 1283–1300.

Cochrane, J. (2001), *Asset Pricing*, Princeton University Press, Princeton, N.J.

Cochrane, J. & Saá Requejo, J. (2000), 'Beyond arbitrage: Good-deal asset price bounds in incomplete markets', *Journal of Political Economy* **108**, 79–119.

Cohen, S. & Elliott, R. (2015), *Stochastic Calculus and Applications*, Birkhäuser, New York, Heidelberg, London.

Cont, R. & Tankov, P. (2003), *Finanical Modelling with Jump Processes*, Chapman and Hall /CRC.

Cox, J., Ingersoll, J. & Ross, S. (1985), 'A theory of the term structure of interest rates', *Econometrica* **53**, 385–408.

Dana, R. & Jeanblanc, M. (2003), *Financial Markets in Continuous Time*, Springer Verlag, Berlin, Heidelberg, New York.

David, A. (1997), 'Fluctuating confidence in stock markets: Implications for returns and volatility', *Journal of Financial and Quantitative Analysis* **32**, 427–462.

David, A. & Veronesi, P. (2013), 'What ties return volatilities to price valuations and fundamentals?', *Journal of Political Economy* **121**, 682–746.

Davis, M. (1997), Option pricing in incomplete markets, *in* M. Dempster & S. Pliska, eds, 'Mathematics of Derivative Securities', Cambridge University Press, Cambridge, pp. 216–266.

de Donno, M. (2004), 'A note on completeness in large financial markets', *Mathematical Finance* **14**, 295–315.

Delbaen, F. & Schachermayer, W. (1994), 'A general version of the fundamental theorem of asset pricing', *Matematische Annalen* **300**, 215–250.

Dellacherie, C. & Meyer, P. (1972), *Probabilités et Potentiel.*, Hermann, Paris.

Detemple, J. & Zapatero, F. (1991), 'Asset prices in an exchange economy with habit formation', *Econometrica* **59**, 1633–1657.

Dothan, M. (1978), 'On the term structure of interest rates', *Journal of Financial Economics* **6**, 59–69.

Duffie, D. (2001), *Dynamic Asset Pricing Theory, 3rd ed*, Princeton University Press.

Duffie, D. & Epstein, L. (1992), 'Stochastic differential utility', *Econometrica* **60**, 353–394.

Duffie, D., Filipovic, D. & Schachermayer, W. (2003), 'Affine processes and applications in finance', *Annals of Applied Probability* **13**, 984–1053.

Duffie, D. & Huang, C. (1986), 'Multiperiod securities markets with differential information', *Journal of Mathematical Economics* **15**, 283–303.

Duffie, D. & Kan, R. (1996), 'A yield factor model of interest rates', *Mathematical Finance* **6**(4), 379–406.

Durrett, R. (1996), *Probability*, Duxbury Press, Belmont.

Epstein, L. & Zin, S. (1989), 'Substitution, risk aversion, and the temporal behavior of consumption and asset returns', *Econometrica* **57**, 937–969.

Esscher, F. (1932), 'On the probability function in the collective theory of risk', *Skandinavisk Aktuarietidskrift* **15**, 175–195.

Fabozzi, F. (2004), *Bond markets, Analysis, and Strategies*, Prentice Hall.

Föllmer, H. & Sondermann, D. (1986), Hedging of non-redundant contingent claims under incomplete information., *in* W. Hildenbrand & A. Mas-Colell, eds, 'Contributions to Mathematical Economics', North-Holland, Amsterdam.

Frittelli, M. (2000), 'The minimal entropy martingale measure and the valuation problem in incomplete markets', *Mathematical Finance* **10**, 215–225.

Fujisaki, M., Kallinapur, G. & Kunita, H. (1972), 'Stochastic differential equations of the nonlinear filtering problem', *Osaka Journal of Mathematics* **9**, 19–40.

Gabaix, X. (2012), 'Variable rare disasters: An exactly solved framework for ten puzzles in macro-finance', *Quarterly Journal of Economics* **127**, 645–700.

Gerber, H. & Shiu, E. (1994), 'Option pricing by esscher transforms', *Transactions of the Society of Actuaries* **46**, 51–92.

Goll, T. & Rüschendorff, L. (2001), 'Minimax and minimal distance martingale measures and their relationship to portfolio optimization', *Finance and Stochastics* **5**, 557–581.

Hansen, L. & Jagannathan, R. (1991), 'Implications of security market data for models of dynamic economies', *Journal of Political Economy* **99**, 225–262.

Harrison, J. & Kreps, J. (1979), 'Martingales and arbitrage in multiperiod markets', *Journal of Economic Theory* **11**, 418–443.

Harrison, J. & Pliska, S. (1981), 'Martingales and stochastic integrals in the theory of continuous trading', *Stochastic Processes & Applications* **11**, 215–260.

Heath, D., Jarrow, R. & Morton, A. (1992), 'Bond pricing and the term structure of interest rates: a new methodology for contingent claims valuation', *Econometrica* **60**, 77–105.

Ho, T. & Lee, S. (1986), 'Term structure movements and pricing interest rate contingent claims', *Journal of Finance* **41**, 1011–1029.

Huang, C. (1987), 'An intertemporal general equilibrium asset pricing model: The case of diffusion information', *Econometrica* **55**, 117–142.

Hubermann, G. (1982), 'A simple approach to arbitrage pricing theory', *Journal of Economic Theory.* **28**, 183–191.

Hull, J. & White, A. (1990), 'Pricing interest-rate-derivative securities', *Review of Financial Studies* **3**, 573–592.

Hunt, P. & Kennedy, J. (2000), *Options, Futures, and Other Derivatives*, Pearson, New York.

Jacod, J. & Shiryaev, A. (1987), *Limit Theorems for Stochastic Processes*, Springer Verlag, Berlin.

Kabanov, Y. & Kramkov, D. (1994), 'Probability theory and its applications.', *Review of Financial Studies* **39**, 222–229.

Kallsen, J. & Shiryayev, A. (2002), 'The cumulant process and Esscher's change of measure', *Finance and Stochastics* **6**(2), 313–338.

Karatzas, I., Lehoczky, J. & Shreve, S. (1990), 'Existence and uniqueness of multi-agent equilibrium in a stochastic dynamic consumption/investment model.', *Mathematics of Operations Research* **15**, 90–128.

Karatzas, I. & Shreve, S. (1998), *Methods of Mathematical Finance*, Springer.

Klein, I. & Schachermayer, W. (1996), 'Asymptotic arbitrage in non-complete large financial markets', *Theory Probab. Appl* **41**, 927–934.

Klein, I. & Schachermayer, W. (2000), 'A fundamental theorem of asset pricing for large financial markets', *Mathematical Finance* **10**, 443–458.

Kreps, D. (1981), 'Arbitrage and equilibrium in economies with infinitely many commodities', *Journal of Mathematical Economics* **8**, 15–35.

Kreps, D. & Porteus, E. (1978), 'Temporal resolution of uncertainty and dynamic choice theory', *Econometrica* **46**, 185–200.

Lando, D. (1998), 'On cox processes and credit risky securities', *Review of Derivatives Research* **2**, 99–120.

Lando, D. (2004), *Credit Risk Modeling*, Princeton University Press, Princeton, N.J.

Last, G. & Brandt, A. (1995), *Marked Point Processes on the Real Line*, Springer Verlag, New York.

Liptser, R. & Shiryayev, A. (2004), *Statistics of Random Processes*, Vol. I, 2 edn, Springer Verlag, Berlin.

Lucas, R. (1978), 'Asset prices in an exchange economy', *Econometrica* **46**, 1429–1445.

Merton, R. (1973), 'The theory of rational option pricing', *Bell Journal of Economics and Management Science* **4**, 141–183.

Merton, R. (1976), 'Option pricing when the underlying stock returns are discontinuous', *Journal of Financial Economics* **5**, 125–144.

Miyahara, Y. (1976), Canonical martingale measures of incomplete assets markets, *in* S. Watanabe, ed., 'Proceedings of the Seventh Japan–Russia Symposium', World Scientific, Singapore.

Miyahara, Y. (2011), *Option Pricing In Incomplete Markets*, Imperial College Press, London.

Moore, B. & Schaller, H. (1996), 'Learning, regime switches, and equilibrium asset pricing dynamics', *Journal of Economic Dynamics and Control* **20**, 979–1006.

Musiela, M. (1993), Stochastic PDE:s and term structure models. Preprint.

Øksendal, B. (2004), *Stochastic Differential Equations*, 5 edn, Springer-Verlag, Berlin.

Øksendal, B. & Sulem, A. (2007), *Appplied Stochastic Control of Jump Diffusions*, Springer-Verlag, Berlin.

Pham, H. (2010), *Continuous-time Stochastic Control and Optimization with Financial Applications*, Springer, Heidelberg.

Protter, P. (2004), *Stochastic integration and Differential Equations*, 2 edn, Springer-Verlag, Berlin.

Reisman, H. (1992), 'Intertemporal arbitrage pricing theoryfrietz', *The Review of Financial Studies* **5**(9), 105–122.

Rietz, T. (1988), 'The equity premium: A solution', *Journal of Monetary Economics* **22**(1), 117–131.

Rodriguez, I. (2000), 'A simple linear programming approach to gain, loss and asset pricing', *Topics in Theoretical Economics* **2**.

Ross, S. (1976), 'The recovery theorem', *Journal of Economic Theory* **13**, 341–360.

Schweizer, M. (1991), 'Option hedging for semimartingales', *Stochastic Processes and Their Applications* **37**, 339–363.

Schweizer, M. (2001), A guided tour through quadratic hedging approaches, *in* E. Jouini, ed., 'Option Pricing, Interest Rates and Risk Mangement', Cambridge University Press, Cambridge.

Sundaresan, S. (2009), *Fixed Income Markets and Their Derivatives*, 3 edn, Academic Press.

Tsai, T. & Wachter, J. (2015), 'Disaster risk and its implications for asset pricing', *Annual Review of Financial Economics* **7**, 219–252.

Vasiček, O. (1977), 'An equilibrium characterization of the term structure', *Journal of Financial Economics* **5**(3), 177–188.

Černý, A. (2003), 'Generalised sharpe ratios and asset pricing in incomplete markets', *European Finance Review* **7**, 191–233.

Černý, A. & Hodges, S. (2002), The theory of good deal pricing in financial markets, *in* H. Geman, D. Madan, S. Pliska & T. Vorst, eds, 'Mathematical Finance – Bachelier Congress 2000', Cambridge University Press, Cambridge.

Index of Symbols

Subject Index